INFORMATION LAW SERIES – 15

CONTROLLING ACCESS TO CONTENT – Regulating Conditional Access in Digital Broadcasting

Natali Helberger

KLUWER LAW INTERNATIONAL

A C.I.P. Catalogue record for this book is available from the Library of Congress.

ISBN 90-411-2345-8

Published by
Kluwer Law International
P.O. Box 85889
2508 CN The Hague
The Netherlands
Tel. +31-70-308-1500
Fax +31-70-308-1515
E-mail sales@kluwerlaw.com
http://www.kluwerlaw.com

Sold and distributed in North, Central and South America by:
Aspen Publishers, Inc.
7201 McKinney Circle
Frederick, MD 21704
United States of America
Tel. +1-877-529-5427 (Customer care)
Fax +1-301-698-7159
E-mail customercare@aspenpubl.com
http://www.kluwerlaw.com

Sold and distributed in all other countries by:
Extenza-Turpin Distribution Services
Stratton Business Park
Pegasus Drive
Biggleswade
Bedfordshire SG18 8TQ
United Kingdom
Tel. +44-1767-604958
Fax +44-1767-601640
E-mail kluwerlaw@extenza-turpin.com
http://www.extenza-turpin.com

Printed on acid-free paper

Printed and bound in The Netherlands.

Contents

Acknowledgements

Finally an author! To tell the truth, I have waited a long time for this day. But little did I know that my first book would not be a riveting adventure about smugglers in the Scottish Highlands, a breathtaking travel book or a mind-boggling science fiction story. It is a doctoral thesis in law.

I would like to express my gratitude to everyone who supported me with their comments, expertise and ideas. First of all, my thanks go to Professor Egbert Dommering and Professor Nico van Eijk, my promoters, who accompanied me during this study with their excellence, inspiration and advice. I would also like to thank Professor Bernt Hugenholtz, Professor Pierre Larouche, Dr Wolfgang Schulz, Dr Aernout Nieuwenhuis, Professor Tom Gibbons, Professor Pamela Samuelson, Dr Lucie Guibault, Dr Wouter Hins, Professor Jules Theuwes, Dr Peggy Valcke and, last but definitely not least, Sjoerd van Geffen. The conversations and discussions with these scholars have convinced me that living in an Information Society that propagates life-long learning is a privilege and a challenge I cherish.

Thanks to Hanna Hedström, Sari Galapo and Brigitte Spiegeler are in place in order for their generous help transforming a thesis into a book that is actually publishable. A word of gratitude also to my language editor, Diane McCartney, for her dedicated and excellent work. The next time that a friend or colleague gives me a piece to read or comment on, I will think of you all. 'Put it forward' will be my motto.

I would like to thank my colleagues. The list is so long that I have decided to restrict it to seven people. First, there's Nirmala, my best roommate, followed by Sabina, Rosanne, Tarlach, Wilfred and Anton. Anja, of course and the other colleagues that make IViR, the Institute for Information Law, the inspiring and cheerful place it is. Cheers to all of you. Cheers also to the Indi-team. And another one to the architect of the house at Rokin—it is still standing!

One thing I learned from the years of research I did for this thesis, is that I have some astonishingly good and enduring friends. They accompanied me patiently, sometimes even stoically, through all the ups and downs. They did not complain about unreturned phone calls, and even accepted the eccentricities of someone who is so deeply immersed in a subject. First of all I must thank my brother Dominik. Special thanks also go to Annette, Anke, Arno, Bart, Claudia, Claudine, Christoph, Dorothee, Ellen, Eva, Jane, Maren, Martine, Paelvi, Stuart and Tanja, to name those whose patience I probably challenged most.

The place of honour is reserved for my parents. I would like to tell them how grateful I am and how proud I am of them. This book may not make for a great read in front of the fireplace on long winter evenings, but it is a book, and I hope it makes my parents a bit proud of me, too. This place is also reserved for Remko. If there is a man with whom one can conquer unknown territory, get stuck somewhere at the second stop of a climb, discuss literature, politics and nonsense alike, cope with snow storms as well as a good bottle of wine, he is the man!

Preface

The time when John Vincent Atanasoff and Clifford Berry were building their digital computer (1937-42) was also the time when the history of the distribution of access-controlled electronic services began. The pioneer was the Musak Corporation in New York, which offered fee-based music services for business customers, and later music programmes to New York's households. The service was distributed via telephone lines and allowed radio services to be received in the connected households by means of a specially designed 'injector box'. Later, in the mid-1950s, Paramount Pictures came up with the Telemeter subscription system, a wired distribution system that required users to insert the payment directly into a coin box that was attached to the viewer's television set.[1] Having said that, the history of controlling access to electronic content with the intention of collecting remuneration dates back even earlier in media history, namely to 19th century Paris, when in 1895 the Lumière brothers showed the first cinematographic performance to a paying audience. A semi-electronic cash machine and a piece of paper were the forerunners of sophisticated architecture of control that electronic access control has become today.

It is an ironic turn of history that it was cinema owners who brought about the fall of the first serious commercial pay-TV service. From the very beginning, electronic control over access to content was subject to fierce controversy. There was a bitter struggle between at least three parties: prospective pay-TV providers who believed they had found a lucrative and alternative way to offer and collect remuneration for electronic services independently of advertisement money or

---

[1] An overview of the history of US subscription television is provided by Gunzerath 2000, pp. 655-671.

public fees; competing providers of electronic services, such as cinema owners and free-TV broadcasters who perceived pay-TV as a competitive threat; and consumers who had gotten used to accessing broadcasting content free-of-charge, but who were suddenly required to pay for access, similar to the purchase of cinema tickets or newspapers. Often, the alleged interests of the consumers were used to bring forward the interests of the competitors. Officially, the competitors' resistance to pay-TV was disguised as a fight against tolls or fees for broadcasting. This was the story of the 'Californian Crusade for Free TV' against Subscription Television, Inc., the company that was formed by a colourful group of investors, publishers, electronics manufacturers, millionaires, brokers and, finally, the Dodgers and Giants baseball organisations. The goal of Subscription Television, Inc. was to launch pay-TV services in Los Angeles and San Francisco in the summer of 1964. The declared enemy of Subscription Television, Inc. was the Association of Theater Owners of America (TOA), which would later form the Citizens Committee for Free TV (CCFTV). The movie theatre owners feared that delivering top sporting events and first-run motion pictures directly into consumers' homes would put the business of movie theatre owners at risk: production companies would prefer to distribute their movies via the new subscription service instead of showing them in movie theatres. The initiatives of the 'primitives, popcorn vendors'[2] found broad support and finally led to the end of Subscription Television, Inc. The company was financially exhausted and no longer able to bear the high investments that were needed to launch the service successfully.

---

2 Pat Weaver, President of Subscription Television, Inc., speech to the American Association of Advertising Agencies, 31 March 1964, Beverly Wilshire Hotel, Los Angeles.

Figure 1—Darn that Pay-TV. 'Darn that pay-TV! Pop says he don't have any more Dollar and a halfs for me to watch each ball game!'

The Citizens Committee for Free TV used highly emotional attacks to mobilize newspapers, broadcasters, and, most importantly, the public against the pay-TV initiative.[3] An example of such an attack is a print ad of an unhappy boy in full football wear crying in front of a blank TV screen with the headline: 'Darn that Pay TV! Pop says he don't have any more Dollar and a halfs for me to watch each ball game!' Below the picture was the text:

> 'What kind of a monster would do this to your child—would come into your home and put a padlock on his TV fun? What kind of a monster would force you to feed your TV set bucketfuls of dollars—or suffer the humiliation of being labelled a "cheapskate" in the eyes of your children? There is such a monster. It's a greedy thing called Pay TV'.[4]

As the argument went, the audience had the privilege of receiving broadcasting free of charge. Behind the scenes, however, the commercial concerns of the competitors prevailed then as now.

Is it 'monstrous' to demand payment for access to broadcasting content? Or is pay-TV a desirable supplement to the traditional broadcasting world as we know it? The fact is, the Californian Crusade for Free TV emerged as the loser. In the wake of technological and market developments, electronic access control has established

---

3 For a thorough overview on the Californian Crusade for Free TV, see Gunzerath 2000.

4 'Darn that pay-TV!', Advertisement in Los Angeles Times, 12 October 1964, sec. 3, p. 5.

itself as part of a technical distribution infrastructure to make access to electronic information conditional upon compliance with the terms and provisions of the service provider. In the language of the European Commission, pay-TV is not a 'greedy thing', but a viable business model. So-called conditional access systems form the technical basis for a considerable number of new convergent services, content management and business models. Access-controlled pay-TV platforms offer access to broadcasting services, cinema, the internet, telecommunications services and e-commerce. Moreover, conditional access can be part of a wider scheme to establish the far-reaching protection and control of the distribution of electronic content, for example, within the context of Digital Rights Management solutions. Alongside these developments, the arguments in favour and against pay-TV have evolved towards a more substantiated discussion. At the heart of the matter are some crucial questions that concern the role electronic access control is supposed to play in media markets and the role of the state in (not) regulating it. The pay-TV discussion revolves around the relationship between electronic access control in broadcasting and freedom of expression, which is still a debated issue, as well as other rationales behind media regulation. The same is true for the concerns of competitors, be it free-TV providers or competing providers of access-controlled services, about the possible anti-competitive practices of pay-TV operators. In response, the European legislator has issued a number of initiatives in European broadcasting and telecommunications law. This is the starting point of this study, which examines the compatibility of electronic access control in pay-TV with guiding rationales behind the regulation of broadcasting content and telecommunications infrastructure. It critically analyses the current European legal framework, and examines whether this framework is an adequate answer to electronic access control in digital broadcasting.

List of Abbreviations

AGCOM	Autorità per le Garanzie nelle Communicazioni
API	Application Programme Interface
ARD	Arbeitsgemeinschaft der Rundfunkanstalten Deutschlands
B2C	Business-to-consumer
CENELEC	European Committee for Electrotechnical Standardization
DRM	Digital Rights Management
DTH	Direct to Home Broadcasting
DTI	Department for Trade and Industry
DTT	Digital Terrestrial Television
DVB	Digital Video Broadcasting Project
EBU	European Broadcasting Union
EC Treaty	Treaty Establishing the European Community
ECHR	European Convention of Human Rights
ECTT	European Convention on Transfrontier Television
ECM	Entitlement Control Message
EEA	European Economic Area
EEM	Entitlement Management Message
EPG	Electronic Programme Guide
ETSI	European Telecommunications Standards Institute
EU	European Union
FCC	Federal Communications Commission
GSM	Global System for Mobile Communications
INDICARE	Informed Dialogue about Consumer Acceptability of DRM Solutions in Europe
ITC	Independent Television Commission

MHP	Multi Media Home Platform
MP3	MPEG Audio Layer 3
MPEG	Moving Picture Experts Group
NRA	National Regulatory Authority
Ofcom	Office of Communications
OFT	Office of Fair Trade
ONP	Open Network Provisions
P2P	Peer-to-peer
RSS	Really Simple Syndication
SAS	Subscriber Authorisation System
SMS	Subscriber Management System
TWF	Television Without Frontiers Directive
WIPO	World Intellectual Property Organization
ZDF	Zweites Deutsches Fernsehen

Chapter 1

Controlling Access to Content

1.1. Introduction

In the electronic world, control over content is control over access to content. Conditional access is a technical solution for controlling access to electronic content. The operator of a conditional access system can determine who has access to which content under which conditions and terms. It enables the recipients' access to be managed.

Conditional access has been widely welcomed as a vital aspect of the business model of many modern broadcasting and online services, and as a driving factor behind the prospering of the 'information economy'. The saturation of advertising markets has resulted in an intensified search for alternative forms of commercializing electronic content. New business models that make the provision of electronic services a profitable undertaking, and solutions to 'packaging' and marketing content were needed. To stimulate these developments, legislation promoting the implementation of electronic content control technologies was passed at national and international levels. The provisions of the Conditional Access Directive,[5] together with rules on the protection of technical measures in the World Intellectual Property Organization (WIPO) treaties[6] and the European Copyright Directive,[7] form a framework that firmly protects the electronic control of content.

Electronic access control, however, also has the potential to fundamentally change the conditions under which electronic content is delivered to the consumer.

---

5 Council Directive 98/84/EC of the European Parliament and of the Council of 20 November 1998 on the legal protection of services based on, or consisting of, conditional access, 28 November 1998, OJ L 320, p. 54 [hereinafter 'Conditional Access Directive'], Articles 4 and 5.

6 World Intellectual Property Organization (WIPO), WIPO Copyright Treaty, adopted by the Diplomatic Conference on 20 December 1996, Geneva [hereinafter 'WIPO Copyright Treaty'], Article 11. WIPO Performances and Phonograms Treaty, adopted by the Diplomatic Conference on 20 December 1996, Geneva [hereinafter 'WIPO Performances and Phonograms Treaty'], Article 18.

7 Council Directive 2001/29/EC of the European Parliament and of the Council of 22 May 2001 on the harmonization of certain aspects of copyright and related rights in the information society, Brussels, 22 June 2001, OJ L 167, p. 10 [hereinafter 'Copyright Directive'], Article 6. Articles 4 and 5 of the Conditional Access Directive, Article 6 of the Copyright Directive, Article 11 of the WIPO Copyright Treaty and Article 18 of the WIPO Performances and Phonograms Treaty provide for the so-called anti-circumvention rules, banning the unauthorized circumvention of technological measures that protect electronic content.

Using electronic access control, service providers can send targeted services to consumers based on geographical location, market segment or personal preferences.[8] Conditional access systems implement structures of individualized control over the subscriber base. This also means, however, that operators of electronic access control can exercise considerable influence over market conditions, competition and individual access to content.

This study examines how conditional access is regulated in European media, telecommunications and competition law. More specifically, it studies regulatory implications of conditional access in pay-TV. The pay-TV example was chosen for several reasons: traditionally, broadcasting was perceived as a medium of great economic consequence and as an effective and intrusive means of mass communication. This is due to the broad reach and universal accessibility of conventional broadcasting. Everybody who owns a TV set and is within reach of a transmission network can receive broadcasting. The individualized and private control of access to broadcasting triggers particularly strong conflicts between economic and public policy rationales. On the one hand, there is a political wish to promote pay-TV as a new business model and as a powerful motor for the digital switchover, meaning the transition from analogue to digital broadcasting. On the other hand, private monopoly control over access to broadcasting does not fit in well with the traditional goals of broadcasting policy, which consist of restricting the influence single players have on the broadcasting sector and on what is often referred to as 'the right to information'.

With the digitization of broadcasting, convergence plays a particularly prominent role in pay-TV, which is another reason why the pay-TV example was chosen. Convergence enables pay-TV operators to offer a wide range of broadcasting and non-broadcasting services.[9] Convergence is also a motive and driving factor behind European Commission policy for the digital broadcasting sector. This is to promote the so-called multi-platform approach, meaning that consumers can access broadcasting, information society services and telecommunications services from multiple platforms.[10] Digital pay-TV platforms are considered to be important

---

8 O'Driscoll 2000, p. 14.

9 Convergence, meaning the amalgamation of previously distinct media, results from the ability of different network platforms to carry essentially similar kinds of services, but also the combination of consumer devices such as telephone, TV set and computer. See European Commission, Green Paper on the Convergence of the Telecommunications, Media and Information Technology Sectors and the Implications for Regulation. Towards an Information Society Approach, Brussels, 3 December 1997, COM(1997)623 final [hereinafter 'Green Paper on Convergence'], p. 7.

10 European Commission, Communication from the Commission to the European Parliament, the Council, the European Economic and Social Committee and the Committee of the Regions on barriers to widespread access to new services and applications of the information society through open platforms in digital television and third generation mobile communications, Brussels, 9 July 2003, COM(2003) 410 final [hereinafter 'Communication on Barriers to Widespread Access'], p. 7. European Commission, eEurope 2005: An information society for all, An Action Plan to be presented in view of the Sevilla European Council—21 and 22 June 2002, Brussels, 28 May 2002, COM(2002)263 final [hereinafter 'Action plan eEurope 2005'], p. 7.

gateways for consumers to access all sorts of services. The analysis of the current legal framework for pay-TV will show why Europe has failed to translate the multi-platform approach in European regulation for this sector.

This study will also illustrate that pay-TV challenges the traditional distinction between broadcasting and telecommunications law. While broadcasting law focuses on content-related aspects, telecommunications law focuses on transport and the technical infrastructure. Electronic access control in pay-TV does not respect this distinction because pay-TV operates at the interface between technology and content.

Finally, the fact that the pay-TV sector is better documented than the internet sector is helpful for illustration purposes. It should be noted, however, that electronic access control is not reserved to the broadcasting sector. Many of the questions that will be addressed also play a role in other media sectors such as the distribution of paid-for content via the internet or mobile platforms and internet download services using Digital Rights Management (DRM) technologies. Although this study focuses on the broadcasting sector, some results of the investigation are worth discussing within the context of other access-controlled services.

In this context, it is not the intention of this study to discuss the general regulatory environment (licensing, access to infrastructure, access to programme material, applicable laws) for digital broadcasting services, neither will the study analyze the legal situation in the European Member States. Instead, this study examines how European broadcasting, competition and telecommunications law have approached the specific subject of access-controlled digital broadcasting platforms and their impact on competition and consumer's access to broadcasting content.

This first chapter is an introductory chapter. It starts with a short general introduction of the notion of conditional access as used in this study (1.2.), and its place in the communications model (1.3.). It also presents an overview of the legal environment to provide a global overview of the setting and current situation and to establish the framework for the analysis in the subsequent chapters. The chapter then explains what electronic access control in pay-TV entails and how it works (1.4.). The next section points out some of the challenges of electronic access control for the process of (mass) media distribution as we know it (1.5.). In so doing, it addresses pivotal regulatory questions of this study and sets the general framework for dealing with them. Chapter 1 concludes with a section outlining some conclusions (1.6.) and each of the chapters in this study (1.7.).

1.2. Conditional Access—Definitions and Scope of the Study

Conditional access is a technical solution for controlling access to electronic services.[11] More specifically, conditional access refers to a combination of hardware and software devices that, combined, enable access to a service that is transmitted electronically to be blocked and subject access to an automated authorization process. The transmitted service can be a content service, a transaction service or a telecommunications service that does not provide content but a means to communicate with another person.

Conditional access automatically identifies the requester, compares his or her identity, grants privileges according to predefined access conditions, so-called 'business rules', and initializes or rejects the processing of the requested content in an intelligible form. Typically, conditional access solutions have three basic components: electronic content protection, identification/authorization and enforcement. Electronic content protection uses encryption or similar technologies[12] to protect content against unauthorized access. The element of identification and authorization consists of feeding the system with personal data, defining business rules, managing subscribers, identifying and verifying users and processes. Finally, enforcement is the technical realization of compliance with the business rules in the form of an automated decision to grant or deny access.

Conditional access systems are certainly not the only way to exercise control over access to content; plug-ins, browsers, operating systems, navigation devices, proprietary standards, network control and so forth, can all play a similarly important role when it comes to providing and controlling access. What makes conditional access systems particularly interesting for the purpose of this study is that they are technical solutions specifically designed to manage consumer access to content. Therefore, they are particularly well-suited to illustrate the complex relationship between control over the technical transport architecture and the accessibility of content, or more precisely, a service carrying content. Conditional access builds an 'architecture of identification'[13] in which compliance with set conditions is an integral part of the authorization process. Because it is technically possible to identify users before they are granted access, conditional access becomes a means of controlling and monitoring consumers. Unlike other monitoring technologies such as cookies and metadata, common user identifiers and technologies that map IP addresses to computers, conditional access technologies enable their operators to relate data directly to a person with whom they may maintain a contractual relationship, for example as a subscriber to a service.

The reasons for using conditional access vary from legal and contractual obligations (for example, the protection of minors and intellectual property rights,

---

[11] Definitions without further references are definitions of the author for the purpose of this study.

[12] More about this in section 1.4.1.

[13] Lessig 1999, p. 34.

the security of data and communication) to business reasons (differentiation, personalization, marketing and advertising strategies) and confidentiality.[14] Perhaps the main reason to use conditional access, however, is to secure the operators' remuneration.[15] Some authors speak in this context of a 'new approach of doing business based on access to services rather than the sale of products'.[16] They argue that in the new 'experience economy', markets and market relations will shift towards an access regime based on securing the time-limited use of assets—the 'age of access'. This development will go hand in hand with a shift from industrial production to cultural production in which the focus lies on the marketing of all kinds of cultural resources in the form of 'paid-for personal entertainment'. Indeed, we can observe a shift towards the exclusive marketing of intangible information products and services. This is best illustrated by the aforementioned legal initiatives to protect technical solutions used to commercially exploit this exclusive control over content.[17] This study primarily examines the last group of conditional access users, namely those who use conditional access to sell and distribute services on an exclusive basis, and here more specifically, in a digital broadcasting environment (see figures 2 and 3).

Conditional access systems can be used to control access to broadcasting, telecommunications or information society services. 'Communications services', as defined under European telecommunications law, refers to services whose aim is the transport of signals, such as telephony services, internet access, the transport of broadcasting signals, etc.[18] 'Communications services' can be broadly interpreted and is sometimes misunderstood to include electronic communications in general, including services providing content or exercising editorial control over content. Because this interpretation does not reflect the intention of the European Communications Framework (see section 1.3.2.), this study has chosen to use the more precise notion of 'telecommunications services.'

European law commonly refers to 'information society services' in the sense of 'any service normally provided for remuneration, at a distance, by electronic means

14 For more details, see the study by Helberger/Van Eijk 2000.

15 It should be noted that the term 'remuneration' can be understood in a narrow sense (meaning the direct financial payment by the recipient in return for the provision of a service by the service provider) or in a broad sense, in which it includes the transfer of other goods of commercial value, and in particular, information or return-services in kind. Note that certain information, for example, about the consumer, consumer behaviour, etc., is increasingly gaining its own market value. The same may apply to certain return-services in kind. More about the distinction in Helberger/Van Eijk 2000, section 1.4.1. For the purpose of this study, 'remuneration' is usually used in a narrow sense.

16 Rifkin 2000, p. 90.

17 Conditional Access Directive, Articles 4 and 5; Copyright Directive, Article 6; WIPO Copyright Treaty, Article 11; WIPO Performances and Phonograms Treaty, Article 18.

18 Council Directive 2002/21/EC of the European Parliament and of the Council of 7 March 2002 on a common regulatory framework for electronic communications networks and services, Brussels, 24 April 2002, OJ L 108, p. 33 [hereinafter 'Framework Directive'], Article 2 (c).

and at the individual request of a consumer of services'.[19] Examples of information society services are most internet services, such as e-commerce services, the provision of search tools, such as search engines, access to and the retrieval of data, the provision of internet access, on-demand, syndication[20] and subscription services, hosting services, web casting, games, etc. Information society services are not reserved to the internet; they can be also electronic services that are delivered via a mobile handset (for example, horoscopes, chats, games, information services) or via the TV set top box (interactive applications, e-commerce services).

'Broadcasting' is defined in Article 1 (a) of the Television Without Frontiers (TWF) Directive as 'the initial transmission by wire or over the air, including that by satellite, in unencoded or encoded form, of television programmes intended for reception by the public'.[21] Broadcasting services can be transmitted via cable, satellite or terrestrial networks. They can be also distributed via IP networks or mobile platforms. Another, still controversial question is if the same service can still be called a broadcasting service.

---

19 Council Directive 98/48 of the European Parliament and of the Council of 20 July 1998 amending Directive 98/34/EC laying down a procedure for the provision of information in the field of technical standards and regulations, Brussels, 5 August 1998, OJ L 217, p. 18 [hereinafter 'Directive on the Provision of Information in the Field of Technical Standards'], Article 1.

20 Syndication services enable consumers to personalize the content they are consuming and to retrieve it from different sources. Examples are Really Simple Syndication (RSS) feeds, peer-to-peer (P2P) networks or 'podcasting'.

21 Directive 89/552/EEC of 3 October 1989 on the coordination of certain provisions laid down by law, regulation or administrative action in Member States concerning the pursuit of television broadcasting activities, Brussels, 17 October 1989, OJ L 298, p. 23 [hereinafter 'Television Without Frontiers Directive'] and Directive 97/36/EC of the European Parliament and of the Council of 30 June 1997 amending Council Directive 89/552/EEC on the coordination of certain provisions laid down by law, regulation or administrative action in Member States concerning the pursuit of television broadcasting activities, Brussels, 30 July 1997, OJ L 202 , p. 60 [hereinafter 'Directive 97/36 Amending the Television Without Frontiers Directive']. In the following, references to the Television Without Frontiers (TWF) Directive refer to the directive in the form as amended by Directive 97/36 Amending the Television Without Frontiers Directive.

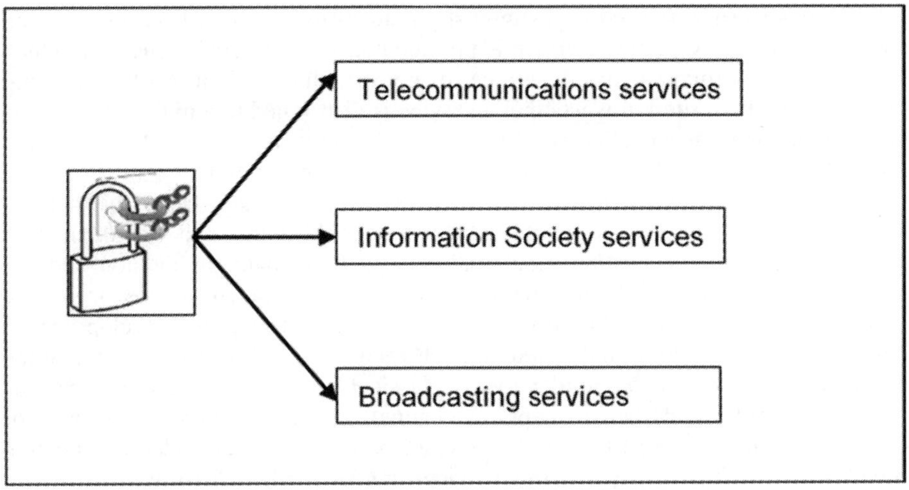

Figure 2—Sectors of conditional access application. The figure provides an overview of the different kinds of services conditional access can be applied to: telecommunications services, information society services and broadcasting services. This study focuses on the latter, and here more specifically on digital broadcasting services.

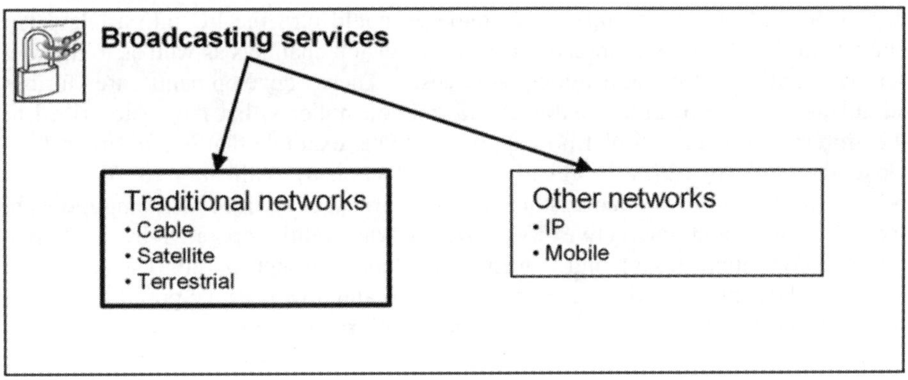

Figure 3—Overview of the different broadcasting environments. Figure 3 illustrates the notion of broadcasting services as used in this study. Digital broadcasting content can be transmitted via different platforms, such as traditional broadcasting networks (cable, satellite and terrestrial networks) or other transmission routes (IP protocol, mobile platforms). The analysis focuses on digital broadcasting services that are transmitted via traditional broadcasting networks.

One major criterion used to distinguish broadcasting services from information society services is whether a service is provided on an individual request (in which case it is an information society service) or not (in which case it is a broadcasting service). Another criterion is whether a service is distributed to a multitude of users at the same time, the so-called point-to-multi-point transmission (in which case it is a broadcasting service), or to an individual user, meaning point-to-point (in which case it is an information society service). However, it must be noted that the distinction between point-to-point and point-to-multi-point, and the distinction between individual request/no individual request blurs with digitization and the resulting convergence of the media. The increased independence from the boundaries of platforms and transmission capacity has favoured the development of new forms of presenting and marketing television and radio broadcasting. Some services still appear to be similar to the classical broadcasting services, such as near-video-on-demand, home shopping channels or subscription television. For other services, and in particular the interactive ones such as video-on-demand, broadcasters' web pages, syndication services, interactive broadcasting[22] and 'Portal TV', it is difficult to assess whether they still fit the definition of broadcasting in its traditional sense. The same is true for cases in which a broadcasting programme that was first transmitted via classical broadcasting channels, such as cable and satellite, is streamed via the internet from the service provider to an individual consumer. Moreover, the internet and mobile platforms are striving to offer (streamed) 'broadcasting' content together with other information society services (for example, news sites increasingly offer moving picture sequences). The deployment of broadband internet technology could turn the PC into a television equivalent that offers widespread access to television channels as well as to internet services and telecommunications services.[23] These developments are further stimulated by European and national information policies that have subscribed to the aforementioned multi-platform approach.[24] One example that demonstrates how close the multi-platform is already to its (technical) realization is Vizzavi, a personalized portal that is accessible across a variety of devices including the web, mobile phones and interactive TV. Vizzavi was jointly operated by Vodafone (mobile communications) and Vivendi (content, internet portal, Canal+) until Vodafone bought Vizzavi's stakes from Vivendi and now uses the portal to offer to its mobile customers exclusive access to news, information and games.

Accordingly, modern pay-TV platforms are not confined to the marketing of digital broadcasting programmes; they also offer additional information society services, telecommunications services and e-commerce services, which are also

---

22 According to the European Audiovisual Observatory 2003 '"[t]rue" interactivity presupposes that an individual message is sent via a return channel, to which the service provider reacts by transmitting the data and services requested by the individual alongside the main television programme', p. 2.

23 European Commission, Communication on Barriers to Widespread Access, p. 2-3.

24 See section 1.1. Note another influential factor for the development of the broadcasting market is national broadcasting regulatory policy.

referred to as 'T-commerce'.[25] This is also why when referring to the services carried via pay-TV platforms the study handles the more neutral notion of '(access-controlled) services'. Within the context of access-controlled services offered via pay-TV platforms, a further distinction can be made between services that focus on the provision of informational input, meaning content, and other services, mostly transactional services or telecommunications services (see figure 4). This study is interested in the former, access-controlled content services, for the simple reason that the use of conditional access within the context of content services can raise interesting public information policy questions about the accessibility and availability of such content. These issues are discussed in Chapter 2.

The reason why the validity of the distinction between broadcasting, telecommunications and information society services is important from a legal point of view is that different regulatory frameworks apply to each category of services. This is explained in the next section. It would go far beyond the scope of this study to discuss the difficult delineation of broadcasting and non-broadcasting services in terms of convergence.[26] This study explains why the delineation does not make much sense in a convergent environment and what the consequences of the distinction are for effective regulation, competition and innovation.

---

25 'T-commerce 2007', Infosat 11/2002, p. 34.

26 For more information see, for example, DLM (Direktorenkonferenz der Landesmedienanstalten), Third Structural Paper on the Distinction of Broadcast Services, available at <www.alm.de/english/english1.htm> (last visited on 15 March 2005).

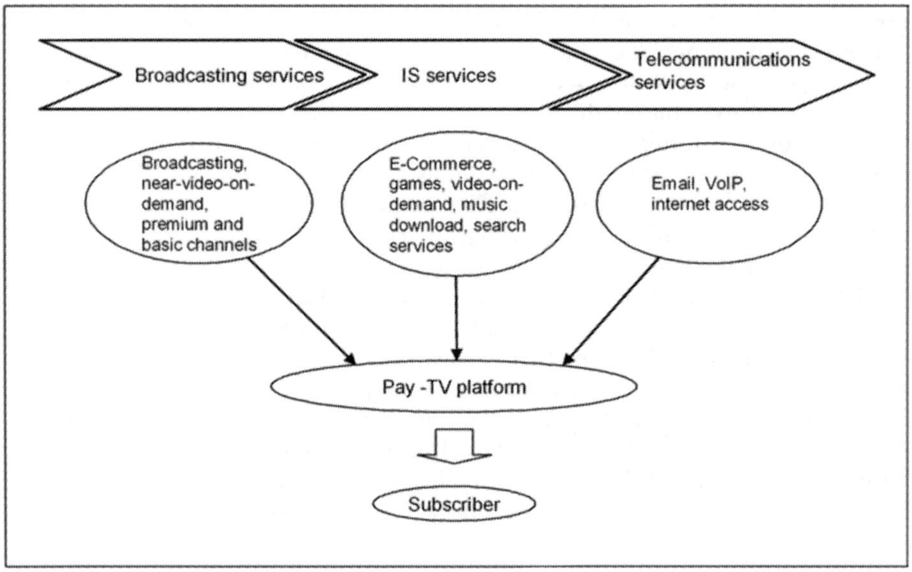

Figure 4—Converging pay-TV environment. Figure 4 illustrates the phenomenon of convergence in digital broadcasting, or more specifically in pay-TV. The boundaries between the different media and services are blurring. As a consequence, subscribers to a digital pay-TV platform can access via this platform not only what was commonly known as broadcasting, but also e-commerce services, interactive services, download services, email and Voice over IP, to name but a few.

1.3. The Place of Conditional Access in the Communications model

The provision of electronic services can be described as a conceptual model composed of four layers stacked on top of each other. As figure 5 shows, each level depicts a specific network function. The first three layers—the physical transmission layer, the network and carrier service layer, and the teleservice layer where the more intelligent applications of the infrastructure are placed—are, for the purpose of this study, also referred to as the 'transport level'. The fourth layer is the 'service level', meaning the level where services are offered to end users. Within the service level, an additional distinction can be made between the actual access-controlled service and the marketing platform (hereinafter 'service platform') via which access-controlled services are offered to consumers. Content is passed down the model, beginning at the service provider end, from one level to the next until it is transmitted over a network. At the remote end, the content is passed up the model to the subscriber's application. The transport and service level are subject to different regulatory environments.

1.3.1. FOUR-LAYER MODEL

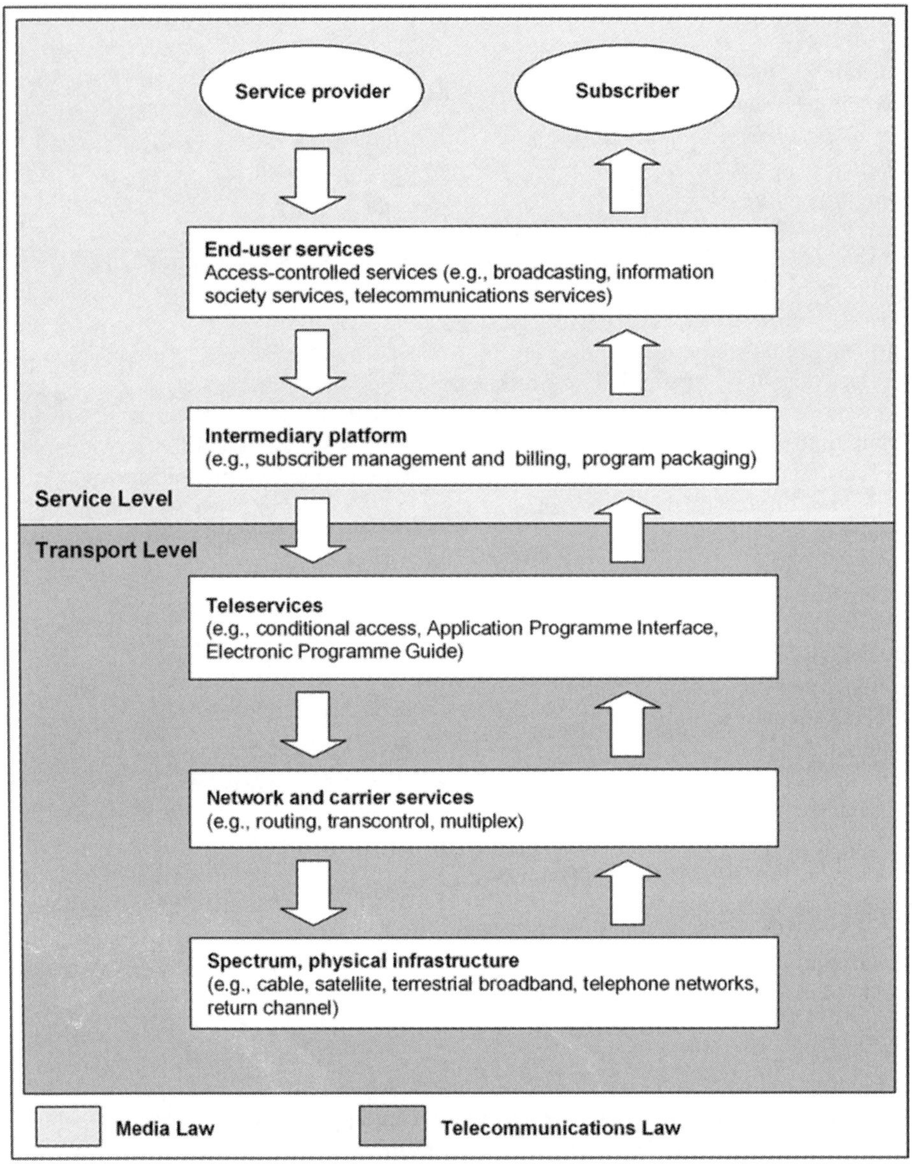

Figure 5—Communications Model. Figure 5 provides a schematic overview of the different layers in the communications chain of access-controlled services. The

figure is an adaptation of a model in Dommering 2000, p. 266. The figure depicts the distinction European telecommunications and broadcasting law makes between the transport level, which is also referred to as the technical level in this study, and the service level, meaning the level at which content services are offered to consumers. While telecommunications law is applicable at the transport level, content services that are transported via the distribution infrastructure fall outside the scope of telecommunications law. The technical conditional access system is situated at the transport level, while the marketing platform and access-controlled content services are part of the service level.

The conditional access is situated at the transport level, and here more specifically, at the upper layer in the distribution chain, meaning the one that is the closest to the service level. This layer is called the 'teleservice layer', or for the purpose of this study, the 'technical platform' of a pay-TV service. Teleservices are services that store, manipulate or present and make content accessible to consumers. Teleservices can also add value to the carrier's services at the lower layers of the Communications model.[27] The 'teleservice layer' is the layer that provides the services/facilities that subscriber applications need to communicate over the respective digital TV network (for example, the decryption of the signals and the processing of the content service). The encryption of service signals relates to the form in which signals are presented to the consumer (meaning encrypted or decrypted). Conditional access control provides a kind of intelligent package for signals that are transmitted via the three transport layers, and is referred to as 'intelligent' because it establishes the automated dialogue between the service provider and the recipient. Subject to this dialogue is the (commercial) exchange of selected content to subscribers. Conditional access control can include additional features, depending on the intelligence and sophistication of the middleware and the software used. Such additional features can be the way in which content is presented to consumers, for example a personalized service offer, the ability of the service provider to terminate the service after a certain period of time, anti-copy protection, the provision of the service in the consumer's preferred language, etc. Figure 5 demonstrates that conditional access is only one element of the technical pay-TV platform. Other elements that facilitate the distribution of access-controlled services are the Application Programme Interface (API), the Electronic Programme Guide (EPG), the operating system, encryption modules, the set top box, etc.[28] This study is interested in the technical platform and its different elements, although conditional access plays a particularly prominent but not isolated role.

Figure 5 shows that the functions of the technical platform are independent of the other transport layers. A service provider can apply electronic access control irrespective of the infrastructure used to transmit the service (cable, satellite, IP-

---

27 See the explanation in Arnbak/Van Cuilenburg/Dommering 1990, p. 135.

28 See for a more detailed description, section 1.4.1.

based carrier services). Conditional access systems for digital services also function irrespective of the kind of content service offered (broadcasting services via internet, films via telecommunications lines, satellite transmitted internet services, etc.). In other words, service providers can apply conditional access technologies within the context of digital service platforms that offer broadcasting services, information society and telecommunications services at the same time. The technical platform provides an independent interface that interacts with the service and transport layers of the model and is the basis for processing (access-controlled) services from the service provider to the consumer.

The same is true for the service platform, which functions independently of the services that are transmitted through it (meaning that a service platform can carry broadcasting, information society and telecommunications services, and offer them to consumers at the same time). This marketing platform belongs to the service level.

The services at the service level are services to which access is controlled, meaning services that are delivered to consumers. As already mentioned, within the service level, the study distinguishes between the service platform and the actual access-controlled services that are offered via this platform. The focus of this study is on the service platform, which is the equivalent of the technical platform at the teleservice layer. The service platform is the marketing and distribution platform for access-controlled services. It describes the entity of content and subscriber-related activities that are needed to distribute access-controlled services (see figure 6). From the consumer perspective, service platforms are the 'portals' to a (digital) world to which they must subscribe before they can enter. From the service provider perspective, providing access-controlled services means providing access to both the service and the technical platform. Frequently, the technical and service platform are integrated with the pay-TV platform and controlled by one and the same operator. Both the service and the technical platform, or elements thereof, can also be offered separately. Moreover, as is shown in section 4.3.2., it is not always easy to ascribe facilities to either the service or the technical part of the pay-TV platform. Often, facilities carry both technical and content or subscriber-related aspects.

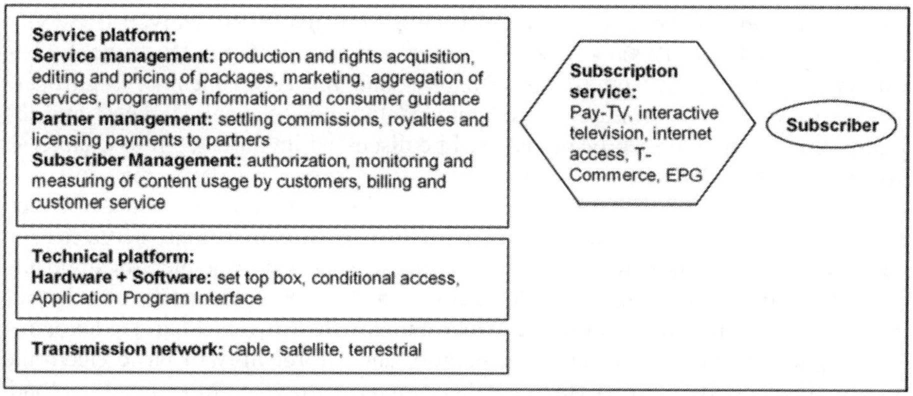

Figure 6—Marketing and distribution scheme for access-controlled services. Figure 6 provides a schematic overview of a pay-TV platform. It demonstrates the distinction that the study makes between the service platform and the technical platform, and follows the distinction that was repeatedly made by the European Commission and European law. The figure shows that the technical platform and the marketing platform are closely connected although they are separate entities. The hardware and the software of the conditional access solution are situated at the technical platform level. The service platform comprises facilitative services, such as aggregation, programme acquisition and subscriber management. Subscribers subscribe to the marketing platform. Some elements of the technical platform are also implemented in the consumer hardware.

1.3.2. THE FIRST THREE LAYERS: TELECOMMUNICATIONS LAW

The first three levels of the transport level fall under telecommunications law. The subject matter of telecommunications law is the transmission of signals via wire or wireless telecommunications networks. Telecommunications law deals with the technical, economic and public information policy aspects of the transport level.[29] This is a field of law that has been widely harmonized by European law.

The European Communications Framework was launched in 2003 and comprises a series of directives and associated measures with the goal to encourage competition in the electronic telecommunications markets, improve the functioning of the Internal market and guarantee basic user interests that would not be guaranteed by market forces. The main instruments of the Communications Framework are five directives (the Framework Directive,[30] the Access and

[29] Dommering 2000, p. 262.

[30] Council Directive 2002/21/EC of the European Parliament and of the Council of 7 March 2002 on a common regulatory framework for electronic communications networks and services, Brussels, 24 April 2002, OJ L 108, p. 33 [hereinafter 'Framework Directive'].

Interconnection Directive,[31] the Authorisation Directive,[32] the Universal Service Directive[33] and the Privacy and Electronic Communications Directive[34]). The Commission Guidelines on Market Analysis and the Assessment of Significant Market Power,[35] the Commission Recommendation on Relevant Markets[36] and the List of Standards and Specifications[37] will be discussed in more depth in Chapter 4. For the purpose of this study, three directives are particularly relevant. The Framework Directive sets out the main principles, definitions, objectives and procedures for the regulation of electronic telecommunications services and networks. The Access Directive lays down procedures and principles for imposing pro-competitive obligations regarding access to and the interconnection of networks on operators with significant market power. Finally, the Universal Service Directive first requires a minimum level of availability and affordability of basic electronic telecommunications services (the so-called universal service obligations). Second, and most relevant for this study, it guarantees a set of consumer rights and consumer protection rules for the sector for users and consumers of electronic telecommunications services.

One important aspect of telecommunications law is the problem of monopoly control over services or facilities if they are crucial gateways for market access. Depending on the circumstances of the individual case, this can be access to transmission networks as well as conditional access or other elements of the

---

31 Council Directive 2002/19/EC of the European Parliament and of the Council of 7 March 2002 on access to, and interconnection of, electronic communications networks and associated facilities, Brussels, 24 April 2002, OJ L 108, p. 7 [hereinafter 'Access Directive'].

32 Council Directive 2002/20/EC of the European Parliament and of the Council of 7 March 2002 on the authorisation of electronic communications networks and services, Brussels, 24 April 2002, OJ L 108, p. 21 [hereinafter 'Authorisation Directive'].

33 Council Directive 2002/22/EC of the European Parliament and of the Council of 7 March 2002 on universal service and users' rights relating to electronic communications networks and services, Brussels, 24 April 2002, OJ L 108, p. 51 [hereinafter 'Universal Service Directive'].

34 Council Directive 2002/58/EC of the European Parliament and of the Council of 12 July 2002 concerning the processing of personal data and the protection of privacy in the electronic communications sector, Brussels, 31 July 2002, OJ L 201, p. 37 [hereinafter 'Data Protection Directive'].

35 European Commission, Commission Guidelines on Market Analysis and the Assessment of Significant Market Power under the Community regulatory framework for electronic communications networks and services, Brussels, 11 July 2002, OJ C 165, p. 6 [hereinafter 'Guidelines on Market Analysis and the Assessment of Significant Market Power'].

36 European Commission, Commission recommendation on relevant product and service markets within the electronic communications sector susceptible to ex ante regulation in accordance with Directive 2002/21/EC of the European Parliament and of the Council on a common regulatory framework for electronic communications networks and services, Brussels, 8 May 2003, OJ L 114, p. 45 [hereinafter 'Recommendation on Relevant Markets'].

37 European Commission, List of standards and/or specifications for electronic communications networks, services and associated facilities and services (interim issue), Brussels, 31 December 2002, OJ C 331, p. 32 [hereinafter 'List of Standards and Specifications'].

technical pay-TV platform. In this context, Article 6 of the Access Directive[38] specifically addresses conditional access questions.

European telecommunications law is based on the idea that the sector-specific regulation of media services is divided into transport, or technical aspects, and content-related aspects. Both aspects fall under very distinct regulatory frameworks.[39] Recital 5 of the Framework Directive leaves little doubt that telecommunications law

> 'does not therefore cover the content of services delivered over electronic telecommunications networks using electronic telecommunications services such as broadcasting content, financial services and certain information society services, and is therefore without prejudice to measures taken at Community or national level in respect of such services, in compliance with Community law, in order to promote cultural and linguistic diversity and to ensure the defence of media pluralism'.[40]

1.3.3. THE SERVICE LAYER: MEDIA LAW

The content-related aspects of the services that are delivered via pay-TV platforms fall, among others, under media law. Such aspects can be the general accessibility and availability of content as well as the quality of content, in particular if it reflects a wide range of different subjects and ideas (pluralism) and if it comes from different sources (diversity). Content-related aspects can also be the way in which content is presented, meaning objectively and in accordance with journalistic principles, and which content is presented, for example, content containing hate speech, content which is dangerous for the development of minors, etc.

Media law is traditionally a product of both public information policy and economically motivated norms that have a variety of goals. In media law, economic goals are to promote functioning competition between a variety of services with the purpose of creating, as the European Commission calls it, 'the world's most competitive, knowledge-based economy'.[41] Economic goals also contribute towards the realization of European Internal Market (Internal Market) principles such as the free flow of services across national borders. The control of economic monopoly power can also be an important public information policy issue.[42] Media ownership law is a form of specialized ex ante structural control designed to limit excessive

---

38 Article 6 of the Access Directive replaces its predecessor, Article 4c of the Council Directive 95/47 of 24 October 1995 on the use of standards for the transmission of television signals, Brussels, 23 November 1995, OJ L 281, p. 23 [hereinafter ' 95/47 Standards Directive']. See also section 4.2.1.

39 European Commission, Towards a new framework for electronic communications infrastructure and associated services, The 1999 Communications Review, Brussels, 10 November 1999, COM(1999) 539 final [hereinafter '1999 Communications Review'], pp. vi-vii.

40 See also Framework Directive, Article 1 (3).

41 European Commission, Action plan eEurope 2005, p. 6.

42 Gibbons 2000, p. 313.

economic and journalistic influences and to prevent the overall media output from being controlled by one or few dominant voices.[43] Central to public information policy is also the availability and accessibility of content for consumers, whereby, in this context, consumers are seen in their role as citizens.

In media law, two major regulatory models are distinguished: the first consists of a concept of elaborate sector-specific content and market control, the other pursues the principle of little interference by regulation and a tendency to leave matters to the market, general competition and consumer protection law or to self-regulation. An example of the former is broadcasting regulation, and of the latter press law. In addition to these two models, there are mixed models such as the regulation of information society services for the internet that can have the character of sector-specific competition law and consumer protection law, notably e-commerce law, leaving it to European Union Member States (Member States) to adopt additional content-related rules in areas in which information society services show a certain degree of journalistic involvement. The broadcasting sector is subject to a significantly greater degree of public intervention than the sectors of the press or of information society services. The regulation of broadcasting is characterized by an alleged ongoing collision of commercial interests, market mechanisms and public information policy. However, broadcasting law, which is at the focus of this study, is undergoing a process of rethinking, as convergence and the invasion of market mechanisms into the media raises a number of existential questions as to the preferable scope and justification of public intervention.

Unlike telecommunications law, European broadcasting law is only partly harmonized in the TWF Directive, while other aspects of broadcasting policy and regulation remain under the authority of Member States. In addition, the Council of Europe has adopted the European Convention on Transfrontier Television[44] that is relevant to all Member States of the Council of Europe. In particular the regulations for pay-TV, however, are fairly harmonized. European regulation exists in the form of Article 3a of the TWF Directive and the Conditional Access Directive, although the latter is not strictly spoken broadcasting law. The Conditional Access Directive provides rules against the unauthorized circumvention of conditional access systems for broadcasting and information society services.

Broadcasting law seeks to 'provide conditions that are most favourable towards pluralism' through rules on licensing, media ownership and programme obligations.[45] In so doing, broadcasting law seeks to protect, among other things, the position of the consumer, or in this context rather the citizen, by creating the conditions for the broad availability and accessibility of multiple content from

---

43 Gibbons 1999, p. 159. Gibbons 2000, p. 313. For an international overview, see the paper by Van Loon 1993.

44 Council of Europe, European Convention on Transfrontier Television, Strasbourg, 5 May 1989, Text amended according to the provisions of the Protocol (ETS No. 171) which entered into force, on 1 March 2002 [hereinafter 'European Convention on Transfrontier Television—ECTT'].

45 Gibbons 1999, p. 159.

different sources. The protection provided is indirect: it does not address consumers or give them specific rights to claim certain behaviour from broadcasters. Instead, broadcasting law addresses broadcasters and imposes programme and other obligations that are (also) in the interest of consumers. Another question that the study will bring to the fore is whether implementing electronic access control in the distribution process will change the assumption of a the audience as passive receivers and hence the paternalistic approach of broadcasting regulation.

Pay-TV arguably still falls under European broadcasting law, as do the broadcasting services that are transported via such platforms.[46] But consumers also enter into a direct contractual relationship with pay-TV providers who set prices and contractual conditions. The relationship between service providers and consumers in the broadcasting sector is thus no longer just a matter of (public) broadcasting law, but is also shaped by contracts. In general, these are standard term contracts and consumers have very limited negotiating power.

Commonly, the legal position and the balance in a commercial relationship between consumers and service providers fall under consumer protection law. The underlying idea of consumer protection law is to balance the negotiating power of contracting parties and provide guidelines that safeguard the interests of the parties. The European provisions on consumer protection are spread over a number of different regulations.[47] Some sectors have specific consumer protection provisions, such as telecommunications law in the Universal Service Directive, the goal of which is to ensure fair and adequate access to, and use of telecommunications services. The Universal Service Directive imposes a number of specific behavioural rules on service providers that have the realization of consumer interests in mind such as pricing principles, data protection, security and secrecy of communication, universal services, etc. For the time being, broadcasting law does not have direct rules on consumer protection.

---

46 See Articles 1(a) and 3(a) of the Television Without Frontiers Directive.

47 Council Directive 84/450/EEC of 10 September 1984 relating to the approximation of the laws, regulations and administrative provisions of the Member States concerning misleading advertising, Brussels, 19 September 1984, OJ L 250, p. 17 [hereinafter 'Misleading Advertising Directive']; Council Directive 93/13/EEC of 5 April 1993 of the European Parliament and of the Council on unfair terms in consumer contracts, Brussels, 21 April 1993, OJ L 95, p. 29 [hereinafter 'Unfair Terms in Consumer Contracts Directive']; Council Directive 97/7/EC of the European Parliament and of the Council of 20 May 1997 on the protection of consumers in respect of distance contracts, Brussels, 4 June 1997, OJ L 144, p. 19 [hereinafter 'Directive on Distant Contracts']; Council Directive 1999/44/EC of the European Parliament and of the Council of 25 May 1999 on certain aspects of the sale of consumer goods and associated guarantees, Brussels, 7 July 1999, OJ L 171, p. 12 [hereinafter 'Sale of Goods and Associated Guarantees Directive']; Council Directive 2000/31/EC of the European Parliament and of the Council of 8 June 2000 on certain legal aspects of information society services, in particular electronic commerce, in the Internal Market, Brussels, 17 July 2000, OJ L 178, p. 1 [hereinafter 'E-Commerce Directive']; Proposal for a Directive of the European Parliament and of the Council concerning unfair business-to-consumer commercial practices in the Internal Market and amending directives 84/450/EEC, 97/7/EC and 98/27/EC, Brussels, 18 June 2003, COM(2003)356 final [hereinafter 'Proposal Unfair Commercial Practices Directive'].

1.4. Electronic Access Control in Pay-TV

Conditional access is used in this study in a broad sense. The conditional access system was defined as a combination of hardware devices (set top box, smart card) and software. Software and devices directly support electronic access control and other facilitative elements of the technical platform that are needed to provide access-controlled services, such as the Application Programme Interface, the memory of the set top box and the return channel or Electronic Programme Guide. Together, they make up the technical platform for pay-TV. Using the example of a fictive conditional access solution for a pay-TV platform, the next section explains how a technical platform for access-controlled services is composed and functions. The description is purely schematic and does not claim to be technically complete or state-of-the-art. However, it does introduce and explain the main notions that are used throughout this study.

1.4.1. FUNCTIONAL CONDITIONAL ACCESS MODEL

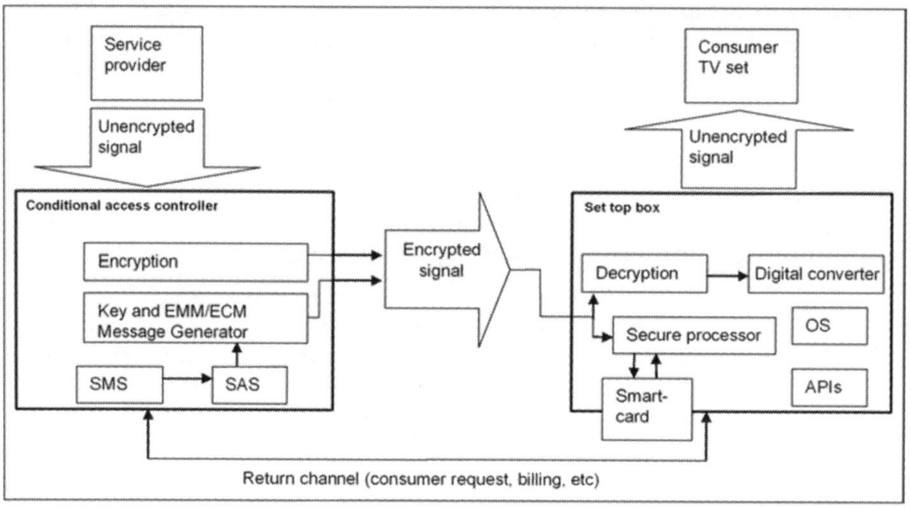

Figure 7— Simplified model of a conditional access system for pay-TV. Figure 7 is a simplified representation of a conditional access environment for pay-TV. The different elements and how they are related to each other is described in the text.

Figure 7 provides an overview of the main components of conditional access: an encryption algorithm, control word, Entitlement Management Message (EMM), Entitlement Control Message (ECM), smart card, Subscriber Management System (SMS), Subscriber Authorisation System (SAS, an operating system and Application Programme Interfaces (API) that coordinate the performance of the different tasks, and finally, an Electronic Programme Guide (EPG). Modern conditional access solutions also implement a return channel (for example, telephone or cable network, internet) that allows service providers to interact with each individual subscriber and vice-versa.

Encryption/Password
Restricting access to a particular service can be accomplished by using, for example, encryption technologies. Once the signal is encrypted, it can only be decrypted by authorized users. Other technologies, notably password systems, biometrical systems and evaluating and filtering devices, can be implemented in addition to encryption technologies. Which particular technology is used depends largely on the kind of information service offered. Broadcasting and other streaming media services usually rely on encryption technologies because they are one way of protecting electronically transmitted signals during distribution. For areas in which the main interest is to prevent unauthorized access to information within the sphere of the service provider, for example, database services or portals, password protection or similar technologies that require prior registration and the entry of a password and user identification (or credit card number, date of birth, etc.) may be the most appropriate.

Control Word
In environments in which conditional access is based on encryption technologies, the signal can be encrypted and decrypted by means of a control word. The control word consists of electronic keys; random strings of bits that are used to initialize the decryption process in the decoder or computer. The final control word is composed of a number of such keys, which also include specific information about the services the consumer is entitled to obtain, access conditions, detailed control information to operate the descrambler, ECM and EMM. All of the information is encrypted together and sent with the content service (for example, broadcasting) signal to the decoder. Alternatively, the necessary decryption and authorization information can be stored on an electronic smart card that is plugged into the set top box. The conditional access operator may also choose to embed only the keys that are necessary to decrypt the ECM and the EMM, etc., while the control word itself is sent together with the broadcasting signal to the set top box.[48]

Entitlement Management Message and Entitlement Control Message
An Entitlement Management Message (EMM) carries the authorization details and is subscriber-specific. The Entitlement Control Message (ECM) carries programme and service-specific information including access conditions and control words that

[48] The security of a crypto-system does not necessarily depend on keeping the crypto-algorithm secret. The security can also depend on keeping the key secret, according to the so-called 'Kerckhoffs' principle'.

are used by the set top box to decrypt the service signal. The decryption process is only performed if the specific subscriber is entitled to receive the specific content.

Smart Card
Smart cards are plastic cards that contain a microprocessor. They are used to store and process information about the cardholder. The information stored on the smart card is needed to initialize the decryption process within the decoder. In addition, modern smart cards incorporate a range of features that make them suitable for a number of transactions such as e-commerce, secure email, personal identification, the storage of customer preferences and personal information and, finally, as an electronic wallet for e-payment solutions. Smart-card technology means a high level of security; only the authorized owner of a smart card can access the services he or she is entitled to receive. Meanwhile, there are smart card readers for PCs that are used, for example, in conjunction with online banking services.

Subscriber Management System
The Subscriber Management System (SMS) is the management centre for the type of conditional access that handles the exchange of services and the return of services between the service provider and the consumer. It initializes personalized smart cards/passwords and their distribution to the respective subscriber. Moreover, the SMS sends out bills and receives payments from subscribers. An important element of the SMS is the customer database, where information about subscribers, including the serial number or IP number of the decoders and information about the services to which they have subscribed, is stored. Functions typically provided by the SMS software applications include:[49]

– Registering, modifying and cancelling subscriber records.
– Carrying out targeted marketing campaigns.
– Managing the set top box and smart card inventory.
– Tracking customer experiences.
– Interfacing with banks and credit card companies.
– Preparing and formatting electronic bills.
– Presenting electronic bills to customers.
– Accounting and auditing facilities, etc.

The SMS is an essential marketing tool because it not only contains billing information, but also specific data about individual consumer behaviour, viewing preferences, etc. Furthermore, it provides contact addresses and other data that have been collected in the course of the subscription process. Obviously, consumer data cannot be accessed without access to the SMS data. Data and addresses are necessary for direct marketing and consumer-tailored initiatives (the SMS controller

[49] O'Driscoll 2000, p. 17.

can analyze subscriber data and adapt its services to respond optimally to consumer demand).

Subscriber Authorisation System

The Subscriber Authorisation System (SAS) is responsible for the technical side of conditional access, the decryption, or making the information accessible. In a password-based system, the SAS initializes the access routine once the correct password has been entered in the system. For SMS-driven encryption-based systems, the SAS sends EMMs and ECMs to the decoding device on the consumer side. For subscription services, the SAS sends EMMs and ECMs to the computer on a regular basis (for example, each month) to renew the subscription rights. For pay-per-view or pay-per-access modes, the SAS sends the information required to access one particular event. The SAS contains databases that can store different kinds of information such as:

- Service information.
- Smart card identification numbers.
- Customer profiles.
- Scheduling data.

Return Channel

The return channel is a direct link between the set top box and the SMS using a modem and a telephone, a cable network or the internet. The return channel establishes a two-way communications mode and allows the broadcaster to directly contact the consumer and vice-versa. The return channel can be used for a number of reasons, including interactivity and audience participation, direct marketing activities, billing (pay-per-view), and the provision of consumer requests (on-demand films). It can be also used to monitor individual consumer's viewing behaviour.

Set Top Box

Originally, the primary task of the set top box, which is also the hardware that enables conditional access, was to host the security functions and provide a platform for the decryption procedure at the consumer end. However, the hardware itself is now increasingly developing its own functionality. The set top box not only plays an important role for access-controlled services, but for digital television in general, because it converts digital signals to analogue signals that can be viewed through an ordinary TV set, and because it provides the necessary technical environment to proceed with enhanced broadcasting or information society services.[50]

[50] Today already, consumers can choose between set top boxes with and without conditional access. The latter are exclusively designed for the reception of free-to-air digital services.

In environments in which access-controlled signals are distributed to a computer, authentication, identification and decryption take place using the computer's own operating system and memory as well as a some player software and plug-ins such as RealPlayer, Microsoft's Music Player, Winamp or QuickTime, etc. Unlike set top box technology, most players can be downloaded from the internet for free. The player software must be interoperable with the file's compression technology, and, if used, the DRM and encryption technology. The more technologies the player software supports, the more different content it will be able to play.

For signals that are transmitted to a TV set, the set top box will usually be the platform used to process all of the functions required to receive and decrypt the signal at the remote end.[51] The set top box incorporates the hardware and software required to receive and decrypt access-controlled broadcasting signals. The main components of a set top box are a decryption chip (holds the algorithm section for conditional access), a secure processor (stores all of the information required for the authorization process (can also be implemented in the smart card), a demodulation unit (in case the received signals are digital signals that have to be translated for an analogue TV set), a central processor (manages the functions and operations in the set top box), memory, modems, an operating system, a set top box browser[52] and a search engine.

Application Programme Interface

As well as having their own operating system, modern set top boxes or other reception devices, such as the X-Box, also have Application Programme Interfaces (APIs) (middleware similar to that of a computer) that handle the encryption and decryption process as well as the increasingly complex processing of services and interactive applications. The middleware is largely independent of the underlying hardware and network components. The advantage is that advanced applications can be written for a common API that can process the application irrespective of a manufacturer's individual specifications for a particular set top box. An independently operating middle level serves application portability and uniformity in the look and feel of interactive applications. Thus, the API functions as a communication bridge (interface) between the operating system of the set top box and the application. The API is used to run advanced service applications such as EPGs, video-on-demand, e-commerce applications and online games that exceed the processing power of the general set top box operating system. In other words, an API is a kind of specialized operating system that can process advanced applications while communicating with the operating system of the set top box and activating the functions and resources required. Depending on the manufacturer and the specifications, an API supports a limited number of application programming languages. Due to the relatively large number of (proprietary) APIs for set top

[51] An alternative are TV sets with integrated set top boxes.

[52] The set top box browser can organize web content in a format that is also viewable on a TV screen.

boxes and the lack of a common standard, applications may not be transferable from one set top box to the other. At the time of writing, discussions are still underway to reach an agreement on the so-called MHP standard for set top box APIs. MHP was developed by the DVB and has been ratified by the European Telecommunications Standards Institute (ETSI). MHP is based on open standards and is a standardized API that allows third parties to develop their own interactive applications.[53]

Electronic Program Guide
The Electronic Programme Guide (EPG) is one example of an interactive programme application that is processed via APIs. The task of the EPG is to lead the consumer through the available choices. In an online environment, their equivalent may be a combination of a portal and search engine. EPGs can be designed with varying degrees of journalistic input and interactivity. Strictly speaking, the EPG is not embedded in a conditional access system, and it can belong to an unencrypted service platform (for example, the EPG of the digital service platform of public broadcasters). Having said that, EPGs will be key elements of most of the modern access-controlled digital service platforms.

When an EPG application is started, the first thing the user might see is the EPG's 'Welcome Page'. The welcome page and the following page(s) could include a collection of icons for all of the available programmes and other services such as internet access, email, games and other interactive applications. The EPG is based on additional information about the programme and other applications (EPG data) that the service operator usually sends together with the broadcasting signal. Based on these data, the EPG can be directly generated by the decoder (the so-called Basic Navigator EPG). The Basic Navigator EPG could become a standard function of future set top boxes.[54] Alternatively, broadcasters or service platform operators could choose to develop their own EPG ('Broadcaster's EPG') and offer it to consumers as part of their services. In this case, the broadcasters (or third parties) would send a specific application with the EPG data which would generate the Broadcaster's EPG.[55]

General Remarks
The previous section shows that the technical platform for access-controlled services is not one device, but a combination of various services and facilities. The combination of elements can differ and single elements can develop their own functionality (as in the case of the set top box or the API). The different components of a conditional access system do not necessarily have to be provided by the pay-TV operator: it is also possible to commission an independent operator.

---

53 See <www.itvdictionary.com> and <www.mhp.org> (last visited on 14 March 2005).
54 In this case, the EPG is pre-determined by the set top box manufacturer.
55 For more details, see the article by Weiss/Wood 1998.

Eventually, functions can be spread across different specialized operators or service providers. For example, broadcasters can choose to implement their own EPG or SMS service while sharing the conditional access system with another service provider. Moreover, a number of independent set top box manufacturers such as Philips, Motorola, Nokia and Pace are developing set top boxes and software solutions for providers of (access-controlled) digital service platforms.

With digitization, it should no longer matter in the mid-term future which transmission medium (cable, satellite, broadband) is used to deliver the signal carrying the content or control word; the conditional access process is transmission-medium independent. Digitized TV signals can be combined with enhanced signals such as video, graphics, texts and web links—so-called enhanced television[56]—and received via satellite, cable, internet[57] or mobile platforms. For digitized signals, it does not matter what the content of the signal is (written word, sound, video signal).

Having said that, conditional access solutions for pay-TV platforms were originally developed separately from conditional access devices for other services and in particular for information society services delivered via the internet. This is often explained by the different structure of communications in these sectors. In the sector of information society services, server-based applications (pull services) such as databases and websites were predominant. Here, the consumer (meaning any web surfer) is largely unknown to the provider. Asymmetric solutions are therefore required, for example, to identify the web surfer. In the area of broadcasting, a consumer first has to subscribe and is hence not anonymous. Consequently, simpler solutions such as encryption technologies may be sufficient.

With the convergence of transmission means and methods, the borderlines are blurring and technical development is overcoming other traditional differences between set top box-based services and PC-based services. Examples are factors such as the resolution of PC and TV screens, the navigation device (keyboard/remote control), font size, memory and transmission capacities. Over the last years, the set top box has evolved into a PC-like device, or vice-versa. It can be a ‘computer for the average consumer’ who wants to send email, surf the web, download content and use e-commerce applications, but who does not want to be bothered with the complexities of the computer itself. Both the set top boxes and PCs have high-speed data interfaces, a lot of memory, powerful processors, a high-speed return channel and APIs that can process all kinds of TV or internet-centric applications. More and more, the reception device used to receive a specific service

---

56 ’Enhanced TV’ refers to technology that enables consumers to receive both television broadcasting and access the internet from the same screen at the same time. Enhanced TV services can vary from simple services that include links to related sites on the internet to highly involved interactivity that merges a TV image with menus, rich multimedia content and supporting text, O’Driscoll 2000, p. 248.

57 For an overview of webcasters, see <http://www.portal.tvoon.de/9715.0.html> (last visited on 14 March 2005).

will be a question of taste.[58] The same conditional access could be used to secure access to broadcasting and/or interactive and personalized information society services.

Digital-broadcasting-service-platform operators continue to drive the development of enhanced digital broadcasting technology. Not surprisingly, however, PC operating system manufacturers are hurrying to secure their position on the market for set top box operating systems (among them Microsoft, Sun and Linux). The other way round, hardware manufacturers such as Philips and Sony are driving the transition towards the PC as a single device for TV reception and alternative for the set top box.[59]

1.4.2. How Pay-TV Works

For consumers, the reception of access-controlled services usually involves entering into a subscription contract that lays down the conditions for the service with the service provider. This 'obligation' generally involves accepting the service provider's conditions and paying a monthly fee for a minimum period of time. Subscribers are often required to sign up for a minimum period of three months, and some providers even issue contracts for up to twenty-four months. The subscription process is also frequently used to collect additional information about the subscriber such as age, sex, profession and interests, how he or she heard of the service offer, etc. The prices for the packages vary, and it can generally be said that the greater the variety of the services, the higher the price. In most cases, the fee will exceed the general broadcasting fee, which the consumer will have to continue paying together with cable subscription fees, etc. Subscription fees and conditions differ widely across Europe.[60] In addition, most providers charge a one-time subscription-opening fee. If the service is provided via a set top box, the consumer will have to purchase or rent the box. A number of providers link the subscription to the purchase/rental of a decoder for which they charge additional fees. Some service providers, however, also subsidize set top boxes or even give them away for free.

The following example of a digital pay-TV service is used to explain (in simplified form) how electronic access control in digital broadcasting works in practice. The service operator must first ensure that the programmes are exclusively transmitted in encrypted or otherwise protected form. At this stage, the consumer receives at best the encrypted, non-readable signal via the TV set. To decrypt the

---

58 In order to receive internet-based services via a set top box, the set top box must be issued with an individual IP address and must be able to download the necessary plug-ins and players to receive web content. In addition, the operating system of the set top box must be able to support a large number of different interactive applications (application portability).

59 For example, the Xbox by Microsoft is connected to the internet and the television, <www.xbox.com> (last visited on 14 March 2005).

60 For an overview see <http://www2.digitalfernsehen.de/Sender/1039791283> (last visited on 14 March 2005); Idate 2000, p. 85.

signal, the consumer must register with the pay-TV provider. This registration is transmitted to the service provider (via mail, email, fax, or the internet). The subscriber's data such as name, age, sex, credit card number or whatever data are required for the subscription process (preferences, special interests, etc.), are stored in the SMS. Based on the information provided, the authorization system updates the SMS database. The SMS creates a personalized smart card that contains the subscriber's personal information, the services he or she is entitled to receive, his or her personal preferences and the keys necessary to decrypt the programme, etc. The smart card is then sent to the consumer. For pay-TV, the SAS subsequently sends to the consumer's set top box, under the management of the SMS, the ECMs and EMMs required to initialize the decryption process. If this information is included on the smart card, the set top box will identify the smart card and thus the consumer, and initialize the decryption process.

The process by which the user requests a specific item, for example, a film from an on-demand service is more complex. The moment the subscriber requests a film, the request is forwarded via a return channel. The SMS checks the data received and decides on the basis of the data stored (including payment of fees, age, etc.) whether the requester is entitled to this access. The service provider is usually free to determine in each individual case the conditions of access, be it the age of the requester, his or her willingness (and ability) to pay the required price, his or her geographic location, etc. If the conditions are fulfilled, the SAS is notified and the ECM and EMM keys used to initialize the decryption process are initiated. The encrypted control word is then sent together with the content to the consumer's set top box. Together with the information stored on the subscriber's smart card and the access code, the encoded signal is correctly decoded and made intelligible via the set top box. Because the capacity of the set top box's operating system is often too limited to process more sophisticated services, the on-demand application will probably be processed via a special API that initializes the process of loading the content from the movie database at the head-end, processes the programme and/or additional information, adapts the presentation's graphic format to suit the platform, checks authorization, etc. Billing is performed either directly via the smart card (electronic purse) or via the return channel.

1.4.3. Main Structure and Lines of Market Development

The initial development of the pay-TV sector was driven by a few relatively strong players. From the beginning, the development of the technical solutions for pay-TV was tightly interwoven due of the economic and technical integration of the technical and service platforms.[61] Very generally, four main lines of development can be identified:

– Market consolidation.

[61] See also section 3.3.1.

- Convergence.
- Vertical integration.
- Intermediary platforms.

Market Consolidation

At the time of writing, the pay-TV sector in the different Member States is generally in the hands of a small number of highly diversified, strong, and international or regional operating enterprises that are also, to a large extent, affiliated. In most cases, one, or at the most three to four major players offer pay-TV services in each country.[62] Pay-TV mainly involves already established national players from the analogue broadcasting world who fight hard to obtain similar positions in the digital world, and in particular in pay-TV.[63]

In addition, the development and operation of technical solutions for pay-TV is in the hands of a small number of enterprises. Moreover, it is usually the larger countries that have two or more competing pay-TV platforms with different conditional access solutions. The solutions are generally proprietary, and the proliferation of interoperability solutions is still rather modest. The result is that consumers with a set top box running one standard are often not able to receive services from a competing platform. A good example of competing pay-TV platforms with different conditional access systems is France with Canal+, TPS and ABSat. Canal+ has a Mediaguard conditional access (Kudelski); TPS and ABSat operate the Viaccess conditional access (France Telecom). Only ABSat, which is the smallest platform, succeeded in concluding an interoperability agreement with both major service platforms so that subscribers to ABSat can also receive ABSat services via the Canal+ set top box.

When a country has several pay-TV platforms, the operators often compete through different transmission platforms, for example, satellite and cable in the UK and The Netherlands or satellite and terrestrial in France. Here, issues of transcontrol can come to the fore and may cause conflicts between cable and pay-TV platform operators (for example, in The Netherlands, where lengthy negotiations were required for Canal+ and UPC to agree on a solution to the decoder problem). However, competition can also take place within one and the same delivery platform such as in the UK where the British pay-TV operator BSKyB distributes a range of third-party services via the Sky platform. Generally, cable operators are considered serious challengers to satellite-centred services. Unlike satellite broadcasters, cable operators already have their own subscriber base and often own the necessary technical equipment and infrastructure to bill individual consumers.

---

62 For a more detailed insight, see European Audiovisual Observatory 2004, pp. 10-12; also: study by Andersen Consulting 2002.

63 Idate estimates that, taking all broadcasting media into account, there are at the most (in the cases of France and the UK) six digital pay-TV operators in one country, Idate 2001, pp. 104, 107. See also Neumann 1998, 225 pp.

Convergence

Established digital broadcasting operators are experiencing competition from new entrants, mostly operators at the network level, and telecommunications and cable operators that are entering the content business. However, banks, sport clubs and video game producers, to give some examples, who are interested in extending their activities to the digital service platforms are also entering the scene. In this context, convergence implies that traditional pay-TV operators are confronted with increasing competition not only from rival pay-TV operators, but also from competitors that are active in internet or mobile markets.

Convergence also implies that pay-TV platforms are not confined to the marketing of digital broadcasting programmes; they also offer additional information society services, enhanced telecommunications services and e-commerce services that could also be offered via the internet. Likewise, traditional internet services are joining pay-TV platforms. And, last but not least, the internet and mobile platforms are striving to offer 'broadcasting' content together with other information society services. For example, news sites increasingly offer moving picture sequences, mobile telephony operators purchase rights for the transmission of soccer events, etc. These developments are further stimulated by European and national public information policies that have subscribed to the aforementioned 'multi-platform approach'.[64]

Vertical Integration

In response to the process of convergence, media markets are seeing many strategic alliances between different 'major' players at the different levels of the pay-TV distribution chain:

– Content production.
– Content acquisition.
– Aggregation and selling to subscribers (service level).
– Transmission via the technical platform (technical level).
– Transmission via physical networks.

In other words, pay-TV operators are often vertically integrated in one form or another (either by ownership or alliances) and thus exercise control over several steps in the distribution chain. Pay-TV operators are often active in the acquisition of content or third-party services (in the form of programmes or whole channels) that are packaged into programme bouquets and sold to consumers, as well as in the operation of the technical distribution platform.[65] Chapter 3 describes a number of

[64] See section 1.1.

[65] IDATE 2001; Amstrong 1999, 258pp.; Neumann 1998, 37pp. and 232 pp.; the papers by Cave/Crandall 2001 (about vertical integration between broadcaster and sports leagues) and Galbiati/Nicita/Nizi 2004, 2pp. (pointing to differences with the US pay-TV market that seems to be much more characterized by vertical separation).

major alliances that involved or would have involved different players in different steps of the distribution chain with the goal to further vertical integration.[66]

Of particular interest to this study is the combination of service and technical platform. The selling of digital content is presently the core business of most conditional access operators. This explains the close links between the service and the technical level, because the technical level is an integrated part of the business model and strategy. This also explains why many service providers are directly involved in the market for conditional access systems or associated facilities. Examples are France Telecom (Viacess), Nethold (Irdeto) and Murdoch (BSkyB) (Videoguard).

Intermediary Platform —A Change in the way Broadcasting Services are Distributed

The trends described above are also a result of the change in the overall distribution strategy for content services. Controlling access to content allows for new, more sophisticated pricing and marketing strategies than the concepts of financing through fees or advertising allow for. Services are no longer 'set off in the air', but marketed individually to subscribers. The possibilities of targeted and personalized marketing allow for differentiation between consumer groups and preferences in the form of price discrimination. Price discrimination and diversification can be a means of attracting the attention of those parts of the audience that have been neglected by advertisers so far. Moreover, a new source of revenue enables the financing of new programmes and services that would otherwise not be possible under conventional advertising models.[67] Market players are now looking for new business models to exploit these potentials. The marketing of digital services via intermediary platforms is one of these models.

Many modern pay-TV platforms function as intermediary platforms, meaning as a kind of portal through which access to a range of services is marketed to consumers.[68] The operators of such pay-TV platforms—modern information brokers—do not necessarily distribute their own content; instead, they are experts at marketing and selling the content of others. The service platform's main task is to aggregate services and deliver packaged and tailored content to subscribers under a common brand name[69] and through a common business platform.

One part of the business strategy consists of selling access to the pay-TV platform to other content service providers, and preferably to those that can add to

---

66 See section 3.3.1.

67 See also Neumann 1998, pp. 148-150.

68 The following section is based on, among others, the papers/studies by Nocke/Peitz/Stahl 2004; Armstrong 1999; Galbiati/Nicita/Nizi 2004; Neumann 1998 145pp., 225pp.; Evans 2003; Wright 2003; Dietl/Franck 2000, 596pp.; Rochet/Tirole 2003 and Bakos/Brynjolfsson 1999. The latter provide an extensive discussion of business models involving aggregation and intermediary platforms using, among other things, the example of site licensing and subscription services, 14pp.

69 For more information on branding, see Breuning 2000, p. 380.

the overall attractiveness of the platform for subscribers. The other part of the business strategy consists of attracting subscribers and selling them access to the platform. Intermediary platforms have a central position in the distribution chain between content service providers and consumers. In other words, pay-TV platforms not only offer third-party content-service providers the technical and marketing services required to sell content to subscribers, they also provide them with another valuable commodity in return: consumer attention. This is an important function of intermediary platforms in more than just the broadcasting sector. It will become even more important as the expected increase in the number of digital channels will make it more and more difficult for broadcasters to attract sufficient consumer attention to sustain the economic viability of a service. And consumer attention is worth money. The amount of attention a particular programme receives determines its economic value for advertisers, providing advertising is permitted on the pay-TV platform, and sponsors, as well as for the retail market in broadcasting content. A pay-TV platform with a large subscriber base can therefore generate positive externalities[70] for the service providers that are carried via the platform. Similarly, the presence of a large number of service providers on a pay-TV platform can generate positive externalities for subscribers because the platform offers them a broad range of services and more choice.[71] The fact that pay-TV platforms serve several sides of the market can influence the decisions the operators make in many ways in terms of pricing, interoperability and the services admitted to the platform.[72] This is because the operator has to get and keep the different sides of a market on board.

Aggregating and selling content via an intermediary platform instead of selling content separately and directly to the consumer can have a number of advantages that make this business model attractive for all parties involved. An important reason is the reduction of administration, transaction, marketing and maintenance costs. The individualized marketing of access-controlled services to consumers is a complex process that requires a lot of organization, from the acquisition and aggregation of content, the journalistic arrangement of marketable units under a common brand name, advertising, subscriber management and service, content protection and monitoring, to the provision of service information and orientation for subscribers (EPG, search index, navigators). The service-platform operator usually performs these tasks centrally. The concentration of all management tasks in one hand allows for more efficient processing. It should be noted, however, that in most cases they can also be offered separately. Control over the intermediary platform does not only allow for the optimized marketing and advertising of one's

---

70 A positive externality can be described as 'something that party A generates for party B but for which party A has no practical way to demand compensation', Evans 2003, p. 2 in footnote 6.

71 Evans 2003, pp. 32-33.

72 See the papers by Armstrong 1999; Nocke/Peitz/Stahl 2004; Evans 2003; Wright 2003; and Rochet/Tirole 2003.

own services, it also favours the efficient distribution of third-party resources,[73] optimizes product variety, increases economic efficiency and reduces redundancy. Platform operators can also benefit from joint control over vertical components of the distribution platform such as the service and technical platform, or from ownership in programme rights, an own distribution platform, etc. This again allows for a more efficient use of the overall distribution platform, including the benefits from the compatibility of the components within the platform. For example, the operator of a pay-TV platform can ensure that all the services that are provided via this particular platform are compatible with each other and with the consumers' set top box. This is also why it can be so important for smaller third-party content providers to have access to a popular platform.

A central element of the pay-TV business strategy is bundling strategies.[74] This is not really new to broadcasting or mass media in general. A traditional broadcasting channel can be understood as a bundle of different content. The difference between traditional channels and the bundling strategies of pay-TV platforms is that the latter bundle a number of services that could also be offered individually. Pay-TV operators can bundle broadcasting with non-broadcasting services, for example broadcasting, telephone and internet services, something that is referred to as 'triple play'. Pay-TV operators can also create pure broadcasting bundles, for example, a lifestyle package that offers cooking channels, health and beauty channels, and travel channels. A package can be offered at a lower price than the individual subscriptions, providing individual subscription is possible. Another example is the bundling of a basic package and a premium package. The former can be made up of a combination of different channels that each subscriber receives, while subscribers must take an additional subscription in order to receive the combination package, which offers particularly popular (premium) content such as sports and movies. As will be explained in the following section 1.5.3., bundling strategies can be profitable and pro-competitive for different reasons. Under certain circumstances, however, they can also be anti-competitive. An assessment of bundling strategies under general competition law is performed in Chapter 3.[75]

There are various examples of bundling strategies used in pay-TV and in the media in general. For example, bundling can take place between pay-TV platform operators and competitors. Platform operators can require competitors who wish to use their conditional access system to use their EPG, or to agree to being marketed via the same service platform and under the same brand name. The example of channel bundling will be discussed in more depth in sections 1.5.3. and 3.4.3. A second example that will be dealt with in more depth requires subscribers to

73 Most of the existing service platforms carry third-party channels and services. These channels are either licensed to the platform operator, or they pay to be carried via the platform.

74 The notions 'packaging' or 'aggregating' are often used synonymously. It should be noted that the notion of 'bundling' used in this context is not necessarily identical with the notion of 'bundling' used in competition law analysis (see section 3.4.3.).

75 See section 3.4.3.

purchase a particular set top box, namely the one that supports the technical standard of the particular pay-TV provider's services.

1.5. The Impact of Conditional Access on the Distribution Chain

Individual control over access to content services introduces new features to the broadcasting world as we know it. To understand them is important for the legal discussion in the chapters to follow. The classic broadcasting model involves the undirected one-way transmission of electronic content towards a multitude of (anonymous) recipients. Once sent, electronic content can be received by anyone who has the necessary technical equipment and is within reach of the respective transmission medium, be it the footprint of a satellite or the local cable or telephone network. Conditional access is changing the general distribution structure for what was commonly known as 'broad-casting'. Pay-TV services address individual consumers separately to authorize access, send them a bill, a specialized advertisement or a specific service they requested.

1.5.1. New Features

Individualization and Interactivity

Conditional access introduces an element of interaction, or interactivity, between the sender and the recipient of the content that was formerly unknown to traditional broadcasting (as opposed to, for example, data retrieval services on the internet where this is common). Because most modern conditional access systems are based on a two-way communications model and include a return channel, they support interactivity between the sender and the recipient.

Contract versus Public Law

Probably one of the most significant changes is that viewers pay directly for broadcasting content. Depending on how much choice the pay-TV platform business model enables, viewers can even specify which content they want to watch and pay for. The distribution of broadcasting content is shifting from a previously public sphere to a more personal sphere where the conditions for access to a service are directly negotiated between the service provider and the requester. This is perhaps not so new to other media such as the press or the internet, but it is new to broadcasting. Access-controlled broadcasting can be responsive to demand, and viewers are consumers.

As long as broadcasting signals were uncontrollable, there was no tangible matter that could be sold to consumers. In free-TV, the contractual relationship exists between the broadcaster and the advertiser; consumers 'pay' for broadcasting content in the form of a public broadcasting fee and their attention. This changed when electronic access control was introduced to the distribution process.

The change towards an individualized, commercialized distribution pattern may explain why traditional broadcasting law has so far been of little use when dealing with electronic access control. Traditional broadcasting law is based on the assumption a) of limited resources and b) that once a programme is transmitted it is generally available. Broadcasting law addresses in the first instance the programme producers, the broadcasters, and imposes on them a number of obligations concerning media concentration, programme quality, pluralism and so forth. Electronic access control introduces access control to the telecommunications process after a programme is composed. Consequently, traditional broadcasting law does not effectively address such control. From the consumers' perspective, consumers depend on access to an access-controlled platform before they can access particular content. For consumers, economic or journalistic influence is thus exercised at an earlier stage, namely when controlling access to the platform. It is the contractual relationship between the consumer, the subscriber and the platform operator that determines which content can be watched at which conditions. When a service provider and a consumer conclude a subscription contract, the relationship between them is no longer governed by broadcasting law alone, but also by the terms and conditions imposed by the platform operator.

Access is no Longer 'Free'

Access to the services that are offered via access-controlled platforms is no longer 'free'. 'Free', in this context, can be interpreted in different ways. 'Free' can mean that consumers must pay an additional fee to access access-controlled content. In this respect, pay-TV resembles other media such as newspapers, cinema or books that are not accessible for free. In a more general sense, 'free' can also mean that consumers must first obtain the platform controller's authorization in order to access the services. Finally, 'free' can also refer to the fact that access to services is not technically free. For example, the transmitted signal is encrypted and requires consumers to have specific decryption equipment such as a set top box to view the content. Hence, in areas in which the majority of consumers do not yet have the necessary decryption equipment, access to access-controlled broadcasting services will not, for technical reasons, be free. As Owen words it:

> 'On a more fundamental level, program choice might be usefully made more responsive to viewer welfare, and less responsive to the notions put forward by philanthropic institutions of what people "ought" to see'.[76]

On the other hand, with the proliferation of digital TV and the encoding of public broadcasting services for copyright reasons,[77] households will, sooner or later, have to install a set top box that converts and decrypts digital signals.

---

[76] Owen 1975, p. 134.

[77] This is already a reality in countries such as Denmark, Switzerland and the UK.

Added Value and Responsiveness
Pay-TV services are still intended for reception by the public providing the consumer agrees to comply with the terms and conditions for access.[78] Moreover, access-controlled services that are marketed via pay-TV platforms can result in more differentiated and specialized service offerings for consumers. Electronic access control can render new services that were hitherto impossible, because of the dependency on advertisers and large audience shares, economically viable. Examples are minority programmes and programming with a high cultural and educational quality, providing there are sufficient interested or wealthy subscribers.

The commercialization and individualization of content services could also provide consumers with a way to express their preferences using a means the market understands: money.[79] In so doing, electronic access control could contribute to making broadcasting more responsive to consumer preferences and interests.[80] Another question is if pay-TV providers will serve the preferences of the largest share of subscribers or offer a real choice. This is a question that will be addressed later in more depth.

1.5.2. Consumer Access to Access-Controlled Services

The ability to manage individual consumer relationships also gives pay-TV operators more power, if not a monopoly position over individual subscribers. Broadcasting viewers are confronted with a new set of problems that are also new to the broadcasting sector. Electronic access control can influence the if and how consumers access content, starting with the question if a consumer can afford to subscribe to a certain service or if he or she has the required reception equipment. This not being the case, he or she is excluded from the service. As Chapter 2 will show, the aspect of exclusion from content that is distributed on exclusive and technically controlled terms has triggered a number of discussions on public information policy.[81] Public information policy in broadcasting is directed at ensuring wide accessibility and availability of broadcasting content. The mission of the broadcasting media is to provide the audience with content and to inform the entire population about newsworthy events 'the public has the right to receive'.[82] In some countries, such as the Netherlands, this mission seems to be carried out primarily by public broadcasting services. In others, for example, in Germany, France and the United Kingdom, commercial broadcasters also participate in this

---

78 In this sense, Dommering 1990, p. 64; Nauheim 2001, p. 131-132; Hins 1999 p. 249.
79 See extensively in section 2.2.1. More generally on the perception of digital television by consumers, Ledoux Book 2004, pp. 123-145.
80 See section 1.3.1.
81 See section 2.2.
82 European Court of Human Rights, Sunday Times, Strasbourg, 26 April 1979, Series A, No. 30 [hereinafter 'Sunday Times'], paragraph 65; European Court of Human Rights, Lingens, Strasbourg, 8 July 1986, series A No. 103 [hereinafter 'Lingens'], paragraph 41.

task It is part of the legislator's positive obligation to provide an environment in which these conditions are fulfilled.[83] Broadcasting threatens to lose its function as an omnipresent and omni-accessible information forum. Both the general availability of and the choice between information sources become questions of access and slip away into private control. More than ever, access to information depends on the pay-TV platform controller and less than ever on the broadcasters or media politicians. For governments, this means that private control of access to information comes with a loss of public control over broadcasting. How serious the issue is taken by regulators is illustrated by the adoption of a set of specific broadcasting rules at the European level. These rules are explained in Chapter 2.[84]

The other side of the coin is that the private management of an individualized relationship between service providers and consumers can influence the chances of rival service providers of gaining access to subscribers, just as it can influence the subscribers' willingness or ability to switch to a more competitive offering.[85] Monopolization of the subscriber base can have considerable impact on competition in the pay-TV sector, and as such will be a matter of competition regulation.

Subscribers are the most precious resource in pay-TV markets. Pay-TV providers depend on subscriptions and willing subscribers are scarce in particular in countries that have a well-functioning free-TV environment. It is not easy to win subscribers when consumers feel that free-TV offers them sufficient programming. Another reason why access to the consumer base is so precious has to do with the economic dynamics in pay-TV markets, namely the existence of first mover advantages, economies of scale[86] and of indirect network effects,[87] and the relatively high individual/collective adaptation costs in this sector. The reception of access-controlled services requires that consumers subscribe to a service platform and make some form of investment to acquire the necessary equipment associated with the conditional access platform (smart card reader, PC, set top box, etc.).[88]

83 Addressed in more detail in section 2.2.

84 See section 2.3.

85 The following section is based, among others, on the papers by Klemperer (1987); Farrell/Shapiro 1988; Shapiro 1999; Besen/Farrell 1994; Evans 2003; Bakos/Brynfolfsson 2000, 2002a and 1999; Galbiati/Nicita/Nizi 2004; Fritsch/Wein/Ewert 1999; Mackaay 1982; Nalebuff 2004.

86 Shapiro and Varian explain economies of scale with the words: 'the more you produce, the lower your average cost of productions', Shapiro/Varian 1999, p. 21. They further distinguish in supply-side economies of scale and demand-side economies of scale. Supply-side economies of scale are, according to Shapiro and Varian, the traditional economies of scale—larger firms tend to have lower unit costs, p. 179. Demand-side economies of scale refer to the fact that a product is widely used and because of that particularly valued by consumers: 'if everybody else uses Microsoft Word, that's even more reason for you to use it too', Shapiro/Varian 1999, p.180. According to the authors, they are the norm in information industries, p. 181. Besen and Farrell use the notion of demand-side economies of scale to describe (direct) network effects, Besen/Farrell 1994, p. 118

87 Clements describes indirect network effects as follows: 'a good becomes more valuable as more consumers use it because there is a greater variety of a complementary good available', Clements 2004, p. 2.

88 See section 1.4.2.

Consumers and content service providers will generally favour the most popular standard that promises the widest coverage. As a result, it is often the leading pay-TV platform that determines the technical standard. When a newcomer enters the market and starts offering services to the installed consumer base, the offering must be sufficiently attractive to justify the often high switching costs, which involve investing in additional consumer equipment, breaking long-term subscription contracts, the prospect of double subscription, etc. The second obstacle is the possible loss of indirect network benefits because if the new conditional access platform is incompatible with the currently available technology and not yet popular, it will have fewer applications and programmes to offer. Content providers may prefer to focus on offering content and writing applications for one dominant platform instead of investing in tailoring their applications to several incompatible conditional access platforms. This can lead to indirect network effects. If indirect network effects are strong, consumers might be reluctant to subscribe to a new system with incompatible technology unless it offers very clear improvements and other subscribers are expected to follow soon, thus creating the critical mass for the new service.[89]

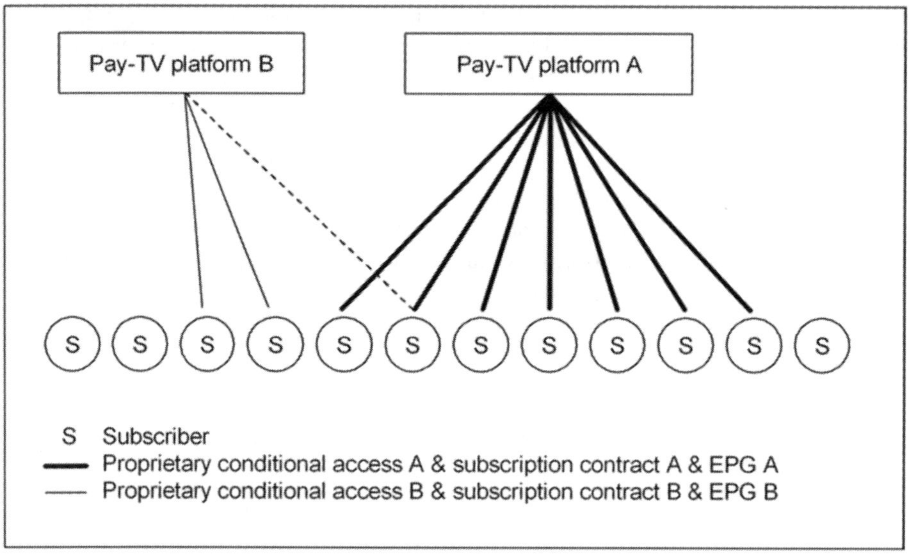

Figure 8—Subscriber Monopolization. Figure 8 illustrates a phenomenon the author refers to as the monopolization of the consumer base. Service provider A has established a lasting relationship with the subscribers to this platform by undergoing a contractual relationship with them, by controlling a proprietary

[89] See Shapiro 1999, p. 5.

standard in their set top box, and by controlling the EPG. In so doing, service provider A can make it more difficult for subscribers to service provider B to access services that are offered on platform A, because their set top box does not support B's standard. Likewise, subscribers to A might find it difficult to switch to services that are offered on platform B.

This is why a pay-TV platform operator can have a gateway position for access to the installed consumer base. Figure 8 shows that access to the installed consumer base is the real bottleneck in pay-TV markets.[90] Exclusive control over the dominant pay-TV-platform is an important means of monopolizing the consumer base, meaning binding subscribers and service providers to a particular platform and excluding other service providers from gaining access to their installed subscriber base. The ease with which consumers can switch between the services offered by a second platform B while they are subscribed to platform A can influence their decision to do so. Factors that may be relevant for them are the compatibility of B's services with A's set top box, whether they are bound to A by a long-term subscription contract or can terminate the contract any time, whether they can afford to subscribe to both A and B's services at the same time and, last but not least, whether they have sufficient information about B's services to make a decision.

Monopolization of the consumer base can be a result of control over a particular facility that is necessary for market entry and for which the operator of that platform holds a monopoly position. More commonly, monopolization of the consumer base will be the result of a combination of control over and the strength of a particular conditional access standard and associated facilities such as the billing system and the information agent (EPG), and the way access-controlled services are marketed through service platforms. The following paragraphs explain in more depth the instruments that pay-TV platforms can use to keep consumers in 'walled gardens' and discourage them from switching to competing providers: bottleneck control, technical and contractual lock-ins, audience fragmentation and, what the study calls, 'the information problem'.

Bottleneck Control

With the ongoing technical and organizational sophistication of electronic telecommunications services, market entry depends on an increasing number of facilities such as control over the conditional access platform or control over other elements of the technical platform, including APIs, Electronic Programme Guides, electronic payment systems, media players, etc. Other facilities required to provide pay-TV services are the marketing platform and programme rights such as those for the Soccer World Championship and new film releases.

[90] See also KEK 2000, 227pp.: Access to the consumers as most important strategic resource in the media.

'Bottleneck' can refer to such services or facilities, and at each level of the distribution chain, be it the transmission network, conditional access, the marketing platform or exclusive rights to content. The term 'bottleneck' is commonly used to describe a situation of monopoly control over a particular facility or service. This can be a temporary or a lasting monopoly situation, and the monopoly position can have its source in legal reasons such as the control over exclusive transmission rights. It can also have its source in practical reasons—the resource cannot be easily duplicated—or in economic reasons that have to do with the degree of market power of the operator of that facility.[91] The precise identification of bottlenecks is thus not straightforward, but depends on a dynamic analysis of a particular situation. This is also why it can be so difficult to identify and remedy bottleneck situations ex ante, as will be demonstrated in Chapter 4.[92]

Access to bottleneck facilities can be very interesting for newcomers to the pay-TV sector. If newcomers lack the financial means to bring in or establish their own technical resources, they will have to seek access to existing resources, possibly under comparably less favourable conditions than the established players arranged for them. The establishment of alternative facilities can be impeded by a number of obstacles such as high irreversible investments for the installation of conditional access and the creation of the necessary distribution and marketing structure, as well as the marketing costs. Other obstacles can be cost asymmetries in accessing necessary resources such as programme rights, and higher production costs or a large subscriber base for the first mover.

From the perspective of the incumbent, exclusive control over bottleneck facilities or the standard technology provides a range of opportunities to impede potential and actual competitors, in particular when exercised by a powerful market player. Possibilities of influencing competition to one's advantage range from plain denial of access to bottleneck facilities or the imposition of unfavourable conditions such as proprietary design, discriminatory behaviour, predatory pricing, bundling strategies, or taking advantage of a stronger negotiating position when purchasing programme rights. The legitimacy of such strategies will be examined in more depth in Chapters 3 and 4.

From a competition policy standpoint, a monopoly situation is not automatically harmful or undesirable. Monopoly control is considered harmful when such control is abused to the detriment of functioning competition. A monopolist that abuses monopoly power might be in conflict with competition law. Chapter 3 explains how competition law can be used to intervene in such cases. Even without abuse, however, monopoly control in broadcasting markets can be undesirable from a general public information policy point of view. In broadcasting law in particular, monopoly positions are regarded with caution because they can conflict with major

[91] According to Poel and Hawkins, a bottleneck exists 'where the availability and/or terms of access to a particular network facility or service environment fall below a benchmark or standard that has been deemed to be in the public interest', Poel/Hawkins 2001, p. 73.

[92] See sections 4.3. and 4.8.6.

broadcasting policy goals such as preventing a player from exercising too much journalistic influence on its audience and negatively affecting pluralism and diversity. This is why bottleneck situations in broadcasting raise concerns in the area of competition law as well as in the area of broadcasting regulation. This latter aspect will be discussed more in depth in Chapter 2.

Technical Lock-ins

In pay-TV, the economic power of a technical platform or elements thereof can be influenced by the popularity of a certain embedded standard. This has to do with the close economic links that are often found between the technical and the service platforms, with the aforementioned dynamics of the market in general, and the influence of indirect network effects and first mover advantages in particular.[93] The consequence is that once an operator of a particular conditional access system has succeeded in establishing a dominant standard, (a) operators of access-controlled services may depend on compatibility with the dominant standard to reach a wide audience, and (b) the success and acceptance of a competing conditional access facility will depend to a considerable extent on whether consumers are able to switch between platforms. This is the argument behind the so-called decoder towers, namely the assumption that consumers are less likely to subscribe to a second pay-TV platform if both platforms require the purchase of different incompatible set top boxes. Arguably, the importance of this argument will vanish if set top boxes are offered at lower prices or are subsidized by the pay-TV platform operator.

Should the first platform use a proprietary standard and the second platform is not compatible, consumers would risk losing indirect network benefits. The newcomer's chances of entering the market would depend on the enterprise's ability to overcome this obstacle.[94] Exclusive control over a technical standard is therefore an important means of binding subscribers and content producers to a particular service platform, and of preventing other service providers from gaining access to the consumer.

Closely related is the aspect of audience fragmentation, which could lead to a reduction of the number of subscribers available to a newcomer. Arguably, digitization will favour the development of more specialized niche channels and hence increase the fragmentation of the consumer base. The use of electronic access control can further contribute to this process by dividing the audience into different zones of incompatible conditional access standards. Audience fragmentation can also take place along national borders. Today's access-controlled services such as pay-TV are often restricted to a national territory and the required smart cards are only sold to residents.[95] This is often due to the licensing practice of content rights,

93 Shapiro 1999, 3pp. About indirect network effects in two-sided markets, see Evans 2003, p. 32.

94 See Shapiro 1999, p. 4; Besen/Farrell 1994, 118pp. and 121pp.

95 See, for example, the subscriber conditions at <www.sky.com/ordersky/home> and <www.canalplus.nl> (last visited on 14 March 2005).

which are often issued on a territorial or language basis. Other reasons are divergent broadcasting laws (for example, in youth protection), and the character of a service as national service for citizens of that state (such as public fee-financed broadcasting). The effect of the use of conditional access systems can be to re-install territorial borders in trans-border media such as satellite distribution. A Danish citizen living in France, for example, may be prevented from accessing the encrypted Danish public service broadcasts of DR1 and DR2, and hence from accessing information from his or her home country and cultural heritage. On the other hand, in countries with no or little own programming such as Luxembourg, using electronic access control in neighbouring countries could prevent those countries' programmes from being broadcast in Luxembourg.[96]

A somewhat grotesque side-effect of using conditional access for exclusive licensing purposes is that consumers in one country can be excluded from the reception of particular events that are transmitted in their own country. This was the case with the satellite transmission of the Soccer World Championship in 2002: Kirch owned the German transmission rights for the Soccer World Championship and sold them to the German public service broadcasting for transmission in Germany and to a pay-TV provider in Spain for transmission in Spain. The Spanish pay-TV provider successfully opposed the German public service's plan to show the games on digital satellite television free-to-air, as this would mean that the games would no longer be exclusively available on Spanish pay-TV.[97]

Finally, another form of technical lock-in is a complex technical design of consumer equipment because it makes switching difficult. For example, research into the available EPGs and set top boxes shows that they, voluntarily or involuntarily, make it very difficult for users to modify predefined settings.[98] Moreover, once consumers have invested in learning how to use one technology, this can be a further reason to prevent them from switching to a different technology.[99]

Contractual Lock-ins

Another strategy used to monopolize the consumer base and to which relatively little attention has been paid in the pay-TV discussion, is that of contractual consumer lock-ins. Contractually locking in consumers describes a situation where subscription contracts are designed to discourage consumers from switching to a competitor.

---

96 Unless broadcasters from neighbouring countries will acquire the additional transmission rights for Luxembourg.

97 German public programming can be received in Spain via satellite.

98 See the study by Jürgens 2002.

99 See also Klemperer 1987 p. 376.

In this context, the duration of the subscription contract is important as well as the ease with which consumers can terminate the agreement.[100] Binding consumers to long-term subscription contracts or making it difficult to terminate the contract is a form of bundling in time. For pay-TV this is usually twelve to twenty-four months. This time frame may have a negative effect on the consumers' mobility and willingness to switch to a competitor before the end of their initial contract.[101] Further research is needed to determine how long the duration of a contract must be before it discourages a consumer from entering into a second contract.[102] Subscription contracts frequently foresee very far-reaching provisions about their automatic extension that are not always easy to detect. [103] Contractual conditions that 'sanction' a termination of the contract can also have a discouraging effect. Examples include an obligation to return a set top box at the end of the contract, or the loss of an email address.[104] Here, terminating the contract has the additional consequence of effectively barring consumers from receiving any digital, access-controlled or other information services before they have invested in new equipment and/or services. Such contractual conditions may be legitimate, reasonable and common in other sectors (for example, mobile phone subscriptions), but they can prevent competing providers of pay-TV services to reach the critical mass of consumers necessary to render their service economically viable.[105] On the other hand, as Shapiro notes, eventually exclusivity provisions can work against the first mover, namely if a second entrant is sufficiently strong and consumers decide to enter into an exclusive relationship with him.[106]

Sub-forms of contractual lock-ins are the aforementioned programme bundling strategies that oblige consumers to subscribe to a whole package of services even if they only wish to access one particular channel, or that make the provision of certain information services (for example, premium channels) conditional upon the subscription to others (such as basic channels).[107] Again, this can discourage

[100] Aghion/Bolton 1985, 389pp. (making a distinction between nominal length and effective length of a contract). See also Farrell/Shapiro 1988, p. 125; Klemperer 1987, p. 376

[101] Differentiating Aghion/Bolton 1985, p. 399.

[102] See also Aghion/Bolton 1985, p. 399, pointing to some difficulties.

[103] See, for example, BBC World Service, Terms and Conditions, No. 2 (Terms): 'The Agreement shall be automatically extended for further periods of twelve months, subject to payment of the Subscription by the Subscriber, unless terminated by either party giving to the other party not less than fifteen days written notice to expire on the last day of the then current term'. Also: Canalplus, Algemene voorwaarden van Canal+ N.V. voor de doorgifte en ontvangst van televisieprogramma's via de kabel en voor de doorgifte en ontvangst van digitale aardse televisiesignalen via de infrastructuur van Digitenne [hereinafter 'Terms and conditions Canal+ Nederland'], No. A4, available at <www.canalplus.nl> (last visited on 14 March 2005). Note in the small print that the contract must be terminated by registered letter.

[104] See Canal+, Terms and conditions Canal+ Nederland, No. C18. See also Shapiro 1999, p. 11.

[105] See Shapiro 1999, p. 8. See also the paper by Bakos/Brynfolfsson 2000; Farell/Shapiro 1989 and Nalebuff 2004, 26pp.

[106] See Shapiro 1999, p. 10.

[107] Section 1.4.3. On the effects for consumer switching costs in case of pay-TV bundling see also Galbiati/Nicita/Nizi, p. 23; Harbord/Szymanski 2005.

consumers from subscribing to additional services that resemble services that they already have in their package.

Information Problem

An important factor for consumer choice is information about what is on offer under which conditions and for which prices.[108] As the example of Google.com or Yahoo.com demonstrates for the internet, control over access to information about content or content services can have an effect that is similar to control over access to content. In the telecommunications sector, it is an acknowledged fact that functioning competition and the consumer's ability to choose between different operators depend on the availability of adequate service information. It is not enough that consumers are technically free to switch between different services and interoperable platforms; they must also be able to access information about the choices available to them.

The significance of access to comparable information about content services is even greater in an access-controlled information environment. In the unencrypted world, consumers can search and choose freely, for example, by flicking through broadcasting channels. When consumers come across channels or programmes that are subject to electronic access control, it will be difficult for them to determine if the content they contain is relevant because it is encrypted or otherwise protected against access. This is even truer in the case of electronic access control to multi-channel service platforms such as pay-TV. Here, consumers find themselves in front of closed doors (knowing that the marketplace lies somewhere behind them). The opposite is also true: how will consumers who subscribe to one service platform know about the services available outside the 'walled garden'? This was, for example, Disney's complaint in a US case against AOL. According to Disney, AOL made it a condition for purchasing placement on the AOL service portal that content providers disable hyperlinks to unaffiliated websites and guarantee that no more than a fixed percentage of traffic at a site within the AOL network be 'diverted' via hyperlinks to sites outside the portal.[109] The less overview consumers have of the marketplace (due to a lack of information), the more they rely on electronic information agents to find what they are looking for. The enormous amount of available information increases the demand for information agents such as EPGs that provide a pre-selection.

---

[108] Fritsch/Wein/Ewert 1999, p. 294.

[109] US Federal Communications Commission (FCC), decision from 11 January 2001 in the matter of applications for consent to the transfer of control of licenses and section 214 authorizations by Time Warner Inc. and America Online, Inc., transferors, to AOL Time Warner Inc., transferee, Memorandum Opinion and order, 22 January, 2001, CS Docket No. 00-30, paragraphs 128-190, available at <http://www.fcc.gov/Bureaus/Cable/Orders/2001/fcc01012.pdf> (last visited on 20 March 2005). See also the statement of Commissioner Michael K. Powell, concurring in part and dissenting in part, from 22 January 2001, available at < http://www.fcc.gov/Speeches/Powell/Statements/2001/stmkp104.doc> (last visited on 20 March 2005).

The power of proprietary electronic information agents lies not only in their ability to present certain programmes more favourably, personally[110] or associatively, but also, and especially, in their ability to generate a biased idea of the available offerings.[111] Unlike, for example, Google.com, most existing EPGs are controlled by the operator of the access-controlled platform and are not open to third parties. EPG controllers can design the EPG in a way that makes it easy to find their own services, but difficult if not impossible to find a rival's service or perform a comparison. This can give EPG controllers enormous potential to manipulate the way consumers access and receive content, particularly where no independent alternatives that would allow consumers to compare different services are available.[112]

1.5.3. CONDITIONAL ACCESS AND COMPETITION

Even if an enterprise has the ability to use monopoly control to the disadvantage of its competitors, it does not say anything about its intention to do so. Obviously, enterprises are not automatically interested in anti-competitive behaviour.

Leverage

One reason that is at the core of many explanations of why an enterprise would engage in potentially anti-competitive practices is leverage. Leverage can be defined as a mechanism whereby a firm with monopoly power in one market can use its power as a lever to foreclose sales in, and thereby monopolize, a second market (hereinafter 'related').[113] Such 'levers' can have different forms such as control over a bottleneck facility, a vertical merger transaction or bundling practices. There could be a case of leverage if the operator of a dominant pay-TV platform refuses to grant a rival access to the conditional access system it is using thereby making it factually impossible for the rival to enter the pay-TV market. Leverage could be the motive behind the same provider making the sale of set top boxes to consumers dependent on a subscription; or if a subscription service provider that offers both premium channels and basic channels makes the subscription to a premium channel conditional on the subscription to a basic channel. The question if, in a concrete case, leverage is the motive behind any of

---

110 Also O'Driscoll 2000, p. 217. For example, one international provider of iTV software solutions recently launched an EPG application that draws viewers into the iTV experience by letting them explore the kind of mood they are in by asking a couple of simple questions such as 'How about leaving on a space flight tomorrow?' or 'Aliens have landed. You're frightened? or You're excited?' The system then makes recommendations to help them find the programmes they 'feel like' watching. The extraordinary potential for gaining more control over consumer decisions is obvious.

111 See also Mackaay 1982, p. 161.

112 Even where alternative EPGs are available, they may be not supported by the technical platform that consumers are subscribed to.

113 Whinston 1990, p. 837.

these practices, or if there are other more likely explanations for the monopolist's behaviour, is for the general competition law authorities or courts to determine.

It was due to the leverage theory that courts and competition authorities in Europe and also the United States tended to treat certain business methods such as tying or vertical mergers particularly harshly. Because the concept of leverage is very relevant to the way electronic access control was treated in the merger decisions of the European Commission and in the sector-specific regulation of conditional access in pay-TV markets, the following paragraphs use two examples to discuss the leverage theory. Although it would clearly exceed the scope of this study to provide a comprehensive picture of the economic and legal arguments accompanying the leverage theory, this is the place to provide at least an overview of the discussion:

Example 1—Pay-TV operator A operates a fairly popular pay-TV platform using its own proprietary conditional access system A, which happens to be the dominant standard for conditional access systems for pay-TV in its region. A makes subscription to its pay-TV platform dependent on the consumers' willingness to purchase its set top box. The set top box is equipped with a technology that can process A's proprietary conditional access technology. There are a few other set top boxes on the market made by independent producers that can process different conditional access standards, but not A's.[114]

In other words, A bundles the subscription to the pay-TV service and the purchase of its set top box technology.[115] If A dominates the market for conditional access solutions and is at the same time active in the market for access-controlled services, A could be in a position to make the life of rivals in the service market more difficult. Rival B also offers access-controlled services in a different, less popular, conditional access standard B. B's standard is, however, supported by a number of independent set top box producers. For B, it could be nevertheless attractive and cost-effective to direct its market strategy at consumers that already own a set top box, namely box A, because of the size of the consumer base. However, conditional access system A is proprietary, and A's set top boxes are not compatible with B's conditional access standard. Subscribers to platform A might, in principle, be interested in subscribing to B too, but that would mean having to buy a set top box that also supports B's technology. Because they already bought set top box A as part of A's bundle, they will probably think twice before subscribing to B, and eventually even decide against it. The case would have looked different if consumers had been free from the beginning to buy set top boxes that support different conditional access standards, namely both A's and B's.

By bundling the subscription and the set top box, A might have leveraged its power over the dominant conditional access standard A to make it more difficult for

[114] For the purpose of this example, the market for pay-TV services and set top box technology are two different markets.

[115] For an economic analysis of bundling strategies, see the papers by Bakos/Brynfolfsson 1999; Liebowitz 1983 (bundling as a form of risk reduction); Hinloopen 2002; Nalebuff 1999; Choi 2004.

B to enter the pay-TV market. B's market entry could be discouraged entirely, or it would leave only a limited customer base to the entrant.[116] Even if it were principally possible for B to remain in the market, A's strategy could substantially lower B's profit expectations.

The strictest opponents of the leverage theory, namely the followers of the Chicago School, would probably warn us from being too quick to assume that the bundling strategy used in this particular case is the result of monopoly power regarding the conditional access standard A and the wish of A to leverage market power over the pay-TV market.[117] At the heart of the Chicago School's critical view of the leverage theory of bundling is the presumption that if an existing degree of monopoly power is legitimately obtained, then so is the profit deriving from it. A could not profitably exercise more monopoly power than it actually had in total. The fact that A has a dominant position for a particular conditional access standard does not imply that it can hinder rivals in the pay-TV market to attract subscribers. A Chicago scholar might argue that it would be more likely for A to adopt this particular strategy because it could maximize its profits by selling subscriptions and set top boxes.[118] According to Posner, if a judge held, on grounds of the leverage theory, that unlawful bundling restricts competition, this would be

> 'tantamount to saying that any time a monopolist decides to handle a step in the production process internally rather than to invite competitive bids, he is guilty of monopolizing because he is unnecessarily restricting competition. This is not the general rule, and it makes no sense to apply it only to the tying context'.[119]

The criticism expressed by the Chicago School was criticized by the so-called 'Post-Chicago School'. Both the Chicago School and the Post-Chicago School critics finally paved the way for a more differentiated view of the leverage theory. One major point of criticism of Post-Chicago scholars of the Chicago School's criticism was that the Chicago School replaced simplistic self-evident arguments—why certain practices should be considered illegal because they triggered leverage—with self-evident arguments that were no less simplistic—why leverage will not occur.[120]

Post-Chicago scholars carried on to develop the Chicago School's criticism to a more differentiated, dynamic and case-by-case-based application of the leverage theory. As different as the arguments are, there seems to be some agreement that a number of market conditions must be fulfilled before one can speak of leverage: namely sufficient market power on the market for set top boxes and for pay-TV

---

116 Nalebuff 1999, 2pp. Instructive also Choi 2004, 87 pp.
117 See Posner 1976, pp. 172-173.
118 Posner 1976, p. 173, 197.
119 Posner 1976, p. 175.
120 Winston 1990, p. 838; Nalebuff 1999, p. 4; Choi 2004, p. 5; Kaplow 1985, 520pp. and Yoo 2002, 201pp.

services, as well as the presence of barriers to market entry in the pay-TV market. Were the pay-TV market fully competitive, it would be difficult to see how A could leverage market power and obtain a monopoly in the pay-TV market. Market entry barriers in the pay-TV sector can be high if newcomers are required to make high irreversible investments for the installation of conditional access systems and the creation of the distribution and marketing structures required to build a reputation and display goodwill, or to obtain the necessary resources and programme rights.[121]

And even if pay-TV platform A had sufficient market power, A would not necessarily employ this bundling strategy to leverage market power if such behaviour were not profitable.[122] Bundling the subscription with the pay-TV platform and the purchase of the set top box could also reduce A's profits, for example, if he were to miss out on a number of subscribers that do not value A's pay-TV platform enough to purchase a set top box that can process only A's services. A would thus have to have additional and probably strategic reasons to engage in anti-competitive bundling. In the provided example, such reasons could be to drive B out of the market or make market entry more difficult.[123] The situation would be different if A produced and sold non-proprietary conditional access solutions. In this case, it would probably be in A's interest to encourage new entries in the pay-TV market because this would result in higher sales of its decoder technology.[124] One must also take the impact of self-correcting market processes into account.[125] If a new pay-TV operator whose platform is more innovative and attractive than A's were to enter the market, A's bundling strategy would not make much sense anymore.

Even if a monopolist has incentives and believes it is profitable to engage in bundling, this does not necessarily mean that such a strategy is undesirable and should be banned. There may be efficiency reasons that justify such behaviour and explain why such behaviour should be permitted.[126] At the beginning of pay-TV deployment in particular, operators face considerable risks and high costs. They need to reach a critical mass of subscribers. A major obstacle to achieving this could be their inability to convince consumers to invest in set top box technology. Offering subscriptions and set top boxes in an attractively priced bundle could make it easier to overcome such obstacles.

Example 2—The importance of efficiency and consumer welfare considerations is demonstrated in the second example, namely that of channel bundling. Pay-TV platform operator C is dominant in the market for retail and wholesale programme content. C sells its programmes to cable operators and other broadcasters who offer

121 For a concise general overview of the discussion, see the paper by Yoo 2002.
122 See also Choi 2004, p. 5, 11.
123 Whinston 1990, p. 840, Yoo 2002, 206pp.
124 Whinston 1990, p. 850.
125 Kaplow 1985, pp. 525-552. See also Klemperer 1987, p. 377, pointing to the effect of intensive competition to attract new subscribers on monopolistic returns.
126 Instructive is the article by Yoo 2002, 200pp.; Larouche 2000, pp. 196-203.

access-controlled services. Besides selling third-party programmes, C also distributes a number of niche channels for sports, films, cooking, gardening and news via its own pay-TV platform. C offers these channels in bundles, in the form of two so-called basic packages: a lifestyle package and a hobby package. C does not offer the channels unbundled to consumers. There are also a number of access-controlled service providers that offer similar niche channels separately.

A well-considered channel bundling strategy would enable C to seize a prominent position in the market in form of packaging the product space and give it a competitive advantage over rivals that offer similar channels separately.[127] The better a platform succeeds in answering audience demand by providing more diverse and broad services, the less room it leaves for alternative offerings from competing information service providers.[128] Service platforms can use such strategies to influence the market chances of competing providers of niche channels, and more widely, those of free-TV providers or providers of E-commerce services.[129] C can use bundling strategies to leverage market power from the wholesale programme market to the retail pay-TV market.

However, large-scale bundling can also be a way to improve the efficiency of C's platform. Bundling can exclude much of the uncertainty about the consumers' valuation for individual channels, which can be a major factor of uncertainty when pricing services and performing transactions.[130] Bundling can, as described previously, also be used to differentiate between different consumer segments that are willing to pay different prices.[131] The role of electronic access control in this context is to enable the cost-effective delivery of services to those consumer segments that have been identified as a particular category of users to which a certain price or service has been allocated.[132] Bundling different channels can also be a means to save production costs and benefit from economies of scales. An important aspect in this context is the way content rights are licensed, namely as presale,[133] package [134] or output[135] deals that are meant to reduce the production risks

---

[127] Wirtz 1994, p. 49. See also Noam 1988, p. 208: differentiation is the rationale of a commercial programming strategy. Neumann 1998, p. 199 pp.

[128] See the paper by Bakos and Brynjolfsson, Bakos/Brynjolfsson 2002a.

[129] The European Commission already observed: 'When a television operator has a leading position in pay-TV and free-TV, and also holds the main programme rights for free-TV and pay-TV, he is in a position to control the interaction between free-TV and pay-TV', European Commission, Commission Decision of 27 May 1998 relating to a proceeding pursuant to Council regulation (EEC) No. 4064/89 (Case IV/M.993–Bertelsmann/Kirch/Premiere), Brussels, 27 February 1999, OJ L 52, p. 1 [hereinafter 'Bertelsmann/Kirch/Premiere'], paragraph 88 and paragraphs 97-99.

[130] See the papers by Bakos and Brynjolfsson, Bakos/Brynjolfsson 1999; Bakos/Brynjolfsson 2000; Nalebuff 1999, p. 19.

[131] Bakos/Brynjolfsson 1999, pp. 15-17.

[132] Differentiating, Nalebuff 1999, p. 16.

[133] Presale refers to the licensing of programmes that are yet to be produced.

[134] The selling of programme rights in packages, where packages are usually composed of both more and less attractive content.

for production companies. For licensees, these deals can be advantageous as they guarantee programme supply while putting such content out of the competitors' reach. Generally, it is the larger operators who can afford to invest the often considerable sums of money in rights packages and take the risks involved—risks they can re-allocate to consumers through differentiated bundling strategies.[136]

Of course, there are limits to the profitability of such bundling strategies, too. Obviously, a platform operator will not sell content as part of a bundle if it can be sold more profitably separately. First film releases or top sports events, for example, are often sold in the form of pay-per-view or premium offers. Because consumers value such content particularly highly, it is also possible to obtain higher revenues by selling it separately.[137] Moreover, the transmission rights of such content are often particularly expensive, which could justify selling them separately and at a higher price than could be asked for if they were offered in a bundle. The costs of programme rights also limit the size of bundles; from a certain moment onwards packages would be too large to offer to consumers at an attractive price. Bundles would then comprise elements that consumers do not value highly, or the bundles would be so large that consumers would have difficulty finding what they are looking for.

Within the context of the leverage theory of bundling, the possible welfare-enhancing effects should be taken into account before banning a particular practice for the sake of functioning competition.[138] It is helpful to distinguish between the effects for consumers and competitors.

Effects for Consumers

Bundling can have both negative and positive effects on consumer welfare. On the one hand, it was explained before that bundling might, from an economic point of view, enhance the efficiency, value and compatibility of a particular service.[139] Subscribers to one of C's basic packages could have access to not only one, but several different channels that fulfil their interests, and access to a package could be cheaper than subscribing separately to similar channels from competitors. This also depends, of course, on C's pricing strategy. To this extent, bundling can eventually lead to more advantageous and larger service offerings for consumers at a lower price and hence increase consumer welfare.

On the other hand, subscribing to a whole bouquet of channels also deprives the consumer of choice: the choice not to take certain channels he or she is not interested in or to take certain channels from third-party providers and so to arrange

[135] Combination of presale and package deal, where licensed packages include the future output or parts thereof, of a production firm. For a general discussion of these practices, see Neumann 1998, 165pp.

[136] Bakos/Brynjolfsson 1999, p. 18.

[137] Bakos/Brynjolfsson 1999, pp. 13-14.

[138] For an in-depth discussion, see the papers of Nalebuff 1999; Bakos/Brynjolfsson 1999; Katz/Shapiro 1998, p. 40pp.; Van Geffen/Nooij/Theeuwes, 95pp., with further references; Bishop/Walker 1999, 209pp. and 292 pp.

[139] Critical Klemperer 1987, p. 877.

his or her own bouquet. For example, a study conducted for the UK Consumers' Association concluded that to watch live Premiere League soccer consumers would have to subscribe to either BSkyB or a cable service. Because of the construction of programming bundles, however, they were forced to purchase a bundle of content including much more than live Premier League soccer.[140] In general, bundling arrangements are less advantageous for consumers with specialized interests, as they might end up paying a proportionally higher price than if the services were offered individually.[141] Bakos and Brynjolfsson explain that the profitability and efficiency benefits of bundling are easiest to quantify when the consumer valuations, meaning preferences for distinct programmes, are identically distributed and not closely correlated for different products. This might mean that access-controlled bundles, similar to television advertising, are again orientated towards mass rather than individual preferences.[142] Bundling different services to large packages in a subscription contract is one means of binding consumers to a particular operator's offer. Providing the package is a) sufficiently broad to cover most of the consumers' interests and b) switching between different service platforms/packages is made difficult because of contractual or technical obstacles, the consumer might find himself or herself in a lock-in situation.

A variation on this theme is a situation in which C does not bundle different channels as basic channels, but offers premium channels with popular sports events and new film releases, and makes subscription to the premium channel dependent on the subscription to a basic channel. Consumers who are only interested in C's premium channel but would like to have a basic channel from another provider must choose between taking both channels from C, although they want only the premium channel, or foregoing the premium channel and taking the competitor's basic channel.[143]

Effects for Competition

The possible effects of bundling for competitors also should be taken into account. The aforementioned efficiency reasons can make it more attractive for smaller third-party service providers with only a small range of content services to include their services in the bundle of a larger platform instead of selling them separately and directly to consumers. This also explains why it can be vital for smaller no-name niche service providers to be included in the package of C's popular platform.[144] In the premium channel variant, participation in C's basic channel could be even more attractive if competitors knew that C's premium channel is so popular that it will attract even more subscribers.

[140] Harbord/Szymanski 2005, pp. 8, 21.
[141] Oftel 1997a, paragraphs 6.1- 6.21.
[142] Bakos/Brynjolfsson 1999, p. 19,
[143] They could, of course, decide to subscribe to both.
[144] KEK 2000, p. 228.

As explained, however, bundling can also directly affect the economic viability of competitors because it can reduce the number of subscribers that are willing to subscribe to their niche channel. One of the controversies in pay-TV is whether a pay-TV operator should be allowed to make the subscription to a premium channel dependent on the subscription to a basic channel from the same operator. Depending on how attractive the premium channel is, such a strategy can make it more difficult for operators of rival basic channels without a similarly attractive premium channel to sell their services to consumers.[145]

It is not the task of this study to decide in favour of one or the other argument—this is a task for the legislators and regulatory and competition authorities. The purpose of this overview is to show that intrinsic statements that certain market practices such as bundling are competitive or anti-competitive should be treated with caution as they may not reflect market realities and in the worst case discourage possible pro-competitive, welfare and efficiency-enhancing behaviour. This makes the regulation in the pay-TV sector, as in many other dynamic sectors, particularly difficult.

More generally, differentiated considerations, especially where the argument of leverage is concerned, are not only necessary in the context of bundling, but also in the context of other possible strategies to leverage market power such as control over bottlenecks or vertical mergers.[146]

Vertical Integration

It is often argued that leverage in the form of bundling or access refusals would be particularly likely to occur where an enterprise is vertically integrated and thus active in two related markets because such enterprises would not only have the ability, but also the incentive to engage in anti-competitive strategies. This argument is often heard in the pay-TV discussion. König, for example, speaks for many when he claims that the risk of anti-competitive behaviour depends on the state of the vertical integration of digital service providers in the distribution chain,[147] and recommends that, in order to approach the 'competition destroying' vertical integration in pay-TV markets, the regulator separate as much as possible the different levels in the distribution chain. The analysis of the early decisions of the European Commission's merger decisions in pay-TV markets in Chapter 3 shows that the European Commission, too, was, in the beginning, quick to ban the emergence of vertical structures in the pay-TV sector. It argued that such structures could invite enterprises to leverage their market power in the form of bottleneck control or other strategies.[148]

---

[145] Breuning 2000, p. 380: 'Vernetzen statt Versparten'.

[146] Whinston 1990, p. 856; Yoo 2002, p. 244.

[147] König 1997, p. 89; Neumann 1998, p. 239; the study of Lemley/Lessig 2000 for the broadband sector. More differentiated, Owen 1975, 888 pp.

[148] See section 3.3.1.

It is important to remember, however, that whether an enterprise that is dominant in one market engages in anti-competitive activities in another market depends on the structure of both markets as well as on the profitability of such behaviour. The previous section sought to illustrate the need for a differentiated view of the argument of leverage using the example of bundling strategies. Furthermore, there can be valid efficiency and welfare arguments that need to be taken into account before making a final decision to step in against particular vertical structures or vertical mergers.[149] Although it would be going too far to provide a comprehensive discussion of the economic aspects of vertical integration this is the place to say at least so much: the discussion of the impact of vertical integration and the resulting threat for leverage is ambivalent.[150] Vertical integration in the form of mergers, internal growth, agreements, etc., can simply flow from a legitimate business strategy. One example would be to aggregate content services and market them to subscribers via a pay-TV platform: here, exercising joint control over the technical and the service platform as well as the content delivered via these platforms could create efficiencies. The control over different elements in the distribution chain could enable the realization of secure cost-efficient access to supply and distribution channels, possibly under more favourable conditions than those generated by the free market. Due to the optimization of internal processes and cost structures, a vertically integrated pay-TV platform could offer similar services at more attractive conditions and so improve its competitive chances. There is nothing wrong with creating efficiencies. This is also why powerful vertically integrated market players can be important drivers behind innovation and progress in national markets. That is to say, vertical integration not only provides opportunities and incentives for anti-competitive behaviour, it can also enhance welfare and economic efficiency. The final decision whether vertically integrated structures are desirable from a public policy point of view must balance the different arguments.

1.6. Conclusion

The introduction of conditional access is changing the traditional 'broadcasting' structure. Conditional access introduces a new gateway or platform to the telecommunications process after a programme is made. Access to media content becomes subject to restrictions and depends on the controller of the pay-TV platform. The pay-TV business model focuses on managing exclusive relationships with consumers so that only subscribers to a platform can access particular content. Because pay-TV subscribers are scarce, platforms are competing for a limited audience. Monopolization of the consumer base in the form of bundling strategies,

---

[149] See the papers by Waterman/Weiss 1996; Speta 2004. See also Nocke/Peitz/Stahl 2004, pp. 14-15 and 18-21.

[150] See the papers by Waterman/Weiss 1996; Litman 1997; Speta 2004; Yoo 2002. See also Nocke/Peitz/Stahl 2004, pp. 14-15 and 18-21; Owen 1989, 888pp.

technical or contractual lock-ins as well as manipulating consumers' search patterns and increasing their search costs are strategies to improve one's competitive position. For the regulators, one of the challenges consists of identifying the situations in which such strategies are anti-competitive, meaning that interference is necessary to protect functioning competition. A second challenge consists of formulating rules that are sufficiently flexible to prevent overregulation and negative effects on innovation, efficiency and welfare. This is the subject of Chapters 3 and 4.

However, the regulation of pay-TV is not only a matter of competition. Public information policy plays an important role, too, and faces different challenges. Existing regulatory measures concerning conditional access in broadcasting are a mixture of competition and public information policy considerations. The fragmentation of the audience into unsubscribed and subscribed consumers could create information lock-ins and lock-outs. States have a positive obligation to create an environment in which consumers can benefit from freedom of expression and are not locked out from matters of interest, and in which democratic principles and other public information policy considerations are safeguarded. Private subscriber monopolies can also raise issues of pluralism and diversity within pay-TV platforms, issues that will be dealt with in more depth in Chapter 2.

Chapters 2, 3 and 4 examine the current European approach to the regulation of pay-TV in general and conditional access in particular within the context of European competition and public information policy. No literature was consulted after 1 December 2004.

1.7. Outline of the Study

Chapter 2—examines the use of electronic access control within the context of public information policy and broadcasting regulation at the European level. The first part of the chapter discusses wider public information policy goals and what the motives are for public intervention. The goal of this section is to identify the parameters for government involvement in conditional access in broadcasting. More specifically, it looks at electronic access control within the context of Article 10 of the European Convention on Human Rights (ECHR), and examines if there is such a thing as a fundamental right of access to information on pay-TV platforms that regulators are obliged to realize. It also examines whether such a claim can be derived from other public policy and democratic principles. The role of public broadcasting as a public interest carrier and the promotion of the free flow of services in the Internal Market are further aspects that are examined. In the second part of Chapter 2, existing instruments used to translate the identified public information policy objectives are described and critically analyzed. In so doing, Chapter 2 focuses on broadcasting regulation and takes a closer look at Article 3a of the Television Without Frontiers Directive (TWF Directive), the right to short reporting in Art. 9 of the European Convention on Transfrontier Television (ECTT) of the Council of Europe and on the 'must-carry' rules in Article 31 of the Universal Service Directive. Throughout the analysis, particular attention is paid to the situation of consumers, alias citizens, in broadcasting law.

Chapter 3—discusses the issue of conditional access control from a general competition law perspective. The chapter starts with a concise overview of European competition law and proceeds with an analysis of the main merger cases in the pay-TV sector over which the European Commission has decided so far. The next section takes a closer look at the applicability of Articles 81 and 82 of the EC Treaty to examples of anti-competitive behaviour that can occur within the context of pay-TV. The examples concern the refusal of access to facilities, discrimination and bundling. The main question is the extent to which general competition law is suitable to respond to the competition problems around electronic access control, which were outlined in Chapter 1. Non-economic aspects in general competition law and the position of consumers will also be dealt with.

Chapter 4—examines the regulation of conditional access in European telecommunications law. This is a critical analysis of Article 6 of the Access Directive, which stipulates mandatory access to conditional access systems. The main question in this chapter is whether Article 6 of the Access Directive can effectively guarantee the functioning of competition in pay-TV markets. Chapter 4 also takes a look at the alternative approach towards bottleneck regulation in Articles 8 to 13 of the Access Directive, which applies to all other telecommunications networks and facilities. The chapter compares both concepts and examines whether the current distinction between rules that apply to broadcasting and non-broadcasting facilities is justified and helpful. It further identifies at a more general level the aspects to consider when identifying critical bottleneck situations in digital pay-TV and explains why the current approach is not adequate to solve the bottleneck problems.

Chapter 5—provides a summary and conclusions. It contains findings of the study, including findings that could be relevant in areas other than pay-TV such as electronic access control in mass media more generally.

Chapter 2

Access-Controlled Broadcasting and the Free Flow of Information

'Information is the oxygen of democracy'. (ARTICLE 19)[151]

2.1. Locks in the Free Flow of Information

The notion of 'free-flowing information' is a picturesque allegory; a landscape in which information flows freely, like the water of a stream, and citizens can satisfy their thirst for information by drawing from an abundance of opinions, ideas and information. This is the environment in which the 'Information Society' finds fertile ground, and where mass media are the natural riverbed through which the information flows.

'The free flow of information' also embodies fundamental principles in media regulation in general and broadcasting in particular, namely by creating the conditions for an environment in which consumers, alias citizens, have access to a variety of informational content from different sources. The accessibility to and availability of information plays a role in every aspect of social life: the economy, the democratic process and control of state power, the functioning of an increasingly multicultural society, work, personal happiness and leisure. To be informed is the Information Society citizens' personal capital, and access to information is an important element that enables them to successfully manage their personal, professional and social lives. The mass media, such as broadcasting, play a vital role in the free flow of information. Broadcasting is one of the major and popular sources of information. Broadcasting, however, is not only a means for citizens to obtain information and inform others; it can also be an instrument for governments to inform citizens and convey their messages to the broad public. This has to do with the wide availability and accessibility of classical broadcasting services; once disseminated everybody who is connected to the distribution network and owns the necessary equipment is a potential recipient.

The introduction of conditional access in broadcasting brought turmoil to the information idyll. Private users of electronic access control established sophisticated systems of dykes and locks to control previously free information

[151] See Article 19 1999.

flows. Broadcasters who wanted access to subscribers of a pay-TV platform suddenly needed more than just a broadcasting licence: they needed access to a conditional access platform and its marketing platform. For consumers, access to access-controlled information is no longer 'free', but a matter of encryption keys and subscription contracts.

What are the consequences for public information policy and the free flow of information? Was there ever any such thing as 'free' (access to) information? And how did public information policy and media legislators respond to the new challenges imposed by electronic access control?

Chapter 2 is divided into two parts. The first part (section 2.2.), lists a number of recurring policy aims that have been put forward in response to the changes in broadcasting that were described in Chapter 1: the privatized, targeted and individualized control over access to content and the resulting lock-in or lock-out situations for consumers that are not subscribed to the pay-TV platform.[152] Probably the most popular, but a very controversial argument, is that national regulators would have to create the conditions in which the public can exercise its 'right to information', which would flow from Article 10 of the ECHR (section 2.2.1.). However, there are also a number of other policy principles that, within the framework of Article 10 of the ECHR or on their own, provide arguments for the government's involvement in electronic access control in broadcasting. Such principles can be democratic and social. They can define the accessibility of information of public importance (section 2.2.2.), the preservation of public broadcasting (section 2.2.3.), as well as the free flow of services in the Internal Market (section 2.2.4.). One aspect that will not be discussed in this chapter but that will be discussed in Chapters 3 (Competition Law) and 4 (Telecommunications Law), is the access of service providers to pay-TV platforms, including access to access-controlled platforms. Chapter 4 also refers to the public information policy considerations involved. Other issues that will not be discussed are the public policy justifications for regulating broadcasting in general, as well as more general aspects of competition law and policy that can become relevant within the framework of exploiting exclusive programme rights. The instruments examined apply to broadcasting services only. It is not the purpose of this study to examine more broadly whether these principles are, or should be, transferable to non-broadcasting services such as online services and DRM-protected content. Because of the need to restrict the scope of the study, Chapter 2 will mention a number of aspects that are relevant when discussing conditional access and broadcasting law and policy but will not claim to provide an exhaustive discussion of all the issues involved.

The second part of this chapter (section 2.3.) analyses how the existing instruments in European broadcasting law that address conditional access translate the information policy concerns discussed in the first part of the chapter. This part

---

[152] See sections 1.4.3. and 1.5.2.

of the chapter discusses instruments issued by the European Communities and/or the Council of Europe. More specifically, it discusses Article 3a of the TWF Directive, which inspired the corresponding obligation in Article 9bis of the European Convention on Transfrontier Television (ECTT) (see section 2.3.1.) about the so-called 'lists of important events'. The ECTT is the equivalent of the TWF Directive and applies to the Member States of the Council of Europe. The chapter continues with a discussion on the right to short reporting that is provided for in Article 9 of the ECTT (section 2.3.2.), and concludes with the must-carry rules in Article 31 of the Universal Service Directive (section 2.3.3). This is a provision from European telecommunications law, but it deals with broadcasting and is drafted to realize public information policy objectives for the broadcasting sector. Chapter 2 concludes with a brief summary of its findings (section 2.4.).

2.2. Conditional Access and Public Information Policy

'Public information policy' is about regulating and realizing the free flow of information, and reconciling the objectives and interests related to it: culture, democracy, economic growth and competition, the free flow of information and, within the context of the European Union: the free flow of services and the realization of Internal Market principles. Public information policy looks at questions related to the technical distribution for the media— telecommunications law—as well as the movement of media services within this infrastructure, such as broadcasting services. The electronic control of access in broadcasting affects both the technical architecture and the circulation of content services.[153] Chapter 2 discusses access control from the broadcasting law perspective.

In Chapter 1, it was explained that conditional access in broadcasting introduces elements of individual authorization and private exclusive control over access to content, while public information policy in broadcasting is still based on the traditional concept of free-to-air broadcasting, as is broadcasting itself. If broadcasting services were no longer widely available because of electronic access control and the resulting fragmentation of the audience, would this mean that the realization of public information policy principles would be endangered? And which public information objectives would then be affected?

These are questions regulators were confronted with while drafting the Conditional Access Directive.[154] The goal of the Conditional Access Directive is to create a universal framework of protection against activities that facilitate illicit

[153] See section 1.5.1.
[154] See also the paper by Helberger 1999a.

access, meaning the piracy of access-controlled broadcasting and information society services.[155]

The downside of technical content protection is that the technology can also affect socially desirable uses of technically protected content. The Council of Europe, after having adopted a recommendation that also protects access-control technologies against unauthorized access,[156] was one of the first to realize that for the broadcasting sector

> '[a]t first sight, the notion of illicit access … is not one that sits comfortably with the principle of freedom of expression and of free access to information enshrined in many national laws and international conventions'. [157]

As far as the Conditional Access Directive of the European Union is concerned, a number of proposals of the European Parliament that were submitted during the drafting of the Conditional Access Directive indicated that the issue of unlimited electronic access control over access to broadcasting content was far more controversial than is reflected in the final version. The proposals illustrate well the whole spectrum of reasons why government involvement with electronic access control could be desirable, if not even necessary.

The Committee on Legal Affairs and Citizens' Rights noticed that conditional access in broadcasting could have a negative effect on the free flow of services in the Internal Market because conditional access operators in one country exclude consumers from another market.[158] The Committee on Legal Affairs and Citizens' Rights proposed to include a recital saying that 'conditional access systems should not be used for the whole purpose of refusing access to citizens in some Member States to services that are freely available in other Member States'.[159]

In the same document, the Committee on Legal Affairs and Citizens' Rights also suggested to include an additional, broader recital on the protection of accessibility and availability of content in general. The proposed Recital 16b required the balancing of, on the one hand, the interests of service providers, and, on the other hand, the interests of the general public not to be excluded from access to

---

[155] See also Council of Europe, European Convention on the Legal Protection of Services based on, or consisting of, Conditional Access, Strasbourg, 24 January 2001 [hereinafter 'Conditional Access Convention'], which resembles the Conditional Access Directive in its major points.

[156] Council of Europe, Recommendation No. R (91)12 of the Committee of Ministers to Member States on the legal protection of encrypted television services, Strasbourg, 27 September 1991 [hereinafter 'Recommendation on the Legal Protection of Encrypted Services'], Explanatory Memorandum, Note 8.

[157] Council of Europe, Recommendation No. R (91)12 on the Legal Protection of Encrypted Services, Explanatory Memorandum, Note 8.

[158] See sections 1.5.2. and 2.2.4.

[159] European Parliament, Committee on Legal Affairs and Citizens′ Rights, Report on the proposal for a European Parliament and Council Directive on the legal protection of services based on, or consisting of, conditional access, Brussels, 21 April 1998, A4-0136/98 [hereinafter 'Committee on Legal Affairs and Citizens' Rights, Report on the Proposed Conditional Access Directive'], Amendment 3.

information about, for example, cultural events. Recital 15a of the same document follows the same line:

> 'Whereas this Directive [the Conditional Access Directive] is without prejudice to the right of the viewer to have access to free-to-air channels within a conditional access service platform without being required to pay an additional fee beyond the normal charge for accessing the platform'.

This is an interesting argument that combines two different aspects. First, it claims that there is a 'right of access to information'. Whether such a right exists or not will be examined in more depth in section 2.2.1. Second, the Committee on Legal Affairs and Citizens' Rights' argument indicates that one characteristic of this 'right of access to information' is the ability to access particular content without having to pay an additional fee. This argument points to the public character of certain broadcasting services, an argument that is commonly made with regard to public broadcasting (section 2.2.3.). Moreover, the Committee on Culture, Youth, Education and the Media proposed to introduce the following recital to the Conditional Access Directive: 'Whereas the encoding of broadcasting services should not unreasonably withhold those services from the general viewer if they were originally offered for non-remunerated use'.[160] In addition to the remuneration argument, the Committee on Culture, Youth, Education and the Media seems to be concerned more generally about the fact that with electronic access control some broadcasting content will no longer be freely accessible.[161] In the end, none of these proposals was incorporated in the final version of the Conditional Access Directive. The Directive does not deal with a potential conflict between electronic access control and information access.[162]

A provision that does so, however, is Article 3a of the TWF Directive, which also stimulated Article 9bis of the ECTT of the Council of Europe. Article 3a of the TWF Directive recognizes the right of Member States to draw up so-called 'lists of important events' that may not be shown exclusively on pay-TV. The list concept is described in detail in section 2.3.1. Listed events must be

> 'outstanding events that are of interest to the general public in the European Union or in a given Member State and which are organized in advance by an event organizer who is legally entitled to sell the rights pertaining to that event'.[163]

---

[160] European Parliament, Committee on Culture, Youth, Education and the Media, Opinion for the Committee on Legal Affairs and Citizen's Rights on the legal protection of services based on, or consisting of conditional access, Brussels, 9 February 1998, A4-0136/98 [hereinafter 'Committee on Culture, Youth Education and the Media, Opinion on the Proposed Conditional Access Directive'], Amendment 2.

[161] A discussion of what 'free' can mean in this context is provided in section 1.5.1.

[162] See explicitly the Conditional Access Directive, Recital 9.

[163] Directive 97/36 Amending the Television Without Frontiers Directive, Recital 21.

Article 3a of the TWF Directive was introduced relatively late in the course of the revision of the directive. The overall goal of the TWF Directive is to establish an Internal Market for broadcasting services and ensure that competition in the common market is not distorted.[164] Part of this strategy is to harmonize national broadcasting legislation and ensure that national rules, including rules that deal with the electronic access control of access to broadcasting content, do not cause obstacles for the free movement of services. The Directive is also bound to the obligations flowing from Article 10 of the ECHR, and seeks to ensure free movement within the context of said provision.[165]

As far as the list of important events is concerned, it is interesting to note that an earlier version of the revised Directive suggested the following wording:

> 'Whereas the approach in Directive 89/552/EEC and this Directive has been adopted to achieve the essential harmonization necessary and sufficient to ensure the free movement of television broadcasts in the Community; whereas Member States remain free to apply to broadcasters under their jurisdiction more detailed or stricter rules [...] including inter alia, rules concerning the achievement of language policy goals, protection of the public interest in terms of television's role as a provider of information, education, culture and entertainment, the need to safeguard pluralism in the information industry and the media and the protection of competition with view to avoiding the abuse of dominant positions, for example securing exclusive rights to major events to the disadvantage of the majority, and/or the establishment or strengthening of dominant positions'[166] (emphasis added by the author).

'Avoiding the abuse of dominant positions' is a notion commonly used in antitrust law rather than broadcasting law. It points towards competition policy considerations rather than public information policy-related reasons for Member States to intervene. The licensing of broadcasting rights has already raised a number of antitrust issues in the broadcasting sector.[167] On the other hand, within the context of this Directive, the notion appears in a list of traditional public information policy goals such as language policy, information, education, culture, entertainment and pluralism. This is not too surprising, as national broadcasting regulation can be a mixture of economically and socially motivated norms. An apt reader could,

---

[164] Television Without Frontiers Directive, Recital 2.

[165] Television Without Frontiers Directive, Recital 8.

[166] European Parliament, Committee on Culture, Youth Education and the Media, Recommendation for second reading on the common position established by the Council with a view to the adoption of a European Parliament and Council Directive amending Council Directive 89/552/EEC on the coordination of certain provisions laid down by law, regulation or administrative action in Member States concerning the pursuit of television broadcasting, Brussels, 31 October 1996, A4-0346/96 [hereinafter `Committee on Culture, Youth Education and the Media, Recommendation on the Directive 97/36 Amending the Television Without Frontiers Directive, Second reading'], Amendment 9.

[167] See Van de Gronden/Mortelmans 2003, 11pp., with further references.

however, interpret this notion to mean that the origin of the list-of-important-events regulation in this early version of Article 3a of the revised TWF Directive is antitrust. Accordingly, an earlier version of Article 3 reads:

> 'Member States shall remain free to require television broadcasters under their jurisdiction to comply with more detailed or stricter rules in the areas covered by this Directive. These rules, which must be compatible with Community law, may concern, among other things:
>
> - […] the taking into account of the public interest in terms of television's role as a provider of information, education, culture and entertainment;
>
> - the safeguarding of pluralism in the information industry and the media;
>
> - the protection of competition with a view to avoiding the abuse of dominant positions and/or the establishment or strengthening of dominant positions by mergers, agreements, acquisitions and similar initiatives'.[168]

Not much of this still very broad and vague wording was taken over in the final version of Article 3a, which reads:

> 'Each Member State may take measures in accordance with Community law to ensure that broadcasters under its jurisdiction do not broadcast on an exclusive basis events which are regarded by that Member State as being of major importance for society in such a way as to deprive a substantial proportion of the public in that Member State of the possibility of following such events via live coverage or deferred coverage on free television. If it does so, the Member State concerned shall draw up a list of designated events, national or non-national, which it considers to be of major importance for society'.

The goal is to regulate the exercise of exclusive rights to particular events by broadcasters in a way that could deprive a substantial part of the broadcasting audience. The reference to competition policy has disappeared. Instead, the final version of the revised Directive points to the protection of citizens' interests[169] and public information policy interests: Recital 18 of the revised TWF Directive

---

[168] European Parliament, Committee on Culture, Youth Education and the Media, Recommendation on the Directive 97/36 Amending the Television Without Frontiers Directive, Second Reading, Amendment 18.

[169] European Parliament, Report on the joint text, approved by the Conciliation Committee, for a European Parliament and Council Directive amending Council Directive 89/552/EEC on the coordination of certain provisions laid down by law, regulation or administrative action in Member States concerning the pursuit of television broadcasting activities, Brussels, 3 June 1997, A4-0201/1997 [hereinafter 'Report on the Joint Text for a Directive amending the Directive 89/552/EEC'], Explanatory Memorandum, paragraph 10.

stipulates that Member States should be able to take measures to protect the public interest, and here more specifically the 'right to information', and to ensure wide access by the public to television coverage of national or non-national events of major importance for society. Examples are provided: The Olympics, the Soccer World Championship and the European Soccer Championship. Member States should be able to take measures to regulate how broadcasters exercise the exclusive broadcasting rights for such events.[170]

Article 3a of the TWF Directive motivated the inclusion of a similar provision in the Council of Europe's ECTT: Article 9bis. Article 9bis of the Convention is called 'Access of the public to events of major importance', and includes an explicit reference to Article 10 of the ECHR.[171] In so doing, the Council's Standing Committee also pointed to the complexity of reconciling freedom of expression principles with property rights and contractual freedom.[172] The final list regulation in Article 9bis of the Convention is modelled after Article 3a of the TWF Directive.

Interestingly, the ECTT distinguishes between the aforementioned 'access of the public to events of major importance' and 'access of the public to information'. The latter is the heading of Article 9 of the Convention that provides for the right to short reporting. The right to short reporting is discussed in more detail in section 2.3.2. The TWF Directive does not have a right to short reporting. In the course of the revision of the directive in 2004, the introduction of such a right was discussed, and if Article 5(3) c of the European Copyright Directive was sufficient.[173]

According to Article 9 of the ECTT:

> 'Each Party shall examine and, where necessary, take legal measures such as introducing the right to short reporting on events of high interest for the public to avoid the right of the public to information being undermined due to the exercise by a broadcaster within its jurisdiction of exclusive rights for the transmission or retransmission … of such an event'.

As the Council of Europe states:

> 'now that television is one of the major sources of information for the public, the exercise of exclusive rights for television broadcasting in a transfrontier

[170] Directive 97/36 Amending the Television Without Frontiers Directive, Recital 18.

[171] ECTT, Article 9bis (2); Explanatory Memorandum, paragraphs 68, 181.

[172] ECTT, Explanatory Memorandum, paragraphs 67, 181.

[173] Meeting of Focus Group 3 on the Revision of the Television Without Frontiers Directive, The right to information and the right to short extracts, Brussels, 23 November 2004. See also European Commission, Communication from the Commission to the Council, the European Parliament, the European Economic and Social Committee and the Committee of the Regions, The Future of European Regulatory Audiovisual Policy, Brussels, 15 December 2003, COM(2003)784 final [hereinafter 'The Future of European Audiovisual Policy'], p. 16.

context of a major event may prove to be detrimental to the right of access of the public to information'.[174]

Also within the context of the right to short reporting, reference is made to Article 10 of the ECHR (freedom of expression) and the 'right to information'.[175] Pluralism is a second important aspect behind the right to short reporting, namely to encourage competition between several broadcasters.[176] Exclusive control over exploitation rights can secure the right holder a competitive advantage. Third-party broadcasters that are not entitled to offer the content to their audience risk losing audience shares. The right to short reporting makes it possible for more than one broadcaster to report on a particular event even if it is 'only' done in the form of a short report. This is an interference with the position of the holder of exclusive rights as complete exclusivity of content can be very valuable. Article 9 of the Convention seeks to strike a balance. The right to short reporting does not guarantee coverage of the full event in free-to-air television, but only the possibility to report about the event by those broadcasters that have not acquired the exclusive rights for full coverage.[177]

The right to short reporting has already found its way into the Convention at the end of the 1980s. It was drafted to protect the public's 'right to information' long before the issue of pay-TV attracted wider attention in Europe. The so-called 'right to short reporting' is also subject to the Council of Europe's Recommendation (91)5 on the right to short reporting on major events where exclusive rights for television broadcast have been acquired in a transfrontier context and to the Draft Recommendation[178] that updates the former. This is also discussed in detail in section 2.3.2. In the Recitals of the aforementioned Draft Recommendation, the Council of Europe goes even further and claims that the public has a right of access to very specific content, namely 'a public's right of access to information on major events', and that this right is worthy of protection.[179] But the 'the public's right of

---

[174] Council of Europe, Recommendation No. R (91)5, of the Committee of Ministers to Member States on the right to short reporting on major events, where exclusive rights for their television broadcast have been acquired in a transfrontier context, Strasbourg, 11 April 1991 [hereinafter `Recommendation No. R(91)5 on the Right to Short Reporting'], Explanatory Memorandum, paragraph 4.

[175] Council of Europe, Recommendation No. R(91)5 on the Right to Short Reporting, Explanatory Memorandum, paragraph 4 and 5.

[176] ECTT, Explanatory Memorandum, paragraph 174; Council of Europe, Recommendation No. R(91)5 on the right to short reporting, Explanatory Memorandum, paragraphs 1-2.

[177] See in detail in section 2.3.2.

[178] Draft Recommendation on the Right to Short Reporting on major events where exclusive rights have been acquired, updating Recommendation No. R(91)5 of the Committee of Ministers to Member States on the right to short reporting on major events where exclusive rights for their television broadcast have been acquired in a transfrontier context, Strasbourg, 16 April 2003, MM-Public(2003)003 [hereinafter 'Draft Recommendation updating Recommendation No. R(91)5 on the Right to Short Reporting'].

[179] Draft Recommendation updating Recommendation No. R(91)5 on the Right to Short Reporting, Recital 6.

access to information' is, unlike Article 9bis of the Convention (list of important events) not restricted to cultural or sports events of 'major importance for society'. It also applies to political and social events of 'only' high public interest such as less important sports and cultural events, and social or political newsworthy events such as a report about an accident, a natural disaster or an armed conflict.[180]

The transfrontier aspect plays a significant role in this context. The Council of Europe expressed its concern about a fragmentation of the European broadcasting landscape into various country or language zones. This could also have a negative impact on cultural diversity throughout Europe, as cultural diversity involves the possibility of transborder exchange and the provision and reception of culturally different information services.[181] Even more dramatic was the wording of the European Committee on Culture, Youth, Education and the Media of the European Communities. The latter warned, referring explicitly to electronic access control, that national services were beginning to vanish into the 'ghettos of encryption':

> '[m]any users may be left with no legitimate access to such services if they are locked away. Others will wish to use their decoder not just in the Member State of purchase, but also in others where its use may be "illicit" because of a discrete copyright arrangement'.[182]

Availability of content and non-exclusion are the ideas that underlie must-carry rules in broadcasting. According to Recital 43 of the Universal Service Directive, Member States should be able to lay down proportionate carriage obligations on enterprises in the interest of legitimate public policy considerations. The Directive leaves open what such general interest obligations are. It only suggests that must-carry obligations may include the transmission of services specifically designed to enable appropriate access by disabled or elderly users.[183] Other general public-

---

[180] More in depth in section 2.3.2.

[181] See Council of Europe, Declaration on cultural diversity, Strasbourg, 7 December 2000 [hereinafter 'Declaration on Cultural Diversity'], paragraph 2.4.

[182] European Parliament, Committee on Culture, Youth Education and the Media, Opinion on the Proposed Conditional Access Directive, section a). See also European Commission, Report from the European Commission on the application of Council Directive 93/83/EEC on the coordination of certain rules concerning copyright and rights related to copyright applicable to satellite broadcasting and cable retransmission, Brussels, 26 July 2002, COM(2002)430 final [hereinafter 'Report on the Application of the Cable and Satellite Directive'], p. 7-8, and section 2.2.4 of this chapter.

[183] Ledoux Book 2004, p. 141. See also European Parliament, Recommendation for second reading on the Common Position adopted by the Council with a view to the adoption of a Directive of the European Parliament and the Council on universal services and users' rights relating to electronic communication networks and services, Brussels, 29 November 2001, A5-0438/2001 [hereinafter 'Recommendation for Second Reading Universal Service Directive], Amendments 6, 26. See also Ofcom, Statement on Code on Electronic Programme Guides, Statement by Ofcom 2004, paragraphs 8-10.

interest objectives can be the promotion of freedom of expression, pluralism and cultural diversity.[184]

Sections 2.2.1. to 2.2.4. take a closer look at the aforementioned arguments used to justify government involvement with the exclusive electronic exploitation of broadcasting rights in pay-TV. They attempt to shed some light on whether there is a 'right to information' in public media, and whether such a right flows from Article 10 of the ECHR (freedom of expression). The first of these four sections, section 2.2.1, takes a closer look at freedom of expression, also within the context of democratic principles and at the pluralism argument. It examines the political and social arguments used to explain why the public might have a protection-worthy interest to access certain events of public importance (section 2.2.2.), the preservation of public broadcasting (section 2.2.3.) and, last but not least, the role that the principle of free movement of services in the Internal Market plays in pay-TV (section 2.2.4.). Section 2.2.4. will also say something about the relationship between EC law and the ECHR.

2.2.1. 'A RIGHT TO INFORMATION'?

The 'right of access to information' or 'right to information' is often used to explain information policy involvement with matters that involve the exclusive exploitation of rights to content. The argument is frequently used in conjunction with pay-TV and in conjunction with the exploitation of intellectual property rights or the control over technical distribution networks such as cable networks.

Article 10 of the ECHR

Many of the rationales put forward to justify the regulation of media in the public interest can be traced back to Article 10 of the ECHR, which has found its way—in one form or another—into all European Constitutions. It provides a standard for national broadcasting legislation. Article 10 of the ECHR[185] protects the 'free flow of information to the public in general'[186], meaning the process of communicating information to the public and/or in public.[187] The participants in the communications

---

[184] European Commission, Communication from the Commission to the Council, the European Parliament, the European Economic and Social Committee and the Committee of the Regions, European Electronic Communications Regulation and Markets 2004, Brussels, 2 December 2004COM(2004)759final [hereinafter 'European Electronic Communications Regulation and Markets 2004], 33pp.

[185] 'Everyone has the right to freedom of expression. This right shall include freedom to hold opinions and to receive and impart information and ideas without interference by public authority and regardless of frontiers. This article shall not prevent States from requiring the licensing of broadcasting, television or cinema enterprises'.

[186] European Commission of Human Rights, Geillustreerde Pers N.V. v the Netherlands, Strasbourg, 6 July 1976, No. 5178 [hereinafter 'Geillustreerde Pers'], paragraph 85.

[187] See also the interpretation of the US Supreme Court in the Red Lion Broadcasting case: 'It is the purpose of the First Amendment to preserve an uninhibited marketplace of ideas in which truth will ultimately prevail, rather than to countenance monopolisation of this market, whether it be by the

process–information providers and recipients—enjoy protection at each stage of the communications chain, starting from the holding of opinions, the imparting of information and ideas to others, the reception of such ideas, to the negative freedom not to receive any information at all.[188] Both recipients and providers of information services benefit from freedom of expression.

Article 10 of the ECHR reflects a wide-reaching international agreement on the particular political and social importance of information and the freedom to exchange it. It is closely linked to the idea of a public dialogue as the foundation for democratic and social life. As the European Court of Human Rights observed,

> 'freedom of expression, as secured in paragraph 1 of Article 10 of the ECHR, constitutes one of the essential foundations of a democratic society, indeed one of the basic conditions for its progress and for the self-fulfilment of the individual'.[189]

The European Court of Human Rights has recognized the role that access to information plays in political processes, as well as in every facet of personal and social life. This is why Article 10 of the ECHR not only covers political content, but also commercial or artistic content, and content of individual interest, learning and entertainment.[190]

The primary role of the state within the context of Article 10 of the ECHR is a passive one, namely to refrain from interference. Article 10(2) of the ECHR leaves room for Member States to make laws that restrict the freedom of expression, providing such laws are a) prescribed by law and b) necessary in a democratic society, that is they are justified by a pressing social need. Whenever Member States issue regulations that restrict the freedom of expression of one party or another and do not comply with Article 10(2) of the ECHR, such law or judgment

---

government itself or a private licensee … It is the right of the public to receive suitable access to social, political, aesthetic, moral and other ideas and experience which is crucial here'. (US Supreme Court, Red Lion Broadcasting Co., Inc. v. Federal Communications Commission, 395 U.S. 367, 1969, Section III A).

188 For a concise overview of Article 10 of the ECHR see Harris/O'Boyle/Warbrick 1995, pp. 373-416; Van Dijk/Van Hoof 1998, pp. 558-585; Van Eijk 1992, pp. 135-164, with an overview on the history; Dommering 2000, pp. 43-45 and pp. 96-101.

189 European Court of Human Rights, Müller and others, Strasbourg, 24 May 1988, Series A No. 133 [hereinafter 'Müller'], Müller, paragraph 33; European Court of Human Rights, Barthold, Strasbourg, 25 March 1985, Series A No. 90 [hereinafter 'Barthold'], paragraph 58; European Court of Human Rights, Sunday Times, Strasbourg, 26 April 1979, Series A, No. 30 [hereinafter 'Sunday Times', paragraph 65.

190 See the often-cited and telling passage in the Handyside-judgement of the European Court of Human Rights: 'Freedom of expression constitutes one of the essential foundations of a [democratic] society, one of the basic conditions for its progress and for the development of every man. Subject to paragraph 2 of article 10, it is applicable not only to "information" or "ideas" that are favourably received or regarded as inoffensive but also to those that offend, shock or disturb the state or any sector of the population. Such are the demands of that pluralism, tolerance and broadmindedness without which there is no "democratic society"', European Court of Human Rights, Handyside, Strasbourg, 7 December 1976, Series A No. 24 [hereinafter 'Handyside'], paragraph 49.

that is based on such a regulation risks being declared in conflict with Article 10 of the ECHR. The consequence is that when governments assess the need for regulatory initiatives regarding electronic access control, be it measures that protect the users of such control or measures that protect the interests of citizens, regulators must take into account the possible negative and positive effects such initiatives have on the realization of freedom of expression. Having said that, Member States do enjoy a certain margin of appreciation in assessing the need for and constitutionality of regulatory initiatives.[191]

The state can also have a positive obligation under Article 10 of the ECHR.[192] Freedom of expression would not be very effective if left to itself or applied in areas in which practical circumstances endanger its realization. The European Court of Human Rights has underlined that Contracting States are responsible for securing the rights and freedoms enshrined in the Convention.[193] It is therefore widely acknowledged and established case-law that Article 10 of the ECHR provides more than a right against the state. Article 10 of the ECHR also imposes on states positive obligations to actively protect and promote the realization of freedom of expression. In Özgür Gündem v. Turkey, the European Court of Human Rights stated explicitly that the

> '[g]enuine effective exercise of this freedom [freedom of expression] does not depend merely on the state's duty not to interfere, but may require positive measures of protection' and, as the Court further said, 'even in the sphere of relations between individuals'.[194]

---

[191] This is demonstrated by the European Court of Human Rights, for example, European Court of Human Rights, Groppera Radio AG and Others, Strasbourg, 28 March 1990, series A No. 173 [hereinafter 'Groppera'], paragraph 72; European Court of Human Rights, Informationsverein Lentia v. Austria, Strasbourg, 28 November 2002, series A No. 276 [hereinafter 'Lentia'], paragraph 35, to name but a few.

[192] See extensively Mowbray 2004, p. 223. See also United Nations, Economic and Social Council, The right to the highest attainable standard of health, E/C. 12/2004/4, CESCR General comment 14, 11 August 2000, p. 382.

[193] European Court of Human Rights, Sunday Times, paragraph 59.

[194] European Court of Human Rights, Özgűr Gűndem v. Turkey, Strasbourg, 16 March 2000, No. 23144/93 [hereinafter 'Özgűr Gűndem v. Turkey'], paragraph 43; European Court of Human Rights, Von Hannover v. Germany, Strasbourg, 24 June 2004, No. 59320/00 [hereinafter 'Von Hannover v. Germany'], paragraph 57. But see also the European Court of Human Rights, Guerra and Others v. Italy, Strasbourg, 19 February 1998, No. 116/1996/735/932 [hereinafter 'Guerra'], paragraph 53 where the court observed that 'the public has a right to receive information as a corollary of the specific function of journalists, which is to impart information and ideas on matters of public interest'. Gersdorf 2000, p. 211; Badura 1989, p. 325; Council of Europe, Declaration on the freedom of expression and information, 29 April 1982 [hereinafter 'Declaration on the Freedom of Expression and Information'], Recital 6. See also Dommering 2000, pp. 4, with further references, and p. 49: 'Ten slotte houdt de communicatievrijheid ook in dat medeburgers het vrije communicatieproces niet verstoren. In die zin dienen medeburgers zich daarvan te onthouden. Maar ook hier ligt er een plicht voor de overheid te zorgen dat de communicatievrijheid in die horizontale verhouding verwezenlijkt wordt'. Harris/O'Boyle/Warbrick 1995, p. 383. Ascher reasons that with the arrival of the information society and the dominance of a few private parties over the information society, the need

The case concerned attacks on journalists, distributors and other associates of the Turkish newspaper Özgür Gündem and the alleged lack of intervention by the government. The European Court of Human Rights did not further define what the boundaries between the state's positive and negative obligations are. It did say, however, that

> 'in determining whether or not a positive obligation exists, regard must be had to the fair balance that has to be struck between the general interest of the community and the interests of the individual, the search for which is inherent throughout the Convention. The scope of this obligation will inevitably vary with regard to the diversity of situations achieved in Contracting States, the difficulties involved in policing modern societies and the choices which must be made in terms of priorities and resources'.[195]

The European Court of Human Rights confirmed its dictum in later cases.[196] In other words, states are obliged to create conditions for the broadcasting sector that are favourable to the realization of freedom of expression and remove public or private obstacles that could hinder its realization.

What this chapter does not discuss is the theory of the horizontal enforcement of human rights or 'Drittwirkung', which is the idea that constitutional freedoms are reflected in legislative provisions and that they therefore also become (indirectly) effective between private parties.[197] According to this theory and on the basis of such laws, claims of private parties against other private parties would also have to be interpreted within the context of constitutional freedoms. The theory of *Drittwirkung* is used to argue that private operators would at least be indirectly obliged to observe freedom of expression because they are subject to legal obligations that are supposed to protect the freedom of expression and hence must be interpreted within this context.[198] An in-depth discussion of *Drittwirkung* might be fruitful were the study to conclude that Article 10 of the ECHR was the basis for an individual 'right to information', meaning the right to claim access to particular privately held information. This study, however, will conclude that this is not the case.

Freedom of Expression and Private Control of Access to Information

Notably within the context of Article 3a of the TWF Directive (list of important events), but also within the context of the right to short reporting in Article 9 of the ECTT, it was argued that the exclusive exploitation of sports events via pay-TV

---

to also protect against private interventions and restrictions becomes even more urgent, Ascher 2002, p. 115. Van Dijk/Van Hoof 1998, p. 26; Birnhack 2004, 39pp., differentiates and points out that, for example, in the US some scholars do not recognize private interference with the free flow of information as a freedom of expression issue. Differentiating Lichtenberg 1990, 114pp.

195 European Court of Human Rights, Özgür Gündem v. Turkey, paragraph 43

196 See, for example, European Court of Human Rights, Von Hannover v. Germany, paragraph 57.

197 See more generally, Van Dijk/Van Hoof 1998 pp. 22-26.

198 De Meij 1996, 79pp.; Urek 1991, 49pp.; Schulz 1998, 164pp.

platforms would conflict with 'a right of the public to information'. This form of exploitation would a) make the programmes only accessible to a limited part of the audience and b) require additional payment for such services.[199] The Council of Europe even claimed that

> '[g]iven the fact that the access of the public to news information about major events can only be fully realized if such access is free, this principle provides that, unless otherwise agreed between them, the primary broadcaster should not be able to charge the secondary broadcaster for the short report'.[200]

Is freedom of expression about gratis access to information?[201] Or is it about access to sports events? Making access to information subject to price negotiations is not new in the media world. Consumers are used to paying for their newspapers. Nobody would seriously argue that the mere fact that one has to pay for a newspaper before the newsagent will hand it over prevents one from receiving information. The same is true for films shown in cinemas or the purchase of CDs and DVDs. Here too, citizens do not usually access the information stored on a CD or DVD without having to pay first. Even the reception of public and commercial broadcasting is far from being 'for free'.[202] To receive public and commercial free-to-air broadcasting, consumers not only have to purchase a television or computer, they must also subscribe to, for example, a cable or satellite network and 'pay' for some programmes in the form of public broadcasting fees. For commercial, advertisement-financed programmes they also pay in non-monetary but money-worth 'assets', such as time and attention.

The production of information goods and services is not for 'free' and often requires substantial investment. Article 10 of the ECHR expressly mentions cinema enterprises, meaning entities that specialize in the screening of films or other kinds of content that are subject to direct remuneration, such as an entrance fee. In its Groppera decision, the European Court of Human Rights found that cable operators fall within Article 10 of the ECHR and as such enjoy its protection[203] irrespective of the fact that cable operators require subscription fees for the showing of a particular programme bundle. In other words, Article 10 of the ECHR protects entities that a) control access to information and b) charge an access fee without assuming automatically that these activities (access control, charging) are intrinsically detrimental to freedom of expression. A decision by the European Commission of

---

[199] For the Council of Europe: Council of Europe, Recommendation 91(5) on the Right to Short Reporting, Explanatory Memorandum, paragraph 47. For the EU: Directive 97/36/EC amending the Television Without Frontiers Directive, Recitals 18-22. See also section 2.2.

[200] Council of Europe, Recommendation 91(5) on the Right to Short Reporting, Explanatory Memorandum, paragraph 47.

[201] See, for example, Hesse 1999, p. 283.

[202] See also Owen 1975, p. 116: 'There is sufficient folk ignorance associated with the Orwellian "freedom" of television, and so strong a public preoccupation with the medium, that politicians would be foolish to seem to tamper with the electronic genie'.

[203] European Court of Human Rights, Groppera, paragraph 55. See also Gersdorf 2000, p. 214.

Human Rights in the Nederlandse Omroepprogramma Stichting case follows a somewhat similar line. The European Commission of Human Rights found that the 'right of freedom of expression' did not provide a right to report about football matches, which the Royal Dutch Football Association organized, without paying a compensation.[204] The selling of broadcasting licenses was one source of financing sports events. The European Commission of Human Rights did not consider this an interference with Article 10 of the ECHR if the organizer of an event limited the right to direct reporting of the match to those with whom the organizer had concluded a contract, including an agreement covering the financial conditions.[205] In conclusion, the argument of a need to preserve 'free-of-charge' access to broadcasting content is a political argument rather than one that is dictated by Article 10 of the ECHR (as to possible other public information policy interests in affordable access to broadcasting, see sections 2.2.2. and 2.2.3.).

Does this mean that cinema owners and pay-TV operators must always be free to control access to information, and that this can never lead to possible tensions with freedom of expression? As already explained, Article 10 of the ECHR commands that a fair balance is struck between the general interest of the community and the interests of the individual.[206] Moreover, whoever exercises the rights and freedoms enshrined in Article 10 of the ECHR undertakes 'duties and responsibilities'.[207] It is the task of the state, as part of its positive obligation, to ensure that a fair balance is achieved.[208] The following section seeks to extract from the interpretation of the European Court of Human Rights of Article of the 10 ECHR an idea of what such a fair balance between the providers of an access-controlled content service and citizens, alias consumers, should be.

That there can be tension between, on the one hand, the exploitation of exclusive rights in content, and, on the other hand, freedom of expression is an issue in the regulation of intellectual property.[209] Copyright law has been described as a

> 'form of public information policy serving the public interest in maximizing the availability of information products by, on the one hand, granting an exclusive right and thereby providing for an incentive to create, and, on the

---

[204] The Royal Dutch Football Association is a private organization, so that insofar only an eventual positive obligation of the Dutch government to protect the freedom of expression of others was at discussion.

[205] See European Commission of Human Rights, Nederlandse Omroepprogramma Stichting v The Netherlands, Strasbourg, 11 July 1991, No. 13920/88, Extract, paragraph 1. See also the paper by Riegel 1988.

[206] European Court of Human Rights, Özgür Gündem v. Turkey, paragraph 43.

[207] European Commission of Human Rights, European Court of Human Rights, Otto Preminger Institut, Strasbourg, 20 September 1994, series A No. 295-A [hereinafter 'Otto Preminger Institut'], paragraph 49.

[208] European Court of Human Rights, Özgür Gündem v. Turkey, paragraph 43; Von Hannover v. Germany, paragraph 57.

[209] Birnhack 2004, 54pp.

other hand, by limiting the scope of the monopoly copyright to ensure information will be widely available and usable'.[210]

Copyright scholars have repeatedly pointed to possible negative implications that exclusive private control over content can have on freedom of expression. The question is whether lessons can be learned from the discussion for the use of electronic access control.

Copyright law grants the author monopoly rights in the exploitation/communication of his or her work. Freedom of expression promotes and protects the freedom to impart or receive information and ideas. As each work is based on ideas and information, there is a potential conflict between information monopolies and freedom of expression.[211] Freedom of expression is the freedom 'to say something in the minimal negative liberty sense if he or she is not liable to be prevented from saying that thing, or to be penalized for saying it'.[212] A holder of intellectual property rights can control how content is used. This gives him or her the power to control what others say using their 'words'. However, the freedom 'to say something' using somebody else's, already published, words, and demanding that others make information accessible that they do not want to make accessible or are only willing to make available under certain conditions, are two very different things.

According to the European Court of Human Rights, Article 10 of the ECHR only protects the reception of content that others 'wish or may be willing to impart'.[213] In other words, Article 10 of the ECHR does protect the right to receive information from publicly available sources. Hence, any further considerations about a possible positive obligation for states to secure access for citizens to the pay-TV platforms would stop here if pay-TV were not already a publicly available source. One could argue that the main motive for pay-TV operators to use electronic access control is to make certain information unavailable to the public; the pay-TV business model is about making content accessible only to selected consumers. Based on this argument, some scholars have rejected the public availability of pay-TV services. They argue that access is deliberately provided only to particular persons with whom the service provider is familiar and has previously concluded a contract.[214] The requirement of prior negotiations would conflict with the assumption of a publicly available service. This follows a line similar to that of the Swiss

---

210 Koelman 2000a, p. 279.

211 Hugenholtz 1989, p. 150: 'Men behoeft geen rechten gestudeerd te hebben om te zien dat er tussen het auteursrecht en de informatievrijheid een zekere spanning bestaat'. See also the papers by Baker 2002, Fraser 1998 and Garfield 2001.

212 Benkler 1999, 390pp.

213 European Court of Human Rights, Leander, Strasbourg, 26 March 1987, series A No. 116 [hereinafter 'Leander'], paragraph 74; European Court of Human Rights, Gaskin, Strasbourg, 7 July 1989 series A No. 160 [hereinafter 'Gaskin'], paragraph 52. See also the decision by the European Commission of Human Rights, Gruppo Interpres S.A. c/Espagna, Strasbourg, 7 April 1997, No. 32849/96 [hereinafter 'Gruppo'].

214 Schwarz-Schilling 1998, p. 491.

Government's argument in the Autronic case. The Swiss Government claimed that television programmes that were transmitted between two fixed points instead of being broadcast free-to-air were not intended for or made accessible to the public with the consequence that the transmission was a matter of telecommunications secrecy, not of Article 10 of the ECHR and the right to receive information.[215]

Nevertheless, one could argue that pay-TV is still principally intended for reception by the public or parts thereof.[216] A person who buys a book or subscribes to a newspaper also concludes a contract. Nobody would claim for this reason that books and newspapers are not publicly available sources of information. That the dedication of the encrypted signal, public or private, plays a decisive role can be concluded from the European Court of Human Rights's Autronic decision. The European Court of Human Rights followed the argumentation that was presented by the European Commission of Human Rights in this case and ruled that Article 10 of the ECHR was also applicable to those (broadcasting) signals that were processed point-to-point providing they were intended for the general public.[217] This interpretation is also reflected in Article 2(a) of the ECTT and Article 1(a) of the TWF Directive both of which define broadcasting as the initial transmission in decoded or encoded form.[218] According to this interpretation, pay-TV is principally a publicly available source.

Assuming that pay-TV is a publicly accessible source, the intention of pay-TV providers is still not to grant access to everyone but only to those who comply with the conditions in the subscription contract. The programmes are encrypted and made accessible only to selected viewers. Can this lead to the conclusion that the right to receive information from publicly available sources also includes a right to access information or, in the case of pay-TV, to receive information in intelligible form?

The question of whether Article 10 EHCR includes a right of access to information has already been the subject of some controversy. In the academic discussion, the notion of a 'right of access to information' is frequently used to plead in favour of government interference with the exclusive exploitation of content. As mentioned in section 2.2., the Council of Europe and the European Parliament of the European Union have repeatedly made reference to a so-called 'right of the viewer to have access'[219], 'a right to information'[220] or even 'a public's

---

215 European Court of Human Rights, European Court of Human Rights, Autronic, Strasbourg, 22 Mai 1990, series A No. 178 [hereinafter 'Autronic'], paragraph 44.

216 In this sense, Dommering 1990, p. 64; Nauheim 2001, p. 131-132; Hins 1991, p. 249.

217 European Court of Human Rights, Auctronic, paragraphs 46 - 48. In this sense also Hins 1991, p. 228 and 248-249. Nauheim 2001, p. 131-132.

218 See extensively on the notion of 'publicly available source', Hins 1991, 244pp. with further references.

219 European Parliament, Committee on Legal Affairs and Citizens' Rights, Report on the Proposed Conditional Access Directive, Recital 15a.

220 Council of Europe, Recommendation No. R(91)5 on the Right to Short Reporting, Explanatory Memorandum, paragraphs 4 and 5; Council of Europe, Recommendation R(91)14 on the legal

right of access to information on major events'.[221] A decision by the European Commission of Human Rights in the case Özkan v. Turkey follows a similar line. Mr Özkan initialized proceedings against the (public) Turkish Radio and Television Institution (TRT) after TRT stopped, without any further explanation, a transmission of the film entitled 'Les Yeux Interdits'. The European Commission of Human Rights indicated that Mr Özkan had a legitimate interest in access to this particular film and even that such an interest was protected by Article 10 of the ECHR. In the end, the European Commission of Human Rights decided that a violation of Article 10 of the ECHR was not given at least 'as long as sufficient alternative sources for that information remain available to the public'.[222] The film in question was not prohibited in Turkey and there were sufficient alternative sources available to the public for watching the film. Consequently, the applicant was not deprived of access to the film in question. The case was settled before the European Court of Human Rights could decide on the question of individual access claims.

One could argue that the freedom to hold opinions and express ideas would be inconsequential if consumers did not have the ability to access information they wish to express their ideas about in the first place. This could be a strong argument in favour of an access right, in particular in relation to information that is not available elsewhere and that the public has a particular interest in receiving. This was the argumentation of the Federal High Court of Justice in the case Lengende.[223] The German magazine Stern signed an exclusive agreement with eleven survivors of the mine disaster of Lengende (Lower-Saxony, Germany), to report about the catastrophe. Bild-Zeitung, a German tabloid, published a report about the event, including frequent citations of remarks that, so claimed Bild-Zeitung, were made by the survivors. Stern sued Bild-Zeitung, without success, for infringing on its exclusive contract with the survivors. The Federal High Court of Justice decided on the case within the framework of the German unfair competition law, and interpreted the relevant provision in the light of Article 5 of German Basic Law (freedom of expression). The court found that exclusive agreements about news events were principally possible and admissible. However, there could be situations in which such agreements foreclose the source of information about newsworthy

---

protection of encrypted television services, Strasbourg, 27 September 1991 [hereinafter Recommendation No. (91)14 on the legal protection of encrypted television services'], Explanatory Memorandum, Note 8

221 Council of Europe, Draft Recommendation updating Recommendation No. R(91)5 on the Right to Short Reporting, Recital 6.

222 European Commission of Human Rights, Z. Noyan Özkan v Turkey, No. 23886/94, Strasbourg, 5 April 1995 [hereinafter 'Özkan'], The Law. See also European Commission of Human Rights, Geillustreerde Pers N.V, paragraph 86. Interesting also a US decision, Board of Education, Island Trees Union Free School District No. 26 v. Pico; 457 US 853, 73 L Ed 2d 435, 1982, and the discussion of this decision by Hins 1991, p. 266-268.

223 German Federal High Court of Justice, Lengende, 27 October 1967, case No. ZR 140/65, published in [1968] 4 Gewerblicher Rechtsschutz und Urheberrecht (GRUR) 209[hereinafter `Lengende'], paragraph I.1.

events. In this case, the agreement is not compatible with the public interest, and access to the source of information must principally remain free for everyone.[224] The argument of alternative sources also played a role in the reasoning of the European Commission of Human Rights in the aforementioned Özkan v. Turkey case. In the specific case of pay-TV, one could make the additional argument that the freedom to receive information is somewhat of a farce if the information is not understandable because it is encrypted. In this context, it is worth remembering that access to access-controlled content is often not so much about access to particular information as it is about access to the platform on which information is offered. The content itself is, in most cases, publicly available, though in encrypted form.

Barendt countered the argument that access to information is a necessary precondition to exercise freedom of expression with the question, providing such a fundamental right to acquire information exists, of how far such a right should go. Was there an obligation to recognize 'implicit constitutional rights to education and travel, equally crucial (it could be said) to the formation of a citizen's ideas and expressions?'[225] Taking the idea further, one may wonder whether the freedom to receive information would also include the freedom to receive it in a language one understands.[226]

Upon closer study, there are more reasons to argue that there is no 'right of access' or 'a right of decryption', or that such a right does at least not flow from Article 10 of the ECHR. One argument why this is not so has to do with the principal character of Article 10 of the ECHR as that of a defence right: it is the task of the state to provide for adequate broadcasting regulation, not of the public to claim individual rights to access particular privately-controlled information. This argument was made, for example, by Barendt, who has examined the question of access rights to broadcasting services in depth[227] and observed that '(v)iewers' interests are, therefore, institutionally, rather than legally, protected through the Parliamentary Commission and the political process'.[228]

The extent to which a party benefits from freedom of expression is a matter of balancing one party's interests with those of the others; freedom of expression of one party does not as such override all other legitimate interests of the other parties. This is inherent to Article 10 of the ECHR, and here more specifically to Article 10 (2) of the ECHR, according to which the exercise of freedom of expression can be subject to restrictions in the interest of national security, for the prevention of disorder or crime, for the protection of health or morals, for the protection of the reputation or rights of others, to name but some. Or, as Mackaay puts it: 'no citizen has an automatic right to information, personal or other, in the hands of others'.[229]

---

224 See German Bundesgerichtshof, Lengende, paragraph I.1.
225 Barendt 1992, p. 21.
226 See Van Dijk/Van Hoof 1998, p. 569, with further references.
227 Barendt 1993, 47 pp. and 145 pp.
228 Barendt 1993, p. 49. See also De Meij/Hins/Nieuwenhuis/Schuijt 2000, pp. 299-303.
229 Mackaay 1992, p. 171

Providing Article 10 of the ECHR conferred such an automatic right of access to information, the restriction in Article 10 (2) of the ECHR would make little sense.

This is to say an individual right of access to information could conflict with the equally valuable protection worthy rights of others. For example, Article 10 of the ECHR can also include the right to remain silent as a negative aspect of the right to freedom of expression.[230] The right to remain silent could be an argument why access to information should not be enforced against the will of others, as well as similar interests, such as privacy,[231] procedural rights or business secrecy. In the pay-TV example, such arguments will be less relevant because, as was observed earlier, pay-TV providers do not aim at remaining silent, keeping the service secret nor do they communicate in private. More relevant is the argument that providers of content services such as pay-TV operators also benefit from protection under Article 10 of the ECHR, namely to impart information as they wish. De Meij, Hins, Nieuwenhuis and Schuijt make this argument, saying that an individual right of access to particular information would impose serious pressure on the media's freedom to determine what information they present to their audience.[232] Barendt further elaborated that in broadcasting regulation the balance between, on the one hand, the broadcaster's programme freedom, and, on the other hand, the interests of viewers and listeners is often struck by imposing programme standards on broadcasters, not individual rights of access to information.[233]

Another important aspect is the need to respect the economic freedoms of providers of access-controlled services, such as the freedom of contract that entitles them to choose with whom they enter into negotiations and conclude agreements, and the freedom to property. Providers of content services benefit from the protection of their property, such as their intellectual property in broadcast material, the broadcasting signal, exclusive transmission rights or the electronic access control technology itself.[234] Rather far-reaching is the statement of the European Commission of Human Rights in the Geillustreerde Pers N.V. case. The applicant Geillustreerde Pers N.V., a publisher of a so-called general interest magazine in the Netherlands, intended to publish in its magazine the complete radio and television programme data for all Dutch broadcasting stations. Geillustreerde Pers N.V. was

---

[230] European Court of Human Rights, Young, James and Webster v The United Kingdom, Strasbourg, 13 August 1981, series A No. 44 [hereinafter 'Young, James and Webster'], paragraphs 57, 66, European Court of Justice, Sigurdur A. Sigurjónsson v. Iceland, Strasbourg, 30 June 1993, series A. No. 264 [hereinafter 'Sigurdur A. Sigurjónsson'], paragraphs 37, 43. See also Van Dijk/Van Hoof 1998, pp. 567-568; Barendt 1992, p. 21; Mackaay 1992, p. 171.

[231] See the paper by Cohen 1996.

[232] De Meij/Hins/Nieuwenhuis/Schuijt 2000, 306.

[233] Barendt 1993, p. 49.

[234] Instructive for the case of a possible conflict between private property and freedom of expression, Hugenholtz 1989, p. 150; Lichtenberg 1990, 115 pp. See also; the paper by Baker 2002 and Grosheide 2000, who emphasizes the economic factor, 217pp. Interesting is also a decision of the Dutch Hoge Raad from 17 December 1993 declaring that the prohibition of selling illegal decoders is not in conflict with freedom of expression, [1994] Nederlandse Juristenblad 274.

prevented from doing so because of a Dutch law that prohibits the reproduction of such information by any party other than the Netherlands Broadcasting Foundation. The European Commission of Human Rights denied that this conflicted with the Geillustreerde Pers N.V.'s freedom of expression and explained that:

> 'the freedom under Art. 10 to impart information of the kind described above is only granted to the person or body who produces, provides or organises it. In other words, the freedom to impart such information is limited to information produced, provided or organised by the person claiming that freedom, being the author, the originator or otherwise the intellectual owner of the information concerned. It follows that any right which the applicant company itself may have under Art. 10 of the Convention has not been interfered with where it is prevented from publishing information not yet in its possession'.[235]

The decision has been seriously criticized. Van Dijk and Van Hoof claimed that the Commission disregarded that the collection of information from any source whatsoever should in principle be free, and that the Commission should have acknowledged this first and then, in a second step, ascertain whether the restriction of freedom of expression was justified on the basis of the 'protection of the … rights of others' in Article 10 (2) of the ECHR.[236] Hugenholtz remarked that the European Commission of Human Rights' decision placed intellectual property uncritically above freedom of expression. It did so without applying Article 10 (2) of the ECHR and acknowledging the need to strike a balance between possible conflicting interests.[237] In other words, as much as Article 10 of the ECHR does not include an automatic right of access to information, it does not include an automatic right to refuse access to information.

The arguments further emphasize the need to weigh differing interests before jumping to the conclusion that one private party should have a right of access to information against another private party. This process of weighing is, arguably, best done in parliament. The wording of Article 10 of the ECHR that restrictions must be prescribed by law and are necessary in a democratic society (Article 10 (2) of the ECHR) further supports this interpretation. This does not exclude that states arrive at the conclusion that there is a need to create a 'right of access to information' against other private parties in the form of statutory rules.[238] If they do so, however, states must carefully balance the interests of all parties concerned according to, among others, Article 10 (2) of the ECHR. From the aforementioned follows that aspects such as the availability of alternative sources of information, the particular public interest character of certain information could play a role[239] and

---

[235] European Commission of Human Rights, Geillustreerde Pers N.V., paragraph 84.
[236] Van Dijk/Van Hoof 1998, p. 562.
[237] Hugenholtz 1989, p. 164; see also Van Dijk/Van Hoof, p. 562.
[238] See also Mackaay 1992, p. 172, referring to information about criminals or hazardous products.
[239] Van Dijk/Van Hoof 1998, pp. 565-566; Mackaay 1992, pp. 168-171. Beers 1992, p. 181.

that the demand of remuneration in exchange for access to information is not as such in conflict with Article 10 of the EHCR.

One situation in which the interest of one party to access information and the interest of the other party to withhold information were weighted repeatedly against each other is the situation of information held by governments.[240] Even in this context, however, the European Court of Human Rights denied an individual right of access to information repeatedly, arguing that Article 10 of the ECHR does not impose an obligation on states to impart information to an individual.[241] It follows that, to the extent that national or regional regulators have acknowledged a right of access to official documents, data, etc. held by public authorities, those initiatives find their rationale in the requirements of a democratic state and the need to enable citizens to control the state, not so much in freedom of expression.[242] Accordingly, the Council of Europe explained that

> 'open access to official documents would allow the public to have an adequate view of and form a critical opinion on the state or the society in which they live and on the authorities that govern them. Information access encourages informed participation by the public in matters of common interest; fosters the efficiency, transparency and effectiveness of administrations and helps maintain their integrity by avoiding the risk of corruption and contributes to affirming the legitimacy of administrations as public services. It strengthens the public's confidence in public authorities'.[243]

---

[240] See Mackaay 1992, pp. 167-171 about the need to distinguish the case of access to government-held information from the case of access to privately held information.

[241] European Court of Human Rights, Leander, paragraph 74. See also European Court of Human Rights, Gaskin, paragraph 52; Guerra and Others, paragraph 53, where a right to certain environmental information was constructed on the basis of Article 8 ECHR; European Court of Human Rights, Sirbu and others v Moldova, Strasbourg, 15 June 2004, No. 73562/01, 73565/01, 73712/01, 73744/01, 73972/01 and 73973/01 [hereinafter 'Sirbu], paragraph 18.

[242] Mackaay 1992, 168pp.; Barendt 1992, pp. 21-22.

[243] Council of Europe (2002), Recommendation (2002)2 of the Committee of Ministers to member states on access to official documents, adopted by the Committee of Ministers, Strasbourg, 21 February 2002, notably paragraph III, with reference to further important recommendations and conventions in this context. For the European Union: Article 255 European Treaty, Treaty on European Union; Regulation (EC) No. 1049/2001 of the European Parliament and of the Council of 30 May 2001 regarding public access to European Parliament, Council and Commission documents, Brussels. 31 May 2001, OJ L 145, p. 43, Recital 2: 'Openness enables citizens to participate more closely in the decision-making process and guarantees that the administration enjoys greater legitimacy and is more effective and more accountable to the citizens in a democratic system. Openness contributes to strengthening the principles of democracy and respect for fundamental rights as laid down in Article 6 of the EU Treaty and in the Charter of Fundamental Rights of the European Union'. European Commission, Communication from the Commission to the European Parliament pursuant the second subparagraph of Article 521 (2) of the EC Treaty concerning the common position of the Council on the adoption of a directive on the re-use and commercial exploitation of public sector information, SEC(2003)627 final, Brussels 28 May 2003, Article 4, to name but a few initiatives. See also M. Mc Gonagle, p. 360pp.; Grosheide 2000, p. 259.

In conclusion, Article 10 of the ECHR does not confer a right of access to information as such.[244] It is already very questionable if such a right is desirable, notably a situation in which one private party could make an enforceable claim against another private party to provide it with certain information it holds. What would this mean for the protection of property, the private sphere and personal autonomy? Even providing such a right was restricted to a right against the media, where to draw the line between private person and media in a time that each consumers can become a 'broadcaster' or 'press service' due to the possibilities of new transmission technologies? In case of pay-TV, a right of access to information might be not even very useful because it is not in the intention of the provider of access-controlled broadcasting to refuse access. His intention is to sell access, although under his own conditions.

But what did the European Court of Human Rights mean in its landmark case Sunday Times? It was in this case that the European Court of Human Rights made its famous claim that 'Article 10 (art. 10) guarantees not only the freedom of the press to inform the public but also the right of the public to be properly informed'.[245] Is there, after all, a right of the public to information? It is not the intention of Chapter 2 to discuss every court case in this area, but two exemplary cases illustrate how this 'right of the public to be informed' is to be understood.

The Sunday Times case concerned the families of thalidomide victims that, as the European Court of Human Rights noted, had a vital interest in knowing all the underlying facts as well as the various solutions and their legal situation. More specifically, the case dealt with the complaint against an injunction by the British High Court to restrain research into the cases of numerous thalidomide victims and the publishing of an article in the Sunday Times that dealt with thalidomide children as well as the settlement of their compensation claims. The European Court of Human Rights found that the families of the victims had a vital interest in knowing all of the underlying facts and that it was 'the right of the public to be properly informed'. The case Plon (Société) v. France follows a similar line. Here, the European Court of Human Rights recognized a legitimate interest of the public to be informed about the state of health of President Mitterrand. The case concerned the publication of a book by Mr Gonod, a journalist and member of several years of the medical staff of President Mitterrand. The book was titled 'Le Grand Secret' and revealed, among other things, information on the state of the President's health. In this case, the European Court of Human Rights found that Article 10 of the ECHR did not form a basis on which the disclosure of certain information could be claimed—unless not disclosing this information would pose a serious danger to

---

[244] In this sense also De Meij/Hins/Nieuwenhuis/Schuijt 2000, p. 306; Hugenholtz 1989, pp. 158-159; Barendt, 1993, pp. 21-23; Mackaay 1992, pp. 167, 171; Badura 1989, p. 322: 'Die Informationsfreiheit gibt dem Informationsinteressenten kein Recht gegen den Staat oder Dritte auf Bereitstellung und Eröffnung von Informationsquellen oder auf richtige und vollständige Informationsdarbietung'; Diesbach 1998, p. 101; Grosheide 2000, p. 223.

[245] European Court of Human Rights, Sunday Times, paragraph 65.

society.[246] The interest that was at stake here was the interest of society in the transparency of political life. The European Court of Human Rights said that 'the more time passed, the more the public interest in President Mitterrand's two seven-year presidential terms prevailed over the requirements of the protection of his rights with regard to medical confidentiality'.[247] In both cases, however, the European Court of Human Rights did not acknowledge an individual right of citizens to seek access to information themselves but of journalists to make the information available.

It should also be noted that in both cases the European Court of Human Rights took no offence to the fact that, in order to receive the information in question (be it on the situation of the thalidomide victims, be it Mr Gonod's book) individuals had to buy the book/newspaper or subscribe to the Sunday Times first. In other words, 'properly' informed does not mean that the public or members of the public must have free-of-charge access to information.

The 'right of the public to be properly informed' must be seen within the context of the role of the media. In the Sunday Times and other cases, the European Court of Human Rights characterized the role of the press as follows:

> 'whilst the mass media must not overstep the bounds imposed on the interests of the proper administration of justice, it is incumbent on them to impart information and ideas concerning matters that come before the courts just as in other areas of public interest. Not only do the media have the task of imparting such information and ideas: the public also has a right to receive them'.[248]

According to this jurisdiction, the mass media play a crucial part in realizing freedom of expression. The media's role is to function as an intermediary—a carrier of the public interest—and to impart information and ideas that the public has a 'right to receive'.[249] It is within the context of this right of the public to receive that the European Court of Human Rights has, in some cases, concluded that the interest in disclosing information must step behind the interest of the media in reporting about it. The European Court of Human Rights stated in Guerra that

> '[i]n cases concerning restrictions on freedom of the press it [the court] has on a number of occasions recognised that the public has a right to receive

246 European Court of Human Rights, Plon (Société) v. France, No. 58148/00, Strasbourg, 18 May 2004 [hereinafter 'Plon (Société) v. France '], paragraph 42.

247 European Court of Human Rights, Plon (Société) v. France, paragraph 53.

248 European Court of Human Rights, Sunday Times, paragraph 65; European Court of Human Rights, Lingens, Strasbourg, 8 July 1986, series A No. 103 [hereinafter 'Lingens'], paragraph 41. Lingens concerned the case of an Austrian journalist who complained about his conviction because of defamation after he had written an article about SS crimes during the Second World War.

249 European Court of Human Rights, Sunday Times, paragraph 65. See also Council of Europe, Recommendation 91(5) on the Right to Short Reporting, Explanatory Memorandum, paragraph 31.

information as a corollary of the specific function of journalists, which is to impart information and ideas on matters of public interest'.[250]

In other words, the media do not only have an important 'watchdog function',[251] their role is also to catch the news and bring it to the 'master' they serve. The role of the individual citizen is a passive one: to receive information and ideas that the media choose to impart.

The Audience—From Receiver to Participant

No smoke without fire—the controversy about the alleged 'right of access to information' points to a more general concern: within the context of electronic access control. The way in which content is (commercially) exploited could result in a situation of electronic exclusion from information. This could have consequences for the functioning of democracy, private, social as well as cultural life in a nation.[252]

The realization that there is an 'other side'—the consumer side—in broadcasting has led some scholars to think more in depth about the consequences of this relationship for the free flow of information. They concluded that, because of the change from traditional broadcasting to an individualized access-controlled distribution pattern, also the challenges for states and its positive obligations concerning freedom of expression change. Access issues come to the fore, and with them the realization that there is a need to pay more attention to the position of the audience. Schulz, Seufert and Holznagel spoke of Zugangschancengerechtigkeit (fair access opportunities).[253] This is the idea that one of the tasks of the media regulators would be to prevent excessive private and exclusionary control over consumers' access to information and to create the conditions that enable citizens and competing broadcasters to benefit from fair opportunities of access to access-controlled content platforms.[254]

---

250 European Court of Human Rights, Guerra, paragraph 53.

251 European Court of Human Rights, Observer and Guardian v. the United Kingdom, No. 51/1990/242/313, Strasbourg, 26 November 1991 [hereinafter 'Observer and Guardian'], paragraph 59.

252 See section 2.2.

253 Schulz/Seufert/Holznagel 1999, p. 93-95.

254 Schulz/Seufert/Holznagel 1990, 95pp. and 108pp. (with explicit reference to conditional access). Also Schulz 1998, pp. 168-190, with reference to Kantian philosophy and the principle of human dignity; Hoffmann-Riem 1990, p. 43, 50pp. and p. 56: 'Deshalb kann und muss die individuelle Rezeptionschancengleichheit abgesichert werden. Dies gilt zum einen in technologischer-räumlicher Hinsicht: Die Einrichtung einer aller Bevölkerungsteile und Gebiete erreichenden Kommunikationsinfrastruktur fällt in den Gewährleistungsauftrag (auch) des Grundrechts der Meinungsfreiheit. Darüber hinaus ist eine den Rezeptionsbedürfnissen entsprechende inhaltliche Vielfalt der Kommunikationsangebote unverzichtbar. Die bloße quantitative Vielfalt unterschiedlicher Rundfunkveranstalter/Programmanbieter bzw. Zeitungen/Zeitschriften reicht dafür nicht'; Birnhack, p. 31 (with view to the public domain); se also the study by Gersdorf 1998 and Gersdorf 2000, p. 210; Larouche 2002a, 19pp. See also the paper by Moglen 1997.

Some Member States have recognized the principle of fair access opportunities more explicitly than others. In Germany, where the above-mentioned argument was made, the German Federal Constitutional Court explicitly recognized the principal of equal opportunity as a constituent element of freedom of expression (here: Article 5 of the German Basic Law).[255] According to the German Federal Constitutional Court, the German media legislator was obliged to prevent a situation of excessively unequal influence by one or more providers.[256] It is worth noting that German broadcasting regulation is based on the so-called dualistic model, where both public and commercial broadcasters participate in the mission of the media to 'inform the audience properly'. The following paragraphs investigate whether there is some broader basis to argue in favour of embracing such a principle in dealing with conditional access inside and outside of Germany's borders.

Communication, meaning the process of receiving and imparting information of which broadcasting is a part, does not usually take place in a social vacuum. It is a social activity that involves at least two parties. In the case of the mass media, an indefinite number of parties may be involved.[257] For the human being as part of a social structure, communications is not only a means to self-development, it is also a means to fit in and maintain social structure(s). Our social structures are largely based on the gathering, processing and disseminating of information, be it as a means to maintain personal contacts (our personal network), to assess and influence the course our society takes, or decisions that are relevant for society. According to this 'social' understanding of communication, the media is a means to step out of one's own individual sphere and enter into, and contribute to the creation of a public forum, a democratic society.

There is a close relationship between participation, freedom of expression and democracy. The free communication of ideas and opinions is what enables intellectual discourse—the competition of ideas and opinions that is so essential to democracy.[258] This is why

> 'The Court's supervisory functions oblige it to pay the utmost attention to the principles characterizing a "democratic society". Freedom of expression constitutes one of the essential foundations of such a society, one of the basic conditions for its progress and for the development of every man'.[259]

---

[255] German Federal Constitutional Court, Blinkfuer, Case No. 1 BvR 619/63, 26 February 1969, Collection E 25, p. 256 [hereinafter 'Blinkfuer'], p. 265.

[256] German Federal Constitutional Court, Niedersächsisches Rundfunkgesetz, Case No. 1 BvF 1/84, 4 November 1986, Collection E 73, p. 118 [hereinafter 'Niedersächsisches Rundfunkgesetz'], p. 118.

[257] Hoffmann-Riem 1990, p. 36 speaks of 'Freiheit auf Gegenseitigkeit'.

[258] See German Federal Constitutional Court, Lüth, Case No. 1 BvR 400/51, 15 January 1958, Collection E 7, p. 198 [hereinafter 'Lüth'], p. 208; German Federal Constitutional Court, Schmid, Case No. 1 BvR 9/57, 25 January 1961, Collection E 12, p. 113 [hereinafter 'Schmid'], p. 125.

[259] European Court of Human Rights, Handyside, paragraph 52; Sunday Times, paragraph 65.

Freedom of expression as one of the essential foundations of a democratic society reveals something about the position of the individual in the process of public communication. Democracy as a form of governance is in its essence a participatory model, meaning a model that it based on the fact that each citizen has the possibility to participate (whether a citizen does so is a totally different question). The central notions of a democracy are the ideas of (political) equality and the opportunity for all members of the society to participate and bring in their different interests, preferences, ideas and opinions.[260] Rawls calls this the principle of 'equal participation'.[261] It entails, so to speak, an element of fairness and non-discrimination.

If the functioning of a democracy is based on the freedom to receive and impart information, and if the role of the media is to inform citizens, then it is consistent to assume that the principle of fair opportunities to participate would also apply to citizens' participation in the media. The German scholar Hoffmann-Riem called this kommunikative Chancengleichheit. Being a social freedom,[262] freedom of expression relies on a balanced form of 'with each other': citizens must be able to freely communicate with each other in a way that each citizen has a fair opportunity to develop himself of herself by exercising this freedom. One party's ability to influence the chances of another party to participate can cause asymmetries.[263] This could lead to inequalities not only in the communicative but also in the democratic process.[264] What is remarkable about this view is that the position of the members of the audience and their relationship towards the media is central as is the acknowledgement that the citizens, alias consumers, play an active role in the communicative process

In traditional broadcasting, fair access opportunities is a fairly well-known principle, even if it is not always referred to by this name. Public service broadcasting is an important example. The public broadcasting programme is a programme that is universally available[265] and that is obliged to carry a mixture of content that reflects the different ideas, opinions, etc. that are present in society. The programme that is delivered to the audience is meant to be a mirror of society; it gives the different groups the opportunity to participate in the media and express their opinions in the public debate. One source of public broadcasting financing is the public broadcasting fee, whereby each member of the audience contributes an equal share and those who cannot afford the fee can usually obtain state support. There are also terrestrial frequencies: citizens who are not subscribed to a cable or satellite operator can still receive a number of programmes, including public

260 Birnhack 2004, p. 13; Rawls 1971, p. 225; Habermas 1979, 268pp.

261 Rawls 1971, p. 221. Instructive also the paper by Birnhack 2004 on the role of copyright law as a means to maximize the citizens' ability to participate in the democratic process.

262 Hoffmann-Riem 1990, p. 38.

263 Hoffman-Riem 1990, p. 38, 40pp. Cohendet 2003, p. 6.

264 Rawls 1971, p. 225.

265 See below section 2.2.3.

broadcasting, via terrestrial frequencies. In these examples, everybody is in principle able to access broadcasting; a problem of unfair access opportunities does not usually arise.

The focus on the role of the media to inform the audience of matters of importance is characteristic for the institutional approach of broadcasting regulation.[266] It focuses on the media's mission. This institutional approach of broadcasting regulation is not designed to address the position of individual members of the audience and their relationship with broadcasters.[267] The institutional approach of broadcasting regulation is strongly reflected in the rather paternalistic approach to national broadcasting laws, the rules on programme standards, quota and, last but not least, public broadcasting.[268]

A reason that was often used to justify the paternalistic approach of broadcasting regulation is the lack of responsiveness of broadcasting, which was said to be 'one of the most difficult problems for media regulation'.[269] 'Responsiveness' refers to the extent to which media can and will respond to the interests and needs of the audience. For example, the responsiveness of the World Wide Web seems to be relatively high because consumers can actively search for content in a global pool of content. It is said that on in the press or on the internet, the mechanism of offer and demand allows for a fairly precise understanding of consumer preferences and interests.[270] In contrast, traditional broadcasting, a 'once-sent-free-access-for-all' medium, has only limited possibilities for learning about the public's needs. Possible instruments are viewing figures, surveys and studies. The following quote is very characteristic of the present concept of broadcasting regulation: 'The basic proposition in conventional PSB [public service broadcasting] thinking is that while broadcasting is designed to benefit viewers and listeners, they neither know what they want nor where their interest lies'.[271] Most broadcasting laws reflect an undisguised belief that consumers do not know what they want or need, an idea that has led to a paternalistic concept in which it is the state that finally decides for consumers.

Arguably, conditional access could bring some changes. Because access-controlled content services are sold directly to the consumer, the service providers are probably the most interested of all players in complying with consumer demand.[272]

---

[266] European Court of Human Rights, Sunday Times, paragraph 65; European Court of Human Rights, Lingens, Strasbourg, paragraph 41. Barendt 1993, p. 49; De Meij/Hins/Nieuwenhuis/Schuijt 2000, p. 299-302. Wentzel 2002, p. 57pp.

[267] See section 1.3.3. See also De Meij/Hins/Nieuwenhuis/Schuijt 2000, p. 304.

[268] See sections 1.3.3. and 2.2.3.

[269] Gibbons 1998, p. 54.

[270] See the range of general and special interest journals and papers that are available at every better newsagent.

[271] Cited in Peacock 1989, p. 53.

[272] Scholz 1995, p. 362, refers to pay-TV as 'im Grunde fast idealtypische Form eines Rundfunks, der strikt den Maßstäben eines vorrangig an der Publikumsnachfrage orientierten Programmangebots

> 'On a more fundamental level, programme choice might be usefully made more responsive to viewer welfare, and less responsive to the notions put forward by philanthropic institutions of what people "ought" to see'.[273]

Conditional access could be a means of improving media responsiveness and, at the same time, an argument against government involvement with media content[274] as well as overly 'elitist patronizing behaviour'.[275] If the offer of a broadcaster were indeed demand-driven, consumers could have a new, powerful means of shaping information markets as a mirror of their preferences and needs. Perhaps 'information on demand' is 'all that is essential to freedom of expression (from a constitutional viewpoint) providing consumers demand the right information about political matters'?[276]

It still remains to be seen whether pay-TV would give consumers what they really want, even if they knew what that was. Pay-TV operators are in a position to differentiate between consumers who are willing and/or able to pay the subscription fee, willing to subscribe to a particular bundle, whose hardware is fit to process the signals of a particular operator, etc. The operator can also distinguish between the consumer's residence or age,[277] to name a few of the possible criteria. Unlike traditional broadcasting, members of the broadcasting audience are not equal in the sense that once a signal is transmitted not everybody in its transmission radius can receive it. Much will depend on the responsiveness of pay-TV and whether it is profitable for pay-TV providers to serve small audience segments. Access-controlled service platforms operate within the boundaries of profitability and audience maximization and therefore may choose to avoid cost-intensive high quality or minority programmes if it is likely that they will attract only small proportions of the population.[278] Moreover, the possibilities for a viable business model are probably restricted. For example, pay-TV still relies on premium programming and T-commerce services, notably in countries where there is still rich free-to-air programming. The viability of new channels finds its natural limits in the amount of money consumers are willing to spend on (additional) services. Likewise, not all right holders will choose to licence their programme rights to pay-TV. Sports event organizers, for example, can also have an interest in the wide distribution of the event. In pay-TV, fair access opportunities become a matter of

---

folgt'. Owen 1975, p. 134; critical: Spence/Owen 1977, p. 164, however also saying that the situation might be worse under advertising-financed television than under pay-TV, because pay-TV prices still reflect the intensity of preferences better than the flat rate paid by advertisers. Also Wentzel 2002, pp. 54, 91; Peacock 1986, section 133; Noam 1988, pp 211-212.

273 Owen 1975, p. 134

274 This argument applies mainly to the public choice model of media regulation, meaning government involvement with the structure of media content. It does not apply to rules that are supposed to ensure the quality and security of media services such as rules on youth protection and against hate speech.

275 Wentzel 2002, p. 57pp.

276 So Owen 1977, p. 27.

277 See section 1.5.2.

278 See section 1.4.3.

the relationship between access controller and consumer and the terms and conditions in the subscription contract. Legislators can respond different ways. One way would be to argue that the realization of public information policy interests in broadcasting, and here in particularly the 'right of the public to be properly informed', is the task of selected broadcasters. This can be free-to-air-broadcasters in general or, more specifically, public broadcasters (see section 2.2.3.). In this case, legislators would want to ensure in the first place that pay-TV does not have a negative impact on the viability of public broadcasting and the realization of this task. The other way is to argue that new technical and economic developments offer new opportunities for the media's mission to inform the audience properly, and for the audience to participate in the public communications process.[279] In the latter case, access issues and the issue of fair access opportunities come to the fore. If fair access opportunities and the broad availability of content is a rationale behind national and European media policy not only for public broadcasting, then there is a need to rethink the current approach. One crucial aspect in terms of regulation of access-controlled broadcasting will then be whether the terms and conditions under which the content is provided are principally set up in such a way that everybody has the opportunity to access content. Consumers should not be excluded arbitrarily[280] or because of insurmountable technical or financial obstacles. This could be the case, for example, if a platform operator were to issue only a limited number of suitable set top boxes or if subscription fees were so high that only few consumers could subscribe, if electronic access control techniques would exclude—without objective justification—disabled or elderly people, people from other Member States or if the way standards are used led to an evident fragmentation of society.

Pluralism and Diversity

Closely related to the principles of democracy, freedom of expression and fair access opportunities, is the idea of pluralism and diversity. This is the idea that viewers should have access to a broad variety of ideas and opinions that are presented to them by the media. Pluralism and diversity are probably the most important quality criteria in broadcasting law. The European Court of Human Rights has frequently stressed that the role of the media in a democratic society

> 'cannot be successfully accomplished unless it is grounded in the principle of pluralism, of which the State is the ultimate guarantor. This observation is

---

[279] See extensively Schulz 1998, 186pp. and 188pp. Looking forward, also the European Court of Human Rights, Lentia, paragraph 39. For example, adaptation to technical and economic circumstances is one reason that has contributed to a development in some national broadcasting regulations, such as in Germany, the UK and France, tha commercial broadcasting participates in the mission of the media to inform the public.

[280] See also Article 14 of the ECHR (Prohibition of discrimination).

especially valid in relation to audio-visual media, whose programmes are often broadcast very widely'.[281]

The court also acknowledged that it can be desirable to not only have a plural programme offer but also an offer that comes from diverse sources.[282] This is also what the Council of Europe says in its Declaration on Freedom of Expression and Information:

> 'Convinced that states have the duty to guard against infringements of the freedom of expression and should adopt policies designed to foster as much as possible a variety of media and a plurality of information sources, thereby allowing a plurality of ideas and opinions'.[283]

The idea behind pluralism and diversity is that the programme that is finally delivered to the audience represents the different groups and opinions in the society. This is meant to offer to all groups, including minority groups, the ability to impart and receive information they are interested in (pluralism), and to protect this process from excessive influence by one party (diversity). A media landscape can lack pluralism if there is a number of competing information service providers that provide more or less the same mass-attractive content. It lacks diversity if the only existing pay-TV platform encompasses fifty channels that are all under the journalistic control of one major operator.

The competition of ideas and opinions that characterizes a pluralistic and diverse media offer must be distinguished from economic competition, although one can influence the other.[284] Monopolizing the consumer base and making it more difficult for subscribers to switch between different services[285] has more than just a detrimental effect on competition. Exclusionary strategies, technical and contractual lock-in situations, and the centralization of large quantities of content in the hand of one or more intermediary platforms can also impact pluralism and diversity inside and outside the platform.[286]

---

281 European Court of Human Rights, Lentia, paragraph 38; European Court of Human Rights, Lingens, paragraph 41

282 European Court of Human Rights, Lentia, paragraph 39. Barendt 1993, pp. 80-81 and 127-128 (including a comparative overview).

283 Council of Europe, Declaration on the Freedom of Expression and Information, Recital 6.

284 Valcke 2003, p. 683-689; Larouche 2002, p. 140-145; Van den Beukel/Nieuwenhuis 2000, pp. 117.

285 See sections 1.5.2. and 1.5.3.

286 For the Council of Europe, see ECTT, Explanatory Memorandum, paragraph 175. For the European Union, see Access Directive, Recital 10: 'National legal or administrative measures that link the competition rules alone may not be sufficient to ensure cultural diversity and media pluralism in the area of digital television'. See also Directive 95/47/EC provided an initial regulatory framework for the nascent digital television industry which should be maintained, including in particular the obligation to provide conditional access on fair, reasonable and non-discriminatory terms, in order to make sure that a wide variety of programming and services are available'. Barendt 1993, 121pp.

Preliminary Conclusion

To draw a preliminary conclusion, although often invoked, there is no such thing as an individual right of access to specific privately controlled content, at least not one that would flow from Article 10 of the ECHR. The 'right of the public to be properly informed' has to be read within the context of the task that the media have to perform. As long as pay-TV is only one additional broadcasting offer alongside free-TV and other media offerings, the realization of freedom of expression is not in acute danger. Perhaps something different would apply in the—not very likely—situation that a pay-TV platform operator controlled access to all broadcasting services, including free-to-air services, or where there is no other way of gathering information about news or events that could contribute to the public debate. In this case, states would be obliged to assess whether the realization of freedom of expression is at stake and eventually interfere.

Still, there is a positive obligation of the media legislator to create the conditions for the realization of freedom of expression in general.[287] Arguably, technological changes call for a fresh look at the media's mission to inform the public properly. And, with technological changes, the demands on the broadcasting regulator to create the conditions for an environment that promotes the realization of expression principles and democratic principles change. Providing pay-TV becomes a more common form of financing broadcasting, the role of pay-TV for the public communications process could gain importance, and with it the idea of fair access opportunities.

Already the European Council of the European Union observed that one of the objectives in the fight against social exclusion is to 'exploit fully the potential of the knowledge-based society and of new information and communications technologies and ensure that no-one is excluded'.[288] And, according to the eEurope 2005 Action Plan, one goal is to give everyone the opportunity to participate in the global information society.[289] In this context, the role of digital TV, of which pay-TV is commonly considered an important driver, has been explicitly acknowledged as a means to have access to content and innovative services.[290] The difference between pay-TV and traditional free-TV is that in the pay-TV business model consumption is a privilege of selected consumers. This can lead to inequalities in the terms and conditions of access to content that did not exist, or do not exist in this form, in traditional free-TV.

---

[287] See section 2.2.1.

[288] European Council, Fight against poverty and social exclusion: common objectives for the second round of national Action Plans, Brussels, 25 November 2002, SOC 470 [hereinafter 'Fights Against Poverty and Social Exclusion'], Annex to Annex 2, objective 2 (a).

[289] See also section 2.2.4.

[290] European Commission, Action plan eEurope 2005, pp. 2, 7.

2.2.2. ACCESS TO CONTENT OF PUBLIC IMPORTANCE

It is interesting to note that neither the list of important events nor the right to short reporting are designed to protect any kind of information other than that of a high or major public importance. States can have distinct national policy interests outside of freedom of expression to preserve the public's access to particular content of public importance and ensure that matters of interest can be received by the whole population rather than by a small circle of privileged subscribers. There are various reasons why content can be of public importance, meaning of importance for society as a whole rather than for single individuals. These can be reasons of public security, such as catastrophe warnings, health-care education and product warnings. Information can be of public importance for political, social or democratic reasons, such as information about major political developments in a country or in the European Union, access to laws and government decisions as well as sports and cultural events. There are also societal reasons to ensure broad access to such events: to give the public the possibility to discuss certain matters of social interest on a wide scale, which requires that the broad public is able to take notice of the matter, and that access to such matter is not restricted to a narrow group. That it can be politically wise to satisfy the citizen's desire for events is something the Roman emperors already understood. The Roman emperors went even a step further and organized the 'games' themselves. However, even in the absence of major national policy reasons, there can be a public interest in avoiding that the popularity and interest that some events generate is exploited for purely commercial reasons, and that such events are not affordable to the majority of the public. After all, popular content is the most prone to being exploited in pay-TV. Such interests can be public welfare arguments such as the promotion of creativity and innovation (for example, the availability of scientific principles and formulas, data, research methods, public domain material or creative works), of democratic processes or of education.[291] It could also be that content is publicly funded, which gives the public a legitimate interest in not being excluded or in not being made to pay a second time for access to this content.[292] This concerns, for example, the highly controversial issue of commercialization of public government information or publicly funded content,[293] public broadcasting content being one example.

2.2.3. PRESERVATION OF PUBLIC BROADCASTING

It is worth noting that the transmission of major sports events, subject to the list of important events and the right to short reporting (see section 2.3.1. and section 2.3.2.), was for a long time the domain of public broadcasters. As public

[291] A thorough discussion of the theoretical foundations of the idea of the commons or the public domain can be found in Ostrom/Hess 2001, 47pp.; Elkin-Koren 2001, 197pp.

[292] More in depth on this problem, Van Eechoud 1998, 178pp.

[293] See the studies by Dommering/Kabel/Hugenholtz 2002 and by Hesse 2002.

broadcasters argue themselves, the transmission of important sports events is one major factor that allows public broadcasters to successfully compete with commercial operators for the citizens' attention. Due to the increasing prices of sports events, public broadcasters are, in the mid-term future, no longer able to compete with pay-TV operators when bidding for transmission rights. Guaranteeing access to the most popular sports events for free-TV is, thus, also a means to improve the chances of public broadcasting.

Traditionally, broadcasting law reserves a major role for public broadcasting. Barendt describes public broadcasting as a descriptive as well as a normative concept and identifies six principal features: universal geographical availability; concern for national identity and culture; independence from state and commercial interests; impartiality; range and variety of programmes and financing by imposing a general charge on consumers.[294] Public broadcasting plays an important role in the realization of public information policy objectives such as the realization of freedom of expression, democracy, pluralism, diversity and fair access opportunities.[295] Public broadcasters were, and still are, a preferred and universally accessible platform for governmental public information policy.[296] Thus public broadcasting is also an important and powerful political tool.

The prominent role of public broadcasting in a multi-channel environment is not self-evident. One could also argue, as some do, that the increasing diversification and the arrival of a multi-channel environment is a reason to curtail the importance of the mission of public broadcasting. If the public had access to a multitude of different niche channels, and if interactivity stimulated demand orientation and diversification,[297] would digitization really 'reinforce the importance of the comprehensive mission of public service broadcasters'? Perhaps it would even diminish it.

For the time being, pay-TV will not render public broadcasting meaningless, and public broadcasting will continue to play a role. On the other hand, it is worth asking what the task of public broadcasting in the digital multi-channel environment is. The Council of Europe recommended maintaining public broadcasting in the new digital environment 'by ensuring universal access by individuals to the programmes of public service broadcasters and giving it, among other things, a central role in the transition to terrestrial digital broadcasting'.[298] This

---

[294] Barendt 1993, p. 52. See also Treaty of Amsterdam, Amending the treaty on European Union, the treaties establishing the European Communities and related acts, Protocol (32) on the system of public broadcasting in the Member States, Brussels 10 November 1997, OJ C 340, p. 308 [hereinafter 'Protocol (32) on the System of Public Broadcasting in the Member States'].

[295] That this is acknowleded at the European level, too, is demonstrated in the Protocol 32 on the System of Public Broadcasting in the Member States.

[296] Van Eijk 1992, 185pp.

[297] See section 1.3.3.

[298] Council of Europe, Recommendation Rec(2003)9 of the Committee of Ministers to Member States on measures to promote the democratic and social contribution of digital broadcasting, Strasbourg, 28

can be a reason to argue, for example, that public broadcasting services should receive a prominent place in an EPG directory.[299] It is interesting to note that the Council of Europe does not describe the role of public broadcasting as a source of high quality content only. In the digital environment, it is not only the content, but also the universal accessibility of public broadcasting that matters. The Council of Europe has even assigned public broadcasting an active role in the roll-out of digital terrestrial networks. In this sense, public broadcasting has been upgraded to not only provide content, but to provide a public forum, or 'virtual arena', that is open to the public. It would exceed the scope of this study to discuss this subject in depth, however, a possible answer could be that, in the future, the public mission of public broadcasting will move away from its (only) task of supplying high-quality content towards providing public access and a forum for social cohesion.

2.2.4. THE FREE MOVEMENT OF SERVICES

National public information policy is also influenced by the policies of the European Union. The realization of the Internal Market and the free movement of services (as laid down in Articles 28-30 and Article 49 of the EC Treaty) is a central objective of the European Communities. Services, in the sense of the Treaty, are also broadcasting services.[300] Article 49 of the EC Treaty entails in the first place a duty for Member States to refrain from or to eliminate any direct or indirect discriminatory public actions (for example, in the form of laws) that are likely to prohibit, impede or render less advantageous cross-border economic activities.[301]

May 2003 [hereinafter 'Recommendation Rec(2003)9 on measures to promote the democratic and social contribution of digital broadcasting'], Appendix, paragraph 21.

299 See section 4.5.

300 See European Court of Justice, Judgment of the Court of 30 April 1974, Giuseppe Sacchi, Reference for a preliminary ruling: Tribunale civile e penale di Biella - Italy, Luxembourg, Case 155-73, European Court Reports 1974, p. 409 [hereinafter 'Sacchi'], paragraph 6; European Court of Justice, Judgment of the Court of 18 March 1980, Procureur du Roi v Marc J.V.C. Debauve and others, Reference for a preliminary ruling: Tribunal de première instance de Liège - Belgium, Luxembourg, Case 52/79, European Court reports 1980, p. 833 [hereinafter `Debauve'], paragraph 8; European Court of Justice, Judgment of the Court of 18 June 1991, Elliniki Radiophonia Tiléorassi AE and Panellinia Omospondia Syllogon Prossopikou v Dimotiki Etairia Pliroforissis and Sotirios Kouvelas and Nicolaos Avdellas and others, Reference for a preliminary ruling: Monomeles Protodikeio Thessalonikis – Greece, Luxembourg, Case C-260/89, European Court reports 1991, p. I-2925 [hereinafter 'ERT'], paragraphs 20 to 25; European Court of Justice Judgement of the Court of 5 October 1994, TV10 SA v Commissariaat voor de Media, Reference for a preliminary ruling: Raad van State - Netherlands, Luxembourg, Case C-23/93, European Court reports 1994, p. I-04795 [hereinafter `TV10'], paragraphs 13 and 16, to name but some.

301 See only European Court of Justice, Judgment of the Court of 11 July 1974, Procureur du Roi v Benoît and Gustave Dassonville, Preliminary ruling requested by the Tribunal de première instance de Bruxelles – Belgium, Luxembourg, Case 8-74, European Court reports 1974, p. 837 [hereinafter 'Dassonville'], paragraph 7; European Court of Justice, Judgment of the Court of 24 November 1993, Criminal proceedings against Bernard Keck and Daniel Mithouard, References for a preliminary ruling: Tribunal de grande instance de Strasbourg - France, Joined cases C-267/91 and C-268/91, European Court reports 1993, p. I-06097 [hereinafter `Keck'], paragraph 11.

But the realization of the Internal Market is more than just a duty for states to respect Article 49 of the EC Treaty. States also have a positive obligation to take every measure that is necessary and appropriate to ensure the realization of the free movement of services:

> 'As an indispensable instrument for the realization of a market without internal frontiers, Article 30 therefore does not prohibit solely measures emanating from the State which, in themselves, create restrictions on trade between Member States. It also applies where a Member State abstains from adopting the measures required in order to deal with obstacles to the free movement of goods which are not caused by the State'.[302]

Similar was also the dictum of the European Court of Justice in the Schmidberger case. The case concerned a complaint by the international transport enterprise Schmidberger against the Republic of Austria regarding the block of the Brenner motorway by a demonstration 'to protect the biosphere'. The Austrian authorities did not ban the demonstration in respect of the demonstrators' freedom of expression and freedom of assembly. Schmidberger argued that the decision not to ban the demonstration amounted to an infringement of Austria's obligations under the EC Treaty to promote the realization of Internal Market freedom (here: free movement of goods). Upon the request of the Austrian Court of Appeals, the European Court of Justice interpreted the principle of free movement of goods under Article 30 et seq., and within the context of Article 5 of the EC Treaty. It found that Article 5 of the EC Treaty required Member States to take every measure that is necessary to fulfil the obligations arising from the Treaty. This obligation would also include the duty to undertake adequate steps to ensure freedoms under the Treaty in situations in which its realization is obstructed as a result of actions taken by private parties.[303]

Chapter 1 explained why the use of electronic access control could prevent consumers in one country from receiving services from another country and that this could lead to a territorial fragmentation of the Internal Market. The way pay-TV services are marketed in some Member States can impede the realization of Internal Market principles. The Schmidberger decision in mind, one can argue that Member States have a positive obligation to make sure that the use of conditional access does not prevent consumers in one country from receiving content services offered by a pay-TV provider in another country. Such action would, of course, also

---

302 European Court of Justice, Judgment of the Court of 9 December 1997, Commission of the European Communities v French Republic, Luxembourg, Luxembourg, Case C-265/95, European Court reports 1997, p. I-06959 [hereinafter `Commission v. France'], paragraph 30. See also Schwartz who argues that this obligation applies not only to restrictions from governments, but also from third parties, Schwartz 1985, p. 116.

303 European Court of Justice, Judgment of the Court of 12 June 2003, Eugen Schmidberger, Internationale Transporte und Planzüge v Republik Österreich, Reference for a preliminary ruling: Oberlandesgericht Innsbruck - Austria, Case C-112/00, European Court reports 2003, p. I-05659 [hereinafter 'Schmidberger'], paragraphs 57-62.

have to take into account the legitimate interests of the services providers and right holders, such as technical reasons (standardization) or obligations towards rights holders.[304] The positive duty to realize Internal Market principles, for example, could also lead to the need to scrutinize more closely the way in which broadcasting rights are licensed on a national basis; a practice that is also responsible for territorial restrictions in pay-TV.

In its Review on the Application of the Cable and Satellite Directive, the European Commission already observed that the practice of transferring rights on a national basis and in combination with the use of electronic access control could run counter to European policies to stimulate the Internal Market. The European Commission criticized in particular the practice of licensing broadcasting rights on a per country basis. It found that only the transfer of rights for a programme for the entire footprint was compatible with Internal Market principles. According to the European Commission, country-based licensing together with the use of encryption technologies in broadcasting was

> 'a problem that affects the European citizens' direct perception of the reality of the Internal Market in their daily life and which thus has an appreciable negative impact in terms of cultural, linguistic, social and economic interpenetration at the intra-community level'.[305]

This clearly indicates that, in Europe, there is already a process of rethinking the desirability of territorial licensing agreements, notably within the context of pay-TV. In the same document, the Commission called for initiatives to encourage a non-national approach to allow the Internal Market to be a genuine market without internal borders for right holders, operators and consumers.

Still, states have to weigh community principles against freedom of expression and property rights of pay-TV providers and right holders. The Schmidberger case is also interesting in this respect as it raises the question of how to reconcile the requirements of the protection of fundamental rights, here of the demonstration, and the freedoms enshrined in the EC Treaty. The European Court of Justice found in the Schmidberger case that the decision not to ban the demonstration was objectively justified. The protection of fundamental rights, so says the European Court of Justice, is a legitimate interest that justifies a restriction of the obligations

[304] See European Court of Justice, Judgment of the Court of 18 March 1980, SA Compagnie générale pour la diffusion de la télévision, Coditel, and others v Ciné Vogt Films and others, Reference for a preliminary ruling: Cour d'appel de Bruxelles – Belgium, Luxembourg, Case 62/79, European Court reports 1980, p. 881 [hereinafter 'Coditel I'], paragraph 16; European Court of Justice, Judgment of the Court of 6 October 1982, Coditel SA, Compagnie générale pour la diffusion de la télévision, and others v Ciné-Vogt Films SA and others, Reference for a preliminary ruling: Cour de cassation – Belgium, Luxembourg, Case 262/81, European Court reports 1982, p. 3381 [hereinafter 'Coditel II'], paragraph 19. See also European Court of Justice, ERT, paragraphs 22-26, concerning the exclusive right to broadcast.

[305] European Commission, Report on the Application of the Cable and Satellite Directive, pp. 7-8.

imposed by Community law.[306] The European Court of Justice, however, has observed that Article 10 of the ECHR could also be subject to restrictions in the general interest. The realization of Internal Market principles is such an interest. Therefore, freedom of expression interests would have to be weighed against the principle of the free movement of services.[307] The competent authorities would enjoy a wide margin of discretion but restrictions placed upon trade within the Community had to be proportionate within the context of the legitimate objective to protect fundamental rights.[308]

2.2.5. CONCLUSION

The motives for regulatory interference with electronic control of access to broadcasting content are various and reach from freedom of expression to national public policy interests and goals that are derived from supranational obligations such as that to ensure the free flow of services within the Internal Market. The information policy considerations discussed within the context of electronic access control pivot around the possible conflicts between, on the one hand, the electronically enforced exclusivity of content, and, on the other hand, the accessibility of such content for the public.

2.3. European Regulatory Instruments that Address Conditional Access

The relationship between electronic access control and the accessibility and availability of information is slowly receiving more attention in the public information policy discussion. The opinions, however, are still very divided on the question of how electronic access control in broadcasting is best approached.

In the second part of Chapter 2, a number of initiatives are discussed that were meant to adapt media law to the 'age of access' and translate the aforementioned public information policy concerns into media law. At the EC level, the leading provision is Article 3a of the TWF Directive (list of important events, section 2.3.1.). Article 3a of the TWF Directive is, so far, the only regulation in European broadcasting law that has been drafted specifically to deal with pay-TV. It has substantially influenced the way Member States deal with the content-related aspects of electronic access control in broadcasting. Other already existing instruments have also been brought into play and apply or were suggested to be applied to pay-TV. These are the right to short reporting (section 2.3.2.), which is provided for in Article 9 of the ECTT of the Council of Europe, and the must-carry

---

306 European Court of Justice, Schmidberger, paragraphs 69, 72-72 and 78-93.

307 See also European Court of Justice, ERT, paragraphs 41-45. See also Van de Gronden/Mortelmans 2003, 16pp.

308 European Court of Justice, Schmidberger, paragraphs 80-82. See also section 3.2.4.

rules in Article 31 of the Universal Service Directive (section 2.3.3.). The goal of the analysis is not so much to describe the present legal situation in detail at the level of the EC or in the Member States, but to identify on a more abstract level the different concepts that have been developed to deal with conditional access.

2.3.1. List of Important Events

Article 3a (1) of the TWF Directive reads:

> 'Each Member State may take measures in accordance with Community law to ensure that broadcasters under its jurisdiction do not broadcast on an exclusive basis events which are regarded by that Member State as being of major importance for society in such a way as to deprive a substantial proportion of the public in that Member State of the possibility of following such events via live coverage or deferred coverage on free television. If it does so, the Member State concerned shall draw up a list of designated events, national or non-national, which it considers to be of major importance for society. In so doing the Member State concerned shall also determine whether these events should be available via whole or partial live coverage, or where necessary or appropriate for objective reasons in the public interest, whole or partial deferred coverage'.

Article 3a recognizes the right of Member States to draw up so-called 'lists of important events'. The lists identify events of particular public importance that should be shown—in their entirety or partially—on free-TV. Pay-TV operators are not banned from showing designated events altogether providing the public has the possibility of following such events on free-TV as well. Note that Article 3a of the TWF Directive applies only to transmission rights for organized events and not to other content, for example, content that is subject to copyright law such as films.

One way to limit exclusivity is to mandate the non-exclusivity of transmission licences for events with the effect that all broadcasters are free to acquire a transmission licence. This is not the course the directive took. The directive does not address event organizers but service providers.

The basic concept behind Article 3a of the TWF Directive and the similarly worded Article 9bis of the ECTT[309] is to limit the exclusive exploitation of transmission rights for the sake of a general public interest in the wide accessibility of certain content.[310] The list-of-important-events concept has been already described as a 'new category of universal service'.[311] Pay-TV operators are not entitled to the exclusionary exploitation of such rights. Unclear is what the scope of

309 In the following, reference is made only to Article 3a of the Television Without Frontiers Directive. Only where the regulation in Article 9bis of the ECTT is substantially different, will it be mentioned explicitly.

310 See also the papers by Hins 1998; Helberger 2002, and sections 2.2.2. and 2.2.3. of this chapter.

311 Bavasso 2003, p. 388.

the limitation on the exclusive exploitation of such events is, what its duration is and under which conditions a partial or deferred coverage is acceptable. Bearing in mind that the rights for deferred or partial coverage are sold separately, this question is important for the future licensing policy of such rights.

Indirectly, the regulation does concern event organizers.[312] The likely effect of the lists is that pay-TV operators stop bidding for such events, are only willing to bid at lower prices, or only bid for events that free-TV broadcasters do not bid for. If pay-TV providers wish to sell access to listed events even though they are available on free-TV, they will need to find new and more attractive forms of presentation. Examples could be the transmission in a High Definition format, added interactive applications or an otherwise enhanced service offering.

In so far as pay-TV operators have already signed for exclusive transmission rights after a national list has entered into force, they will probably also have to offer those rights to free-to-air television providers, for example, in the form of sub-licences. The directive is silent on an eventual duty of event organizers to cooperate and agree on the issuing of sub-licenses to free-TV providers. Moreover, the directive does not provide guidelines on the conditions under which pay-TV operators have to grant such sub-licences to free-TV. Obviously, such conditions have to be reasonable and adequately priced as the provision would otherwise run into a void.

Not regulated in Article 3a of the TWF Directive is the question of what happens if no free-TV broadcaster is willing to acquire the transmission rights for a certain event, for example, because the licences are too costly.[313] This could lead to the event not being transmitted at all. Article 3a of the TWF Directive aims at preventing broadcasters from transmitting such events on an exclusive basis; it does not oblige free-TV broadcasters to broadcast such events on a non-exclusive basis. Strictly speaking, the list concept does not guarantee 'access' to information of major importance, but rather only enables such events to be made 'accessible' to the public in free-TV.

Events should not be transmitted in a way that deprives 'a substantial proportion' of the public rather than individual citizens from accessing such content. In other words, for the application of Article 3a of the TWF Directive, size matters when it comes to deciding whether public information policy interests are at stake or not. This is another indication that one goal behind the list-of-important-events concept is the realization of universal accessibility. The qualification of a 'substantial proportion' is left to the Member States. While in Germany[314] the substantial proportion consists of two-thirds of all households and only seventy per cent in

---

[312] Interesting in this context is the Irish implementation law that foresees that the licensing contract between the pay-TV provider and the event organizer can be declared void (see below).

[313] So far, only the UK has regulated this case (see below).

[314] Article 5(a) of the Fourth Interstate Treaty on Broadcasting, 29 September 2000, OJ C 227, p. 4.

Austria,[315] Ireland considers a free-TV broadcaster qualified if he reaches at least eighty-five per cent of the population.[316] Italy[317] and the UK[318] consider ninety per cent of the population to be substantial. The idea of one broadcasting channel being able to reach seventy to ninety per cent of the population is still a realistic concept in a digital multi-channel environment. One also wonders what the situation would be if, in the future, pay-TV were to prosper and access-controlled platforms succeeded in reaching a substantial proportion of the population. In this case, at least one criterion for the application of the list-of-important-events regulation would not be fulfilled. Article 3a of the TWF Directive, however, does not seem to consider the broad accessibility of certain events on pay-TV a satisfactory option. It makes very clear that the public must have the possibility to follow such events on 'free television'.

This leads to the principal question of what 'free television' really is. The observation made in Chapter 1 of this study should be borne in mind: it cannot be excluded that traditional free-TV services such as public broadcasting services will encrypt their services too, for example, to comply with the licensing conditions of content producers. Recital 22 of the TWF Directive defines that

> 'free television' is the 'broadcasting on a channel, either public or commercial, of programmes that are accessible to the public without payment in addition to the modes of funding of broadcasting that widely prevail in each Member State such as licence fees and/or the basic tier subscription fee to cable network'.

According to this definition, 'free' television primarily means 'free' from additional costs. It should be noted that this notion is not identical to public broadcasting.[319] The definition provided in the Explanatory Memorandum of the ECTT follows the same line, although it does acknowledge that there may be a need for a more specific definition.[320] According to this interpretation, public broadcasters that encrypt their programmes, for example, for reasons of copyright (as is the case in Denmark and Austria) could still be considered 'free' as long as they do not impose an additional access fee. The incongruity of this interpretation is demonstrated by Germany and Italy, which, when implementing the lists, interpreted 'free' differently, meaning in the sense of non-encrypted or broadly

315 Federal Act on the exercise of exclusive television broadcasting rights (Exclusive Television Rights Act (Fernseh-Exklusivrechtgesetz - FERG)), 19 January 2002, OJ, C 16, p. 8, Article 5a.

316 Broadcasting (Major Events Television Coverage) Act 1999, 26 April 2003, OJ C 100, p. 12, Article 3.

317 Decision No. 8/1999 of the Communications Authority adopted on 9 March 1999, Brussels, 30 September 1999, OJ C 277, p. 3.

318 Broadcasting Act 1996, Chapter 55, Part IV, OJ C 382/2, 18.11.2000. The Television Broadcasting Regulations 2000, 18 November 2000, OJ C 328/6. Code on Sports and Other Listed Events, in: ITC Code on Sports and Other Listed Events, revised January 2000, 18 November 2000, OJ C 328, p. 8.

319 Hins 1998, p. 321.

320 ECTT, Explanatory Memorandum, paragraph 183.

available television.[321] Accordingly, plans of the German cable operator Kabel Deutschland to encrypt all signals that it carries, including free-TV programmes such as public service programming, to prevent signal theft (the so-called Grundverschlüsselung), triggered agitated debates on the issue if, under these circumstances, free-TV programming was still 'free'. It was argued that even if consumers did not have to pay additional fees to access free-to-air broadcasting, those who did not have the necessary reception equipment would still be excluded from accessing it.[322]

Article 3a of the TWF Directive also does not give much guidance on the criteria by which Member States can identify events of major importance. Recital 21 of the revised directive only stipulates that events of major importance for society should be outstanding events and of interest to the general public. It is left to Member States to specify what 'outstanding' and 'of interest to the general public' means and determine the procedures they follow. The directive only provides a non-exhaustive list in Recital 18 of Directive 97/36 as an example: the Olympics, the Soccer World Championship and the European Soccer Championship. It is evident from these examples that major events for society point to an interest for society as a whole because of their importance for the forming of a national and/or cultural identity, and not only to the 'the right of the public to be properly informed'. Article 3a of the TWF Directive does not require that the criteria according to which the lists are drawn up be transparent.

On the one hand, the vagueness of the directive respects the authority of Member States in cultural matters. On the other hand, considering that the main goal of the directive is to harmonize public interest initiatives of Member States in this field, one might wonder if such vagueness will achieve this goal. Moreover, the list concept as it is now leaves the concerned parties with considerable legal uncertainty. This was also the opinion of the Council of Europe, who felt the need to provide for more elaborate provisions in Article 9bis (2) and the Explanatory Memorandum of the ECTT. As the Council correctly noted, the list of important events interferes significantly with the contractual freedom and property rights of the event organizers and pay-TV broadcasters.[323] These freedoms are necessary to achieve the public-interest objective pursued with regard to the impact on social and economic activities that are linked to the trade of broadcasting rights. Events must

---

[321] Article 5a of the Fourth Interstate Treaty on Broadcasting reads: '(1) ... In the Federal Republic of Germany, events of major importance for society (major events) may be broadcast in encrypted form on pay-TV only where the broadcaster or a third party makes it possible under appropriate conditions, for the event to be broadcast on a free and generally accessible television channel ... Only channels which can be received by more than two-thirds of all households shall be deemed to be generally accessible'. In Italy, 'free' television is understood as a channel that reaches more than 90% of the population (and they may not incur additional costs for the acquisition of technical equipment), Article 1 Decision No. 8/1999 of the Italian Communications Authority adopted on 9 March 1999.

[322] Digital-TV 2004, 'ARD und ZDF kündigen digitale Kabeleinspeisung', 20 February 2004, available at <www.presseportal.de> (last visited on 14 March 2005).

[323] ECTT, Explanatory Memorandum, paragraph 181.

be clearly defined according to objective criteria to provide for sufficient legal certainty. The list must also not discriminate against other market participants or lead to market foreclosure against broadcasters from other Member States, right holders or other economic operators.[324]

Unlike the TWF Directive, Article 9bis (2) of the ECTT and the Explanatory Memorandum also provide guidelines concerning the list definition procedure. Member States must adopt clear and transparent procedures, and the criteria used to define listed events must be explicit and public. Procedures must also foresee appropriate consultations with interested parties and take possible practical implications into account such as the need to have some planning certainty in advance.[325]

Transparent guidelines on procedures and criteria seem essential considering the sometimes high economic values involved. Against the background of the more elaborate provisions of the ECTT, the lack of such provisions in the TWF Directive is evident and problematic. Reference to public interest is very vague. Striking in both regulations is the lack of involvement of consumer representatives when defining what major events are.

The List Concept in Practice

As shown, Member States have far-reaching authority to design the procedures and determine the criteria according to which national lists are drawn up. Interestingly, the existing national lists prove to be more harmonized than one might expect under these conditions. None of the Member States that have implemented the list concept have sought to implement Article 3a of the TWF as a means to protect content other than the (live) transmission of popular sports and cultural events.

At the time of writing, only a few Member States have implemented Article 3a of the TWF Directive and officially notified the Commission. Among these Member States are Austria, Germany, Ireland, Italy and the United Kingdom, whose lists are discussed in more depth. Denmark has decided to withdraw its list.

In Austria, the lists are drawn up by the federal government. The government must involve a 'cross-section' of television broadcasters, right holders, enterprises, consumers, employees and persons working in the fields of culture and sports. Austria interprets events of 'substantial social' interest as events that meet at least two of the following criteria:

- The event already commands widespread attention in Austria, particularly as a result of reporting in the media.
- The event is an expression of Austria's cultural, artistic or social identity.
- The event is—in particular because of the involvement of top-level sportsmen and women—a sporting event of special national significance or one that

[324] ECTT, Explanatory Memorandum, paragraphs 182 a-d.
[325] ECTT, Explanatory Memorandum, paragraphs 182 g, h.

commands widespread attention among viewers in Austria owing to its international importance.
- The event was previously broadcast on free-TV.[326]

Television broadcasters that have acquired exclusive rights for the transmission of a listed event will allow the event to be viewed on a free-TV channel (Article 3 (1) FERG). This is achieved by means of an amicable agreement in which the conditions of transmission on free-TV are laid down. If such an agreement is not reached, the Federal Communications Senate can interfere and determine adequate conditions and market prices (Articles 3 (3) and (4) FERG). A television operator that does no comply with this obligation can be sued in addition to receiving administrative fines for damages as well as the duty to compensate for loss of earnings (Article 3 (5) FERG).

In Ireland, the Minister for Arts, Heritage, Gaeltacht and the Islands draws up the lists after consulting with the event organizers and broadcasters, and after giving the public an opportunity to comment.[327] Events of 'public interest' are qualified according to criteria similar to those in Austria, namely:
- The extent to which the event has a special general resonance for the people of Ireland.
- The extent to which the event has a generally recognized distinct cultural importance for the people of Ireland.
- Whether the event involves the participation of a national or non-national team or by Irish people.
- Past practice or experience with regard to television coverage of the event or similar events.

The Irish version of the list-of-important-events concept is strict. Broadcasters that do not offer free-to-air television, reach less than eighty-five per cent of the population, and have acquired exclusive rights to broadcast designated events may not broadcast them unless the event was made available to a free-TV broadcaster with more than eighty-five per cent coverage. Should the first broadcaster does not comply with its obligations, the High Court may, upon request, issue remedies, including an obligation to make the event available to a free-TV broadcaster. It may also declare the licensing contract with the event organizer void (Article 6 Broadcasting (Major Events Television Coverage) Act).

In Italy, the Italian Communications Authority, AGCOM, draws up the list according to the following criteria:
- The event and its outcome are of special and widespread interest in Italy, and interests persons other than those who usually watch this type of event on television.
- The event enjoys widespread recognition by the general public, has particular cultural significance and strengthens the Italian cultural identity.

---

326 Exclusive Television Rights Act, Article 5a.

327 Broadcasting (Major Events Television Coverage) Act 1999, Article 3.

- The event involves a national team in a specific sporting discipline in a major international tournament.
- The event has traditionally been broadcast on free television and has enjoyed high viewing figures in Italy.[328]

The United Kingdom probably has the most exhaustive provisions for the list concept.[329] Here, the Secretary of State draws up the lists after consulting with the BBC, the Welsh Authority, the Independent Television Commission (ITC) (now Ofcom) and event organizers (Article 97 of the Broadcasting Act). It applies one main criterion, namely the event must have a special national resonance and not simply a significance to those who ordinarily follow the sport concerned, and it must serve to unite the nation, meaning it is a shared point in the national calendar.

The Broadcasting Act already restricts the acquisition of exclusive rights without the previous consent of the UK broadcasting National Regulatory Authority (NRA) (Article 101 of the Broadcasting Act). Generally, a pay-TV broadcaster may not distribute a listed event unless a designated free-TV broadcaster (the BBC, Channel 3 or Channel 4) has acquired the rights to live coverage. If no free-TV broadcaster has acquired the broadcasting rights, the broadcasting NRA will give its consent if the availability of such rights was generally known and no free-TV broadcasters expressed an interest in acquiring them (section 13 of the Code on Sports and Other Listed Events). The ITC has included extensive guidelines in its Code on Sports and Other Listed Events that determine when a genuine opportunity to acquire the rights on fair and reasonable terms has been offered.

Germany stipulates that events of major importance for society may be broadcast in encrypted form on pay-TV only when the broadcaster or a third party makes it possible under appropriate conditions for the event to be broadcast on free-TV. Should parties fail to reach an agreement, an arbitration procedure is in place.[330] Germany does not further specify the notion of 'major importance for society' to make the criteria that it applies transparent. Judging from the final list, Germany shares the confidence of the other list countries that sport is of major importance to its people: Summer and Winter Olympics; all European Championships and Soccer World Championship matches involving the German national soccer team, as well as the opening match and the semi-finals, irrespective of whether the German team is involved; the semi-finals of the German FA Cup; the German national soccer team's home and away matches; the final of any European soccer club competition involving a German Club.

---

328 Decision No. 8/1999 of the Communications Authority. See also Article 72 (1) c of the Dutch Media Law, which also refers to audience shares and whether a programme was transmitted on free-TV in the past.

329 Broadcasting Act 1996; The Television Broadcasting Regulations 2000; Code on Sports and Other Listed Events, in: ITC Code on Sports and Other Listed Events.

330 Article 5(a) of the fourth Interstate Treaty on Broadcasting, 29 September 2000, OJ C 227, p. 4.

	Olympic Games Summer/Winter		**Soccer World Champion-ship**[331]	**European Soccer Champion-ship**[332]	**Other Sports Events**	**National Cultural and Other Events**
Austria	+	+	+ (only the games for men)	+ (for men)	Austrian Soccer Cup, World Alpine Skiing, World Nordic Skiing	Vienna Philharmonic Orchestra's New Year's Concert, Vienna Opera Ball
Ireland	+	-	+	+	Hurling Finals, All-Ireland Senior Inter-Country Soccer, Rugby World Cup Finals, Irish Grand National, Irish Derby, Nations Cup at Derby Horse Show	-
Italy	+	+	+	+	Tour of Italy cycling competition, Formula One Italian Grand Prix	San Remo Music Festival
UK	+	+	+	+	Scottish FA Final, Grand National, Derby, Wimbledon Tennis Finals, Rugby League Challenge Cup Final, Rugby World Cup Final	-
Germany	+	+	+	+	German national soccer team matches (home and away)	-

Table 1—Table 1 provides an overview of the events that are featured in the lists of important events that were drawn up by Austria, Ireland, Italy, the UK and Germany following the implementation of Article 3a of the TWF Directive. Overview of the implementation of Art. 3a TWF Directive in national broadcasting laws.

There are a number of aspects about the way Member States implemented the list concept that are worth noting (see also the overview in table 1). First of all, all national lists stress the importance of cultural, artistic and societal criteria rather than political or democratic criteria or the informational value of an event in general. Not considering specific local preferences, soccer enjoys an impressive popularity among the citizens of Europe (and probably most of all in Italy); it is unanimously considered to be of utmost public importance. Here, it is the aspect of national identity or the ability of an event to, as the British regulation says, 'unite the nation'. None of the national lists discussed include events other than sports and cultural events, for example, political events. It is also interesting that the social

[331] Usually only to the extent that the national team is involved, and often restricted to finals and semi finals.

[332] Usually only to the extent that the national team is involved, and often restricted to finals and semi finals.

importance of an event is measured using criteria such as the participation of national sportspeople, the importance for the national identity and audience viewing figures. The latter is interesting because viewing figures and their adequacy as an indicator for social relevance and public interest are controversial. There is a general consensus that viewing figures are a measure for popularity, but does this also mean that they are a valid measure for public, political and societal importance?

Remarkable is also the repeated reference to the text on whether an event has been traditionally transmitted on free-TV. Not only the societal importance, but also the fact that the audience is used to following such an event on free-TV plays a role. The reference to traditional coverage is symptomatic for one characteristic feature of the list concept: the attempt to maintain a traditional state in the presence of the arrival of a new player: pay-TV platforms. An apt reader could conclude from this that the lists are not only about the broad accessibility of major events for the audience, but also about the relationship between free-TV and pay-TV. One could even take this a step further and argue that the lists are about competition between public broadcasters, which were traditionally the major bidders for transmission rights for important sports events,[333] and pay-TV operators, for whom sport is a major crowd-puller to sell their services to subscribers.

As far as the procedures of a) drawing up the lists and b) enforcing the lists are concerned, the national implementations do not provide for much more clarity than Article 3a of the TWF Directive. All of the national lists examined are drawn up either by ministries or regulatory authorities for the broadcasting sector. The involvement of third parties such as event organizers, broadcasters or the public is restricted, if provided for at all, to the possibility of being consulted. Only three of the examined five Member States that have implemented the list concept foresee a broader consultation procedure.

Once transmission rights have been acquired, most Member States foresee arbitration procedures between pay-TV and free-TV broadcasters although the national laws differ considerably in detail. To the extent that Member States foresee arbitration procedures, the arbitration procedures do not foresee an active role for event organizers. Only the UK has issued extensive guidelines on when transmission rights are fairly offered to free-to-air broadcasters. And only the UK has adopted a regulation in the event that free-to-air broadcasters are unwilling to acquire licences for transmission rights for listed events.

Assessment

The biggest challenge for the list-of-important-events concept is defining which types of information are of major importance to the public. What are the relevant criteria, and who should be able to define them? The public sphere is not confined to a narrowly defined realm, be it politics, be it sports. The public dialogue includes

---

333 See section 2.2.3.

a multitude of political, social and cultural spheres that affect ideas and understanding, and hence political opinion, about the world, values, concepts and the choices one makes or others have made. Viewed from this angle, the notion of information or content of public interest should be interpreted broadly, broader in any case than important sports events.

Consequently, it is difficult to distinguish which kinds of services and formats are of importance for the public debate. A considerable part of content services contributes indirectly to the process of forming a public opinion, offer identification with and an assessment of one's own values and judgements, form part of a political or social dialogue about actual issues and/or reflect the variety of different opinions and ideas on particular subject matters. Soaps, talk shows and documentaries play their part in the democratic process, as do news reports and political speeches. For example, politicians can be observed taking chances by appearing on more popular talk shows rather than 'high standing' programmes because they acknowledge the impact of 'low-quality' (in terms of intellectually less demanding) programmes for public opinion.[334]

Furthermore, the 'public' is made up of individual members with individual interests and information needs. This makes it difficult to argue in general terms that certain content is more important than other content. It also explains why the repeatedly proposed criterion of 'national importance' or 'identity' is always a political decision, and is not necessarily the most objective criterion when it comes to identifying public interests.

Moreover, were citizens to choose the kind of information that is important to them, the outcome would probably be different to that of a government-driven decision. The question is whose perspective counts. Article 3a grants Member States wide discretion on if and how the provision is implemented, as well as on the event-definition criteria, the body authorized to draw up the list, the extent to which interested parties must be heard, etc. The national lists are developed by ministries and regulatory authorities, possibly in close cooperation with broadcasters and sports clubs.[335] Astonishing is how little the public, or relevant groups in the public, are involved in this process. To this extent, Article 3a is a rather paternalistic instrument. It is primarily the government's perspective that counts. In contrast, for example, the policy approach towards the internet is characterized by a more consumer-oriented approach that acknowledges the autonomy of citizens to determine the kinds of content that are important to them personally. Here, it is the (collective) intelligence of consumers that counts. One possible justification for this difference in treatment could be the higher level of responsiveness of the internet as a medium. This argument, however, might become obsolete with service differentiation and demand orientation in pay-TV: In the future, convergence,

[334] See also the paper by Gleich 2001.
[335] For a critical discussion see also the paper by Helberger 2002.

digitization and electronic access control could contribute to making the broadcasting environment more responsive to individual citizen's preferences.

A problem of a more practical nature is that the national lists reserve certain events for free-TV, but that the lists are no guarantee that such events will be widely available and accessible. Transmission rights to popular sports events are expensive—so expensive that the majority of advertising-financed free-TV broadcasters cannot afford them. Public broadcasters are not only faced with limited financial resources, they are also legally obliged to invest the income from public broadcasting fees not only in sports events but also in a diverse and balanced programming. In the end, this might result in a situation in which no free-TV broadcaster can bid for the rights, while pay-TV operators will be prevented from showing the events on encrypted television. This is a situation that neither Article 3a of the TWF Directive nor most national implementation laws, with the notable exception of the UK,[336] address.

Last but not least, the list concept deals with content that must be publicly accessible. However, the list concept fails to provide a solution for all the other content that is subject to electronic access control, for example sport events that are not listed. As was shown in section 2.2.1., it can be in the interest of realizing public information policy objectives to ensure that content on pay-TV is offered on fair, non-discriminatory and affordable terms. The list concept, however, does not restore the balance between controllers of access to content and those seeking access. With the exception of the listed events, pay-TV providers are entirely free to foreclose electronic access to all kinds of content and content services, and make access to such content subject to their own conditions and requirements.

The European Commission appears to share at least some of these concerns. In a consultation that the European Commission organized within the framework of a review of the TWF Directive, the Commission concluded that the last word in this matter had not yet been spoken:

> 'The Commission considers that the issue of rights to access to newsworthy events needs further attention and will invite experts to further discuss this matter (in a focus group). The issues at stake have to be dealt with in conjunction with copyright regulation. Three main questions have to be answered: to what extent does the Copyright Directive provide an adequate solution through its "fair dealing" provisions? Do the general policy objectives at stake (pluralism of information sources, etc.) require the statutory definition of a "right to access" for broadcasters and news agencies? What type of intervention should be provided for?' [337]

---

[336] ITC, Extracts from the ITC Code on Sports and Other Listed Events, revised January 2000, paragraphs 12 and 15.

[337] European Commission, The Future of European Regulatory Audiovisual Policy, p. 16.

In other words, the European Commission acknowledges that Article 3a of the TWF Directive might not suffice to take care of national public information policy concerns when dealing with content that is subjected to electronic access control. For the European Commission, this means that Member States might wish to adopt additional regulations, and that this could trigger the need for further harmonization measures on the EU side. Interesting is the reference to copyright law. At first sight, this reference may seem confusing because organized events are usually not subject to copyrights but to ownership and contractual rights of the event organizer. On the other hand, the transmission of the broadcast is subject to intellectual-property protection of the broadcasting signal, irrespective of whether the content transported is subject to copyrights or not. As copyright and neighbouring rights are not the subject of this study, this is not the place to deal with this question in depth. However, the following brief explanation of the nature of the discussion is in order: in the above quote, the European Commission addresses the question of the need at the European level for an additional right to short reporting, and whether Article 5 (3)c of the Copyright Directive suffices to satisfy the public information policy concerns of Member States when dealing with electronic access control.[338] Article 5 (3)c of the Copyright Directive reads:

> '[R]eproduction by the press, communication to the public or making available of published articles on current economic, political or religious topics or of broadcast works or other subject-matter of the same character, in cases where such use is not expressly reserved, and as long as the source, including the author's name, is indicated, or use of works or other subject-matter in connection with the reporting of current events, to the extent justified by the informatory purpose and as long as the source, including the author's name, is indicated, unless this turns out to be impossible'.

The following section 2.3.2. discusses the right to short reporting in more depth, and, in this context, also refers where appropriate to Article 5 (3)c of the Copyright Directive. In particular, it highlights some of the important differences between the right to short reporting and the relevant exception to copyright law, and illustrates why applying Article 5 (3)c of the Copyright Directive is probably not a satisfactory solution.

2.3.2. SHORT REPORTING

Article 9 of the ECTT reads:

> 'Each Party shall examine and, where necessary, take legal measures such as introducing the right to short reporting on events of high interest for the public to avoid the right of the public to information being undermined due to the

[338] European Commission, The Future of European Regulatory Audiovisual Policy, p. 16.

exercise by a broadcaster within its jurisdiction of exclusive rights for the transmission or retransmission, within the meaning of Article 3, of such an event'.

Article 9 of the ECTT does not distinguish between the broadcasting of content that is subject to exclusive rights in encoded or decoded form. Providers of pay-TV can be obliged to allow short reporting, too. As far as pay-TV providers are concerned, the ECTT foresees the complementary application of both, the right to short reporting and the list of important events, depending on whether content is of high public interest or major importance to the public.[339]

The Council of Europe characterizes events of 'high public interest' as events that justify short footage in a television programme.[340] This is not a very precise definition but it indicates at least one aspect, namely that the informational character of the event is paramount. In its Recommendation on the Right to Short Reporting, which covers the right to short reporting on major events for which the exclusive rights for their television broadcast have been acquired in a transfrontier context, the Council of Europe explains that it is mainly up to broadcasters to assess which events are of particular interest to their audience. It further distinguishes between events that are and are not organized. To the first category belong any sports, cultural, social or political events that are organized by public or private natural or legal persons. To the second category belong topical, non-organized events such as accidents, natural disasters and armed conflicts. These are events that are normally shown within the framework of news and current affairs programmes. In general, news events will not be subject to exclusive rights—all broadcasters are free to report them. However, as the Council of Europe points out, access to non-organized major events could be limited in some cases because, for example, the owner of the event's premises has so decided, public authorities have restricted access to the venue for reasons of public security, or because the people involved have agreed to give an exclusive report to one particular broadcaster. [341]

A more detailed explanation of what the right to short reporting entails can be found in the Council of Europe's Recommendation on the Right to Short Reporting, too. According to Principle 2(1) of the Recommendation, 'any secondary broadcaster should be entitled to provide information on a major event by means of a short report.[342] Unlike Article 3a of the TWF Directive, the right to short reporting does not aim at full coverage in free-TV, but at the making of excerpts to inform those parts of the audience who are not able to watch the content otherwise (for

---

[339] Council of Europe, ECTT, Explanatory Memorandum, paragraphs 66, 178.

[340] Council of Europe, ECTT, Explanatory Memorandum, paragraph 177.

[341] Council of Europe, Recommendation 91(5) on the Right to Short Reporting, Explanatory Memorandum, paragraphs 11-15.

[342] 'Short report' is defined in the Recommendation as brief sound and picture sequence about a major event such as to enable the public of the secondary broadcaster to have a sufficient overview of the essential aspects of such an event, Council of Europe, Recommendation 91(5) on the Right to Short Reporting, Explanatory Memorandum, Definitions.

example, because they are not subscribed to the respective pay-TV platform). Their purpose is to provide the public with sufficient information about the essential aspects of a major event, not to provide supplementary elements of information or entertainment.[343] 'The Recommendation and its Explanatory Memorandum also provide very detailed provisions on the content of a short report, on when a short report should be broadcast (not before the primary broadcaster has had the opportunity to carry out the broadcast of the main event), on a possible remuneration of the primary broadcaster as well as on the number of short reports per day or event and their duration.

A short report can be made by a) recoding the signal of the primary broadcaster and/or b) by having access to the site to cover the major event (Article 2 of the Draft Recommendation). While a) concerns the recording of a signal that has already been made available, b) goes further to the extent that it provides a real right of access against the event organizer or owner of the premises (for example, the management of a mine that forbids access to a mining disaster). A short report is defined as

> 'brief sound and picture sequences about a major event such as to enable the public in general or the relevant public in a given country to have a sufficient overview of the essential aspects of the event in which a broadcaster from another country holds exclusive rights. Such short reports are intended for inclusion in regularly scheduled news bulletins of secondary broadcasters, but this does not prevent the primary broadcaster from using short reports for its own programmes or for any other use of its choice'.[344]

Conceptually, the right to short reporting in Article 9 of the ECTT is probably what comes the closest to an access right for broadcasters. This is certainly true where it gives a right of access to an event's site. As far as the right to recording a signal for the purpose of using it in a short report is concerned, it resembles an exception to exclusive rights in the public interest, similar to the aforementioned Article 5 (3)c of the Copyright Directive. Unlike Article 3a of the TWF Directive, the right to short reporting does not impose any restrictions on the exclusivity of the transmission—the event can still be broadcast on an exclusive basis. But, similar to an exception in copyright law, it does oblige the entity that carries out the exclusive transmission to allow certain uses, namely the making of short reports. In other words, the right to short reporting makes the exercise of exclusive transmission rights subject to certain limitations.

One difference between an exception such as Article 5 (3)c of the Copyright Directive and the right to short reporting is that the latter is formulated as a right. This conceptual difference can have consequences for the position of the user

---

[343] Council of Europe, Recommendation 91(5) on the Right to Short Reporting, Explanatory Memorandum, paragraphs 39, 41.

[344] Council of Europe, Recommendation 91(5) on the Right to Short Reporting, Explanatory Memorandum, paragraph 27.

invoking a limitation in court. As Guibault points out, a users' 'privilege to use protected works grants no individualised right of redress and is therefore only as strong as the "legitimate interest" or the "public policy objective" behind it'.[345] The consequence is that a judge, when deciding whether an exception applies will first have to weigh the interests of the person benefiting from the limitation against the property rights of the event organizer. Apart from legal uncertainty for the broadcaster, the balancing process may have actually different results depending on the jurisdiction where the dispute is held. Guibault showed that in France, for example, the authors' rights are considered more favourably than users' interests as compared to other countries.[346] The right to short reporting does not include the reservations provided for in Article 5 (3)c of the Copyright Directive. According to Article 5 (3)c of the Copyright Directive, the possibility to reproduce and communicate parts of the transmission is only allowed in cases where such use is not expressly reserved. The right to short reporting has no such reservation; this makes the right to short reporting a stronger instrument than Article 5 (3)c of the Copyright Directive. Moreover, the subject matter of Article 9 of the ECTT and of Article 5 (3)c of the Copyright Directive is different. Article 5 (3)c of the Copyright Directive stipulates a limitation to broadcasters' exclusive neighbouring and copyright rights in the broadcasting signal, irrespective of what the content is and if it is protected by property rights or not. Article 9 of the ECTT concerns contractual exclusive rights and/or property rights to a site.[347] For the purpose of Article 9 of the ECTT, 'exclusive rights' have been defined as

> 'rights acquired contractually by a broadcaster from the organiser of a major event and/or from the owner of the premises where the event is taking place, as well as from the authors and other rights holders, with a view to the exclusive television broadcast of the event by that broadcaster for a given geographical zone'.[348]

Finally, unlike Article 5 (3)c of the Copyright Directive, the scope and modalities of the right to short reporting have been specified in great detail by the Council of Europe, especially in the aforementioned Recommendation R(91)5.

Assessment

The right to short reporting is more flexible than the list-of-important-events concept and does not work with predefined definitions of 'high public interest'. It is up to the broadcasters, and in the last instance the courts, to decide if an event is of 'high public interest' or not. On the other hand, this increases the range of events that can be subject to a right to short reporting and are not restricted to a predefined

---

[345] Guibault 2002, p. 109.

[346] Guibault 2002, pp. 257-161.

[347] Council of Europe, Recommendation 91(5) on the Right to Short Reporting, Explanatory Memorandum, paragraph 16.

[348] Council of Europe, Recommendation 91(5) on the Right to Short Reporting, Definitions.

list. Although one may wonder whether courts are the best place to pass judgement on the rather political question of whether an event is in the public interest or not, the right to short reporting concept provides more ad-hoc flexibility and room for a more audience-oriented decision of the events that are in the public interest. It should also be noted, however, that the audience plays a passing role in the right to short reporting; again, it is the mission of the media to keep the audience informed and to decide which events it needs to known.

The right to short reporting focuses on keeping the public informed more than on ensuring the public's access to the content in question. This is easily explained by the fact that the right to short reporting protects informational interests only. The right to short reporting supports the media in its mission to keep the public 'properly informed' despite private exclusive control over newsworthy content. The public does not need to watch the full event to be properly informed.

The real problem with the right to short reporting and access of the public to information is probably a very practical one, at least where it is invoked against pay-TV operators. The right to record a signal for the purpose of making a short report would not be effective if the primary broadcaster is not obliged to provide the signal in un-encrypted form. The secondary broadcaster would have to have access to the encrypted signal or the event itself first before being able to make a short report. Article 9 of the ECTT, however, does not include a corresponding obligation for the primary broadcaster. In practice, this could lead to lengthy negotiations and the risk that the interest of the public to be informed about a particular event becomes obsolete with the passage of time.

2.3.3. Must-carry Rules

According to Article 31 (1) of the Universal Service Directive:

> 'Member States may impose reasonable must-carry obligations, for the transmission of specified radio and television broadcast channels and services, on undertakings under their jurisdiction providing electronic communications networks used for the distribution of radio or television broadcasts to the public where a significant number of end-users of such networks use them as their principal means to receive radio and television broadcasts. Such obligations shall only be imposed where they are necessary to meet clearly defined general interest objectives and shall be proportionate and transparent. The obligations shall be subject to periodical review'.

So far, media regulators have relied heavily on free-to-air public broadcasting as a medium to realize the public interest to being properly informed and having access to information (see section 2.2.3.). Where technological developments lead to new modes of distribution and new forms of private control over distribution means, media legislators are again confronted with the question if there is a need to seek new ways of fostering the wide accessibility of public broadcasting. One example is access to the EPG, which is addressed by Article 6 (4) of the Access Directive and is discussed in more depth in Chapter 4 (Telecommunications Law).[349] Another, older example, are must-carry obligations. Must-carry obligations are a means to make public broadcasting universally accessible on different platforms, traditionally: cable platforms. Subscribers to one transmission platform, for example a cable platform, will not be excluded from access to services that are originally transported on another platform. The solution that must-carry rules offer is to make selected general interest services available on different platforms.

With the revision of the legislative framework for the telecommunications sector, the must-carry principle has found a place in Article 31 of the Universal Service Directive. The idea behind must-carry is one of mandatory access and carriage obligations for the network operator in favour of selected, predefined programmes, referring to designated public interest carriers. Although regulated in telecommunications law, must-carry rules relate to the content of a service. Must-carry is an example of where the strict approach of separation of content and infrastructure-related aspects in telecommunications regulation has not been maintained.[350]

Not all Member States foresee must-carry rules in their national laws.[351] Moreover, while some Member States restrict must-carry obligations to public broadcasters, others include other services of allegedly public interest in the must-carry package. According to the European Commission, however, most of the Member States that have adopted must-carry rules have done so for very similar general interest objectives such as pluralism, cultural diversity and freedom of expression.[352]

Characteristic of must-carry rules is the aforementioned strong functional view of the media:[353] the must-carry principle is about the universal and equal accessibility of public interest programming. It is not about the individual consumer's access to a certain platform. Access is realized by admitting designated common interest carriers to the distribution platform. The decision which services consumers can hope to receive after they have subscribed to a cable network is usually reserved to governments or regulatory authorities (Article 31(1) of the Universal Service Directive).

Similar to the example of cable television, it was proposed to apply the must-carry idea to digital access-controlled programme platforms. For example, the European Parliament has recommended:

> 'Member States may impose reasonable "must carry" obligations, for the transmission of specified radio and television broadcast channels and services,

---

[349] Section 4.5.

[350] See section 1.3.3. See also the paper by Capiau 2001.

[351] European Commission, European Electronic Communications Regulation and Markets, p. 33.

[352] European Commission, European Electronic Communications Regulation and Markets 2004, p. 34. Eurostrategies 2003, 34pp.

[353] See section 2.2.1.

> on undertakings under their jurisdiction providing either (a) electronic communications networks used for the distribution of radio or television broadcasts to the public, where a significant number of end-users of such networks use them as their principal means to receive radio and television broadcasts, or (b) conditional access systems and other associated facilities. Such obligations shall only be imposed where they are necessary to meet clearly defined general interest objectives and shall be proportionate and transparent. The obligations shall be subject to periodical review'.[354]

The argument is that digital programme platforms, like cable networks, constitute (privately controlled) distribution networks and that the operators of both can exercise control over the accessibility and availability of (particular) information services. In both cases, there are public information policy interests in restricting the degree of private control over access to designated public interest services (see section 2.2.).

The final version of the Universal Service Directive leaves little doubt that the must-carry principle was not written with access-controlled service platforms in mind. According to Recital 45 of the Universal Service Directive, services that provide

> 'content such as the offer for sale of a package of sound or television broadcasting content are not covered by the common regulatory framework for electronic communications networks and services. Providers of such services should not be subject to universal service obligations in respect of these activities'.

Arguably, this would still leave Member States free to implement must-carry-like regulations with regard to the service platform. As far as the technical conditional access platform is concerned, the European Commission has noted that access to platforms was already covered by the Access Directive,[355] and that an application of the must-carry principle was therefore not necessary.

Assessment

The application of must-carry rules to pay-TV platforms might not only not be necessary because Article 6 of the Access Directive is in place, it might also not be particularly helpful. Again, one major problem with the must-carry rule is the question of the kind of content services that should have the status of must-carry

---

354 European Parliament, Recommendation for Second Reading Universal Service Directive, Amendment 26.

355 European Commission, Communication from the Commission to the European Parliament pursuant to the second subparagraph of Article 251 (2) of the EC treaty concerning the common position of the Council on the adoption of a Directive of the European Parliament and of the Council on universal services and users' rights relating to electronic communication networks and services, Brussels, 18 September 2001, SEC/2001/1407 final [hereinafter 'Communication Concerning the Common Position on the Universal Service Directive'], p. 10-11.

services, and who should be entitled to decide this. One possibility would be to restrict the regulation to public broadcasters. An example for another, more sophisticated, concept can be found in The Netherlands. The Dutch media law foresees the possibility for local communities to establish independent programme councils. The task of the programme council is to 'advise' cable operators (the advice is almost of a binding nature) on the composition of a so-called 'basic package' which must reflect the interests and preferences of the local population.[356] What is remarkable about the Dutch approach is that it suggests that the decision about the content that is to be included in the must-carry package should involve, at least indirectly, the consumers who are represented in the programme councils. This is only fair, one might say, considering that it is their interest that is at stake.

However, the most important reason why must-carry rules are probably not the best answer to electronic access control is that the situation of pay-TV subscribers differs from that of cable subscribers on one important point. Unlike cable subscribers, subscribers to a pay-TV platform do not usually face any major obstacles to access services that are distributed outside the platform. The pay-TV platform is not an alternative distribution platform, but is offered in addition to free-TV. This is why, unlike the situation of cable subscribers, subscribers to pay-TV platforms can still receive, in most cases, free-TV, including public broadcasting. Unless this condition is no longer a given, must-carry rules are only of very limited value for this discussion.

2.4. Conclusion

There is something tantalizing about the electronic control of access to information. Public information policy makers and legislators are enthusiastic about the range of new, more innovative, more attractive services that electronic access control can trigger. But there is, of course, the other side of the coin: the foreboding that electronic access control will bring with it a range of undesirable consequences for the promotion of public information policy goals, notably the free flow of information, and the broad accessibility and availability of content from different sources.

Chapter 2 examined how the effects of the proliferation of electronic access control for the individual accessibility and availability of content are treated in the broadcasting regulation discussion and how they are translated into regulatory instruments. One finding of this part of the analysis is that the most popular argument—electronic access control would conflict with an individual right of access to information—is a fallacy. There is no such thing as a right to access-controlled content services, and even less is there a right to have access to such services free of charge. This is at least true where such a right is claimed to flow

[356] See the Dutch Media Law (Mediawet), Article 82(l)-(n).

from Article 10 of the ECHR. As far as freedom of expression is concerned, if the European Court of Human Rights has referred to a 'right of the public to be properly informed', such 'right' must not be understood in the sense of an enforceable 'right' to access certain content against the will of the operator of a pay-TV platform. The 'right of the public to be properly informed' has to be read within the context of the task the media has to perform, which is to make information accessible. According to this interpretation, as long as the media comply with its function, the balance in the sense of Article 10 of the ECHR will not be endangered by pay-TV.

Yet Member States have a positive obligation to watch over and promote the realization of an environment in which citizens can exercise freedom of expression, also within a changing media environment. Moreover, freedom of expression is only one of several public information policy objectives that can be referred to as a justification for government interference with electronic access control in pay-TV. Democratic principles, pluralism, diversity and the broad availability and accessibility of public interest content and services are other valid public information policy interests that are at stake. One conclusion of Chapter 2 is, however, that with the change in the general distribution structure towards access-controlled platforms and commercial, individualized service provider-consumer relationships, the position of the audience and the strong paternalistic approach in broadcasting regulation should be reassessed.

The problem with pay-TV services, if there is one, is not so much that consumers cannot access certain information—this is the core business of pay-TV providers, but that the terms and conditions of access discourage them from doing so. It is also possible that single individuals cannot comply with the terms and conditions for technical, residence or financial reasons. The author concludes that one way of realizing important information policy goals in the regulation of access-controlled broadcasting can be to ensure that terms and conditions of access to pay-TV platforms are fair, affordable and non-discriminatory. This is particularly true in states where the media's mission to inform the audience is not reserved to public broadcasting, but where commercial broadcasters also participate in this task. Government interference must not necessarily take the form of legal initiatives. Non-legal initiatives are also possible solutions to provide for fair access opportunities, such as granting financial support for socially weaker citizens, promoting interoperability and the availability of affordable and user-friendly consumer hardware. This is a route that existing instruments in European broadcasting law do not take.

Chapter 2 then examined how the different policy concerns were translated into regulatory instruments. It is evident that existing instruments such as broadcasting regulation in general, are still characterized by a functional view of the media. It is the media's task to bring information to the consumers. Consumers as active information seekers are not part of this concept. This is true for Article 3a of the TWF Directive and the right to short reporting in Article 9 of the ECTT. It is also true for the must-carry rules discussed in conjunction with the regulation of electronic access control.

The goal of the list of important events and the right to short reporting is to ensure the universal accessibility of content that is subject to exclusive rights in free-TV. Article 3a of the TWF Directive imposes limits on the exclusive exploitation of transmission rights for certain events. The right to short reporting grants broadcasters a right to use content or access events that are subject to exclusive rights. Both instruments have different goals. The right to short reporting primarily protects the interest of citizens of being properly informed. Article 3a of the Television Directive also serves other public interest objectives, to promote access to content of public importance to society in order to foster social cohesion, competition between free-TV and pay-TV and, at least this is the interpretation of the author, also public broadcasting.

There are doubts about the extent to which both tools are adequate and effective in achieving their goal. The author's main point of criticism is, however, that both initiatives are directed towards maintaining a concept of broadcasting that belongs to the analogue past. States have traditionally taken recourse to public law instruments when regulating broadcasting markets. The changed distribution structure in pay-TV and its focus on private, individualized relationships questions this approach. Possibly, and very likely, free-TV broadcasting will not cease to exist in the future. On the other hand, one cannot ignore that access-controlled content services are here to stay. Chapter 1 explained why there are strong indications that they will become an increasingly important part of the modern media landscape. Existing media law is ill-prepared for the resulting changes to the distribution structure, meaning the change from broadcasting to individual access, from citizen to consumer and from public mission to profit-driven services. Where access to broadcasting content becomes a matter of private contracts and control over access to content, traditional concepts do not help much. Ensuring that some events or excerpts thereof remain available on free television may be a means to preserve the importance and competitiveness of free-TV in general, and public broadcasting in specific. But the more pay-TV prospers and is perceived by consumers as a third possible form of financing broadcasting, the more inadequate this approach must seem.

The arrival of electronic access control is not necessarily a threat to a flourishing media landscape; it could also be seen as an opportunity for new, more responsive content services. Precondition is, however, that consumers can access access-controlled services at fair, affordable and non-discriminatory conditions, that they have a real choice between different competing platforms, including platforms from other countries, and that they have reliable information on the services available to them.

The key to finding a solution that takes the changing and increasingly interactive and individualized distribution patterns into account lies in the commercial relationship between content controller and consumer. This is a relationship that, so far, has been ignored in European broadcasting law. Further research should be carried out to define how this relationship can be addressed so that it can promote the wide availability and accessibility of media content. Chapter 4 will show that the answer might lie in telecommunications law.

Chapter 3

Conditional Access and General Competition Law

3.1. Introduction

Chapter 3 takes a look at the relationship between electronic access control and European competition law. Content, has become a profitable commercial commodity. Today, a significant share of content distribution is an economic undertaking. Content is the fuel that keeps the engines of the 'information economy' running. This also means that the shape of today's information markets is the result of the structuring force of economic mechanisms. Determined by fundamental economic considerations and strategic thinking, the activities of the players in these markets are protected by individual economic freedoms.

The idea of freedom of competition, which is part of the individual economic autonomy, includes the freedom to pursue one's own economic advantage in a competitive environment. This is, in its positive sense, the ability to participate in the marketplace, for example, by concluding contracts, by offering access-controlled services according at one's own will and according to one's own economic interests. In its negative sense, freedom of competition refers to the freedom from public influence, which includes the right to determine the way one organizes one's business as well as the freedom not to be forced to promote competitors to one's own disadvantage. On the other hand, the freedom of competition is limited, it's most obvious restrictions being the interests of competitors in providing services themselves and the overall competition. In other words, freedom of competition is relative and reaches only as far as it is tolerated under general competition law and policy.[357]

The study's choice of European law is based on a number of reasons. The first reason derives from the fact that many conflicts in this sector are likely to be relevant for the European Internal Market due to the size of the actual market players, their increasing international involvement, and the widespread distribution of services by means of cross-border transmission techniques such as the use of satellite and IP protocol. Second, European competition law influences, to a considerable extent, existing national competition law and the way it is interpreted.[358] The relevant provisions discussed in this context can probably be

[357] Herdzina 1999, p. 12.
[358] Wyatt/Dashwood 2000, p. 648.

found in similar form in national competition laws. Finally, Articles 81 and 82 of the EC Treaty create direct rights with regard to the individuals concerned—rights that national courts must protect.[359]

This study distinguishes between general European competition law, as discussed in this chapter, and sector-specific rules that are specifically designed for a particular sector and regulate, among other things, aspects of competition. Such sector-specific rules are discussed in Chapter 4 (Telecommunications Law). Among the sector-specific provisions that are relevant to this study are the access rules in the Access Directive, related rules in the Framework Directive and the Universal Service Directive. In addition to competition-related aspects, these sector-specific rules also realize other public interest objectives.

It is worth noting that competition law focuses primarily on the relationship between economic entities that are in competition with each other. Competition law only considers consumer interests indirectly, to the extent that functioning competition is a process of approximation towards the consumers' demands and for the consumers' benefit (see section 3.5. for a detailed discussion). This is also why the question of if and how consumers access pay-TV services will only fall under the scope of competition law in a situation where the way in which and the conditions under which they are granted access to access-controlled services has a broader anti-competitive impact.

The second part of Chapter 3, section 3.2., starts with a brief overview of relevant European competition law (section 3.2.1.). It also looks into the way the European Commission has defined pay-TV and the associated markets (section 3.2.2.). Because market definitions in competition law are created on a case-by-case basis, the purpose is not so much to categorize certain markets, but rather to give an idea of the arguments underlying the European Commission's market definitions and how these arguments were refined in the course of time. The ensuing analysis of European competition law consists of two aspects. The third part of Chapter 3 (section 3.3.) discusses the aspect of merger control as well as some of the major merger decisions made by the European Commission in the pay-TV sector (section 3.3.1.). This is to determine whether it is possible to identify some guiding principles of the European Commission's practice in this field (section 3.3.2.). Part 4 (section 3.4.) deals with antitrust issues and the applicability of European competition law. It investigates the applicability of Article 81 and 82 of the EC Treaty to some of the possible anti-competitive practices that were identified in Chapter 1, namely access refusals (section 3.4.1.), discriminatory practices (section 3.4.2.) and tying strategies (section 3.4.3.). The analysis ends with two brief parts on economic and non-economic considerations in European competition law, namely the position of consumers (section 3.5.) and the role that non-economic

[359] Wyatt/Dashwood 2000, pp. 643-644, who also point out the practical problems that arise during the application of these provisions.

considerations can or should play in competition law procedures (section 3.6.). The last part (section 3.7.) is the conclusion.

3.2. Rules and Markets

3.2.1. OVERVIEW OF RELEVANT EUROPEAN COMPETITION LAW

European competition law distinguishes between three main categories of anti-competitive behaviour. The first two deal with antitrust; the third concerns mergers: Article 81 of the EC Treaty prohibits all collusive agreements between enterprises that may affect trade between the Member States and that aim at or have the effect of preventing, restricting or distorting competition within the common market. This concerns forms of collusive behaviour such as coordinated behaviour, cartels and price control that directly influence the performance of the parties involved. Examples could be (the so far rather theoretical case of) an agreement between leading providers of conditional access systems on access prices for third-party broadcasters or a joint agreement on a particular API standard with the intention of making market entry for other service providers more difficult.

Article 82 of the EC Treaty is concerned with the market power exercised by a company and the extent to which it allows market players to use its market power to inhibit the activities of competitors and thereby influence the normal working of market mechanisms. This provision deals with impeding strategies, namely the contractual or factual limitations on the economic freedom of actual or potential competitors. The denial of access to the conditional access system or components thereof could fall into this category, as would the granting of access under discriminatory terms and conditions. Other possible examples are the use of proprietary standards with the intention of blocking market access and undue bundling practices that restrict the economic freedom of programme providers to choose on which digital service platform they market their content.

The third category involves concentrative structures and the abuse of a dominant position, which may be covered by Articles 81 and/or 82 of the EC Treaty when the conduct involves the weakening of competition. This refers to economic action that aims at reducing the number of independent economic entities, meaning competitors, in a market through mergers or joint ventures at the horizontal or vertical level. Although Article 82 of the EC Treaty does not expressly mention any kind of merger control, the European Court of Justice decided that, under certain circumstances, mergers also constitute an abuse of a dominant position.[360] They can fall under Article 82 of the EC Treaty and the Council's Merger Regulation No.

[360] European Court of Justice, Judgment of the Court of 21 February 1973, Europemballage Corporation and Continental Can Company Inc. v Commission of the European Communities, Case 6-72, European Court reports 1973, p. 215 [hereinafter 'Europemballage'], paragraphs 25-27.

139/2004.[361] Joint ventures can also fall under Article 81 of the EC Treaty providing that agreements that restrict competition are involved.

3.2.2. MARKET DEFINITION

From both an economic and legal perspective, the identification and definition of product markets is a necessary precondition for market analysis because the identification of actual and potential competitors and the relevant economic behaviour are relative and can only be identified within the context of the respective market.[362]

Market definition is not an exact science, and economic theories for defining markets are numerous.[363] One aspect shared by all theories is that market definition involves an element of subjective assessment. Moreover, the definition of a relevant market can—and does—change over time because the characteristics of products and services evolve and the possibilities for demand and supply substitution change.[364] Depending on the dynamics of the market, a definition can rarely be much more than a snapshot of the actual competitive situation in a certain market. This is especially true in fast evolving markets such as in the converging markets for broadcasting, information society and telecommunications services.[365] For this reason, one should refrain from putting too much emphasis on the way the European Commission defines markets in a particular case; in the next case, it might adopt a different market definition. The analysis acknowledges that market

[361] European Council, Council Regulation No. 139/2004 of 20 January 2004 on the control of concentrations between undertakings, 29 January 2004, OJ L 41, p. 1 [hereinafter 'Merger Regulation No. 139/2004'], recasting Council Regulation (EEC) No. 4064/89 of 21 December 1989 on the control of concentrations between undertakings, 30 December 1989, OJ L 395, p. 1 [hereinafter 'Merger Regulation No. 4064/89']. Merger agreements reached, or public bids announced, before 1 May 2004 will continue to be assessed under the old regulation Merger Regulation No. 4064/89. See also European Commission, Commission Regulation No. 802/2004 of 7 April 2004 implementing Council Regulation (EC) No.139/2004 on the control of concentrations between undertakings, 30 April 2004, L 133, p. 1 [hereinafter 'Commission Regulation Implementing the EC Merger Regulation'] and European Commission, Commission Guidelines on the assessment of horizontal mergers under the Council Regulation on the control of concentrations between undertakings, 5 February 2004, OJ C 31, p. 5 [hereinafter 'Commission Guidelines on the Assessment of Horizontal Mergers'].

[362] In the language of the European Commission, a relevant product market comprises 'all those products and/or services which are regarded as interchangeable or substitutable by the consumers, by reasons of the products' characteristics, their prices and their intended use' (this is the so-called substitutability principle), European Commission, Commission Guidelines on Market Analysis and the Assessment of Significant Market Power, paragraph 44.

[363] Herdzina 1999, 73.

[364] European Commission, Recommendation on Relevant Markets, Explanatory Memorandum, p. 4.

[365] See European Commission, Notice on the application of competition rules to access agreements in the telecommunications sector, 22 August 1998, OJ C 265, p. 2 [hereinafter 'Access Notice'], paragraph 47: 'Any attempts to define particular product markets in [the EU] would run risk of rapidly becoming inaccurate or irrelevant. The definition of particular product markets ... is best done within the context of a detailed examination of an individual case'.

definition under general competition law is case-oriented and that the following analysis can be no more than a first and simplified impression of existing practice.[366] Nevertheless, market definition constitutes an important part in the competitive analysis, and it is helpful to understand the criteria according to which the European Commission proceeds. The following analysis of the European Commission's decision practice may provide an indication of its previous practice for the pay-TV sector, the criteria on which it has based its market definitions so far, and how it considers the impact of convergence and change in the economic environment.[367] The below list of recognized markets is therefore indicative rather than exhaustive.

Given the impact of market definition, and bearing in mind the particular difficulties associated with assessing very dynamic markets such as in the area of digital broadcasting, the European Commission has developed additional, sector-specific guidelines to assist national competition authorities in defining markets and assessing the level of competition.[368] Two important elements in the European Commission's market definition procedure in telecommunications markets are demand-side and supply-side substitution.[369] A third element is the existence of or the potential for future competition.[370] Relevant product or service markets comprise all of the products or services that are sufficiently interchangeable or substitutable, not only in terms of their objective characteristics (price, intended use, etc.) but also in terms of the conditions of competition.[371] The Commission, however, acknowledges that 'product substitutability between different electronic telecommunications services will arise increasingly through the convergence of various technologies'.[372] As far as pay-TV is concerned, the Commission makes a general distinction in at last two main types of relevant markets, namely markets of services or facilities provided to end users (retail markets) and markets of facilities necessary to provide such services offered to operators (wholesale markets). Within these two types of markets, the European Commission further differentiates for

---

[366] For a more elaborate analysis, see the study by Bird/Bird (2002).

[367] The other aspects of the decisions will be discussed in more depth in section 3.3.1.

[368] European Commission, Access Notice; Commission Guidelines on Market Analysis and the Assessment of Significant Market Power, to name but some.

[369] See European Commission, Access Notice, paragraph 40; Commission Guidelines on Market Analysis and the Assessment of Significant Market Power, paragraph 48.

[370] See European Commission, Access Notice, paragraph 40.

[371] Demand-side substitutability is used to measure the extent to which consumers are prepared to substitute other services or products for the service or product in question, whereas supply-side substitutability indicates whether suppliers other than those offering the product or service in question would switch in the immediate to short-term their line of production or offer the relevant products or services. See also European Commission, Commission Guidelines on Market Analysis and the Assessment of Significant Market Power, paragraphs 49-54.

[372] European Commission, Commission Guidelines on Market Analysis and the Assessment of Significant Market Power, paragraph 47.

each of the levels in the production chain between demand and supply-side characteristics.[373]

Distinct from market definitions for the purpose of a general competition law analysis are market definitions within the context of the application of telecommunications law, as described more in depth in Chapter 4.[374] To this extent, Articles 14 to 16 of the Framework Directive set out the procedure that must be followed by National Regulatory Authorities (NRAs) in relation to a series of specific regulatory obligations in telecommunications law. For the purpose of the application of telecommunications law, the Commission Guidelines on Market Analysis and the Assessment of Significant Market Power, according to Articles 14 to 16 of the Framework Directive are particularly relevant.[375] The guidelines provide a (sector-specific) list of criteria for the definition of relevant product and service markets at the transport level, as well as an overview of the existing European Commission's decision practice. Within the context of Article 15 of the Framework Directive, the European Commission also issued a Recommendation on Relevant Markets in the telecommunications sector with the intention of updating the list on a regular basis.[376] For Chapter 3, it remains to be said that, as the European Commission indicated itself,[377] market definitions according to the Communications Framework do not necessarily follow the principles of market definition under general competition law. Therefore, general competition law authorities can identify diverging product markets. Moreover, restrictions can be imposed on enterprises by sector-specific NRAs in addition to sanctions imposed by the general competition authorities. Hence, it is important to understand the complementary character of general competition law in areas in which sector-specific rules exist.

Retail Markets

Pay-TV/Free-TV—The European Commission decided repeatedly that access-controlled services, namely pay-TV services, would constitute a product market separate from that of free-TV, meaning advertising or licence-fee financed television.[378] It was argued that, in the case of fee and advertising-financed

---

[373] Access Notice, paragraph 45, Commission Guidelines on Market Analysis and the Assessment of Significant Market Power, paragraph 64.

[374] See section 4.6.2.

[375] European Commission, Commission Guidelines on Market Analysis and the Assessment of Significant Market Power, see above.

[376] European Commission, Recommendation on Relevant Markets, see above.

[377] European Commission, Recommendation on Relevant Markets, Explanatory Memorandum, p. 8.

[378] To name but some: European Commission, Commission Decision of 27 May 1998 relating to a proceeding pursuant to Council regulation (EEC) No. 4064/89 (Case IV/M.993 – Bertelsmann/Kirch/Premiere), 27 February 1999, OJ L 52, p. 1 [hereinafter 'Bertelsmann/Kirch/Premiere'], paragraph 18; European Commission, Commission Decision of 15 September 1999 relating to a proceeding under Article 81 of the EC Treaty (Case IV/36.539 - British Interactive Broadcasting/Open), 6 December 1999, OJ L 312, p. 1 [hereinafter 'BiB'], paragraph 24; European Commission, Commission Decision of 30/04/2002 declaring a concentration to be

television, there is a trade relationship between the programme supplier and the advertising industry, whereas in the case of pay-TV the programme supplier enters a contractual relationship with the subscriber. Because of this, different conditions for competition were given.

Interestingly, the European Commission also indicated that this distinction might blur due to the convergence of the sectors[379] as well as in cases in which programmes are financed from a mixture of sources.[380] This is true in situations in which pay-TV providers have opened their platforms to advertisers and/or in which pay-TV platforms also carry free-TV programming for their subscribers. Consequently, it cannot be excluded that the European Commission will, at some stage, find that pay-TV and free-TV services operate in one and the same market. Were the European Commission to move away from its two-separate-market assumption, this could have consequences for the finding of a dominant position of pay-TV providers and the applicability of competition law. And, indeed, NewsCorp made exactly this argument in the NewsCorp/Telepiù case in order to demonstrate that the suggested merger would not lead to the creation of a dominant position in the Italian broadcasting sector.[381] The Commission did not follow this argument, though, because of the observed differences in the supply-side substitutability (different means of financing, revenue models for channel suppliers, the 'windows' licensing policy of programme producers, etc.) and the demand-side substitutability (different content offered, different financing models and business strategies) between pay-TV and free-TV.[382] On the other hand, and as was remarked by the parties consulted, both pay-TV and free-TV providers were competing for the attention of consumers. To this extent, both providers were competing in the retail market for viewers and in the wholesale market for attractive content.[383] In contrast, consumer organizations pointed to the differences between subscribers to pay-TV offerings and free-TV viewers.[384] It would lead too far to weight all of the arguments mentioned in this decision. The NewsCorp/Telepiù case demonstrated, however, that the existing distinction between pay-TV and free-TV is based on a number of factors that can change with ongoing economic and technological developments, notably those in the wake of digitization and convergence:

---

compatible with the common market (Case IV/M.0057 - TPS) according to Council Regulation (EEC) No. 4064/89, 8 June 2002, OJ C 137, p. 28 [hereinafter 'TPS'], paragraph 14.

379 European Commission, Commission decision of 2 April 2003 declaring a concentration to be compatible with the common market (Case Comp/M.2876 – Newscorp/Telepiù), according to Council Regulation (EEC) No. 4064/89, 16 April 2003, OJ L 110, p. 73 [hereinafter 'Newscorp/Telepiù'], paragraphs 18-28.

380 European Commission, Commission Decision of 9 November 1994 relating to a proceeding pursuant to Council Regulation (EEC) No. 4063/89, (Case IV/M.469 – MSG Media Service), 31 December 1994, OJ L 364, p. 1 [hereinafter 'MSG'], paragraph 32.

381 European Commission, Newscorp/Telepiù, paragraph 18.

382 European Commission, Newscorp/Telepiù, paragraphs 19-25.

383 European Commission, Newscorp/Telepiù, paragraph 26.

384 European Commission, Newscorp/Telepiù, paragraphs 32-33.

'convergence between media and telecommunications on the other hand is likely to bring about an increasing proximity between the different ways in which entertainment and information are brought to consumers, and the ways in which these consumers enjoy them'. [385]

It should be noted that, so far, the European Commission has not seen the need to distinguish between analogue and digital pay-TV. It is argued that digital pay-TV is only a further development of analogue pay-TV and therefore does not constitute a separate relevant product market.[386]

So far, no attempt has been made to claim that instead of defining one joint market for pay-TV that each (proprietary) pay-TV platform constitutes a market of its own. The argument would be particularly plausible in situations where consumers are strictly bound to one particular pay-TV platform due to long-term subscription contracts, proprietary conditional access technology, high switching costs, etc. In such situations, the presence of any additional platform might not constitute a real choice, meaning no real demand substitutability exists. For the time being, however, no cases are known to the author in which the Commission has argued in this way.

Digital Interactive Television Services—In its BiB decision, the European Commission for the first time defined a separate market for digital interactive broadcasting services. According to the European Commission, in the BiB decision, the difference between pay-TV and digital interactive TV was that the latter was 'largely transactional or informational', whereas pay-TV was essentially entertainment oriented.[387] Underlying this distinction is the traditional idea of broadcasting as a point-to-multi-point medium as opposed to information society services that are distributed on an individualized basis. Chapter 1, however, explained why this distinction is increasingly difficult to maintain in practice.[388]

In the BSkyB/Kirch decision, the European Commission further described and differentiated between markets for digital interactive services that are available via television and personal computers.[389] The European Commission explained this to be a result of the relatively small number of households with personal computers and the fact that interactive services intended for reception via the TV set can be integrated into traditional broadcasting entertainment channels. Again, one may

---

[385] European Commission, Newscorp/Telepiù, paragraph 39.

[386] European Commission, TPS, paragraph 14.

[387] European Commission, BiB, paragraph 23.

[388] See section 1.2.

[389] European Commission, Commission decision of 21 March 2000 declaring a concentration to be compatible with the common market (Case IV/M.0037 – BSkyB/Kirch Pay TV) according to Council Regulation (EEC) No. 4064/89, 15 April 2000, OJ C 110, p. 45 [hereinafter 'BSKyB/Kirch'], paragraph 38; European Commission, Commission decision of 4 August 1998 declaring a concentration to be compatible with the common market (Case IV/M.5 – Cegetel/Canal+/AOL/Bertelsmann) according to Council Regulation (EEC) No. 4064/89, 28 January 2000, OJ C 024, p. 4 [hereinafter 'Cegetel/Canal+/AOL/Bertelsmann'], paragraph 12.

wonder to what extent this finding will hold true bearing in mind that it does not take much more than a simple TV card to turn the computer into a television set.

Paid-for Content Provision via the Internet/Portals/Internet Advertising—The European Commission has indicated that there are distinct markets for the provision of internet advertising, portals and particular types of paid-for content and advertising-funded content. This distinction is based on the assumption that these different activities earn revenues in different ways and from different sources, and that they are frequently carried out by different undertakings and require substantially different inputs.[390] As the Commission remarked on another occasion, different content markets will be relevant to each delivery mechanism as long as consumers regard the provision of services across the different access mechanisms as non-substitutable.[391]

Intermediary Platform—It is also worth noting that the European Commission has already acknowledged a distinct role for intermediary platforms. The European Commission repeatedly uses the notion of 'programme platform'[392] or 'direct-to-home distribution platform'[393] to describe a pay-TV marketing platform that is distinct from the technical conditional access platform. To this extent, existing market definitions reflect a strict distinction between the technical and content-related aspects of pay-TV. The fact that there is a demand and a market for particular products or services does not prevent the existence of a separate market for aggregation services such as pay-TV platforms, internet portals and services from traditional industries such as the Yellow Pages or business guides. The European Commission considers the bundling or 'packaging' of services, for example, that are different from the provision of individual services as part of a package. Moreover, the European Commission explains that the promotion and marketing of a service bouquet differs from the marketing of single services because of the economies of scope involved.[394] The demand substitutability of a

---

[390] European Commission, Commission Decision of 27 May 1998 declaring a concentration to be compatible with the common market (Case IV/JV.1 - Telia/Telenor/Schibsted) according to Council Regulation (EEC) No. 4064/89, 31 July 1999, OJ C 220, p. 28 [hereinafter 'Teleia/Telenor/Schibsted'], paragraph 15; European Commission, Commission decision of 11 October 2000 declaring a concentration to be compatible with the common market and the EEA Agreement (Case COMP/M.1845 – AOL/Time Warner), 9 October 2001, OJ L 268, p. 28 [hereinafter 'AOL/Time Warner'], paragraph 35.

[391] European Commission, Commission Decision of 20/07/2000 declaring a concentration to be compatible with the common market (Case IV/M.0048 - Vodafone/Vivendi/Vanal Plus) according to Council Regulation (EEC) No. 4064/89, 20 May 2003, OJ C 118, p. 25 [hereinafter 'Vodafone/Vivendi/Canal+'], paragraph 41.

[392] European Commission, Bertelsmann/Kirch/Premiere, paragraphs 26 and 27.

[393] European Commission, Commission Decision of 19 July 1995 declaring a concentration to be incompatibe with the common market and the functioning of the EEA Agreement (Case IV/M.490 – Nordic Satellite Distribution), 2 March 1996, OJ L 53, p. 20 [hereinafter 'Nordic Satellite Distribution'], paragraph 58.

[394] European Commission, BiB, paragraphs 16-18 (for the bundling of interactive services); European Commission, Newscorp/Telepiù, paragraph 74.

package of interactive services is distinguishable from the demand substitutability of the individual services that form part of the package.[395] The finding in a later decision that access-controlled single channels and access-controlled platforms might constitute one market does not change much in this respect, as in both cases intermediary services, namely the operation of the technical conditional access platform and the marketing platform are needed.[396]

Wholesale Markets

Licensing of Broadcasting Rights—The European Commission identified separate markets for the licensing of broadcasting rights, which it further divided in the course of time into markets for:

- Football events that take place regularly (annually) and in which national teams participate.
- Football events that do not take place regularly and in which national teams participate.
- Other sports.
- TV thematic and generic channels (for example, animation, documentary and fiction).
- TV programmes.
- Premium films.[397]

Interestingly, the European Commission distinguishes for the latter between films produced by major Hollywood studios (Universal, MGM, Paramount, Sony (Columbia), Disney (Buena Vista, Touchstone and Miramax), Twentieth Century Fox and Warner) and those produced by other studios.[398]

This distinction into different markets for broadcasting licenses is still somewhat rough, particularly from the point of view of demand-side substitutability. The majority of consumers do not value sports events or feature films in the same way.[399] According to their personal preferences, the transmission of a soccer game might not be substitutable with the transmission of a table soccer game, a consumer who wants to watch horror movies might not want to watch Disney movies instead, etc. Consequently, even if the leading producer of horror movies does not have dominant market power in an overall national market for premium films, it could still have dominant market power in the national market for horror movies. On the

---

[395] European Commission, Vodafone/ Vivendi/ Canal+, paragraph 47, for internet portals.

[396] European Commission, Newscorp/Telepiù, paragraphs 44-46.

[397] For example: European Commission in Commission Decision of 13 October 2000 declaring a concentration to be compatible with the common market (Case IV/M.2050 - Vivendi/Canal+/Seagram), 31 October 2000, OJ C 311, 31, p. 3 [hereinafter Vivendi/Canal+/Seagram'], paragraphs 17 -18; Newscorp/Telepiù, paragraph 52 et seq., to name but some.

[398] See European Commission, Newscorp/Telepiù, paragraph 57.

[399] See also European Commission, Newscorp/Telepiù, paragraph 51.

other hand, from a broadcaster's or even advertiser's point of view, the genre or popularity of a particular programme might count.[400]

Markets for the Wholesale Supply of Niche Films and Sports Content, and Channels for Pay-TV—The European Commission found that the licensing of broadcasting rights for pay-TV is a distinct market that can be divided into first-window and second-window licensing. It also found that the wholesale supply of premium films and sports a) content and b) channels for pay-TV forms a separate market from the supply of services for free-TV.[401] It is also distinct from thematic and general interest channels.[402] The argument used by the European Commission was that while thematic or general interest channels were supplied to customers as part of a package, film and sports channels are charged on an individual basis. Moreover, the wholesale price of acquiring film and sports channels is far higher than that of other channels. It was not excluded that, in the future, additional separate markets for sports and film channels will be defined.[403] Within this context, the European Commission has so far worked with rather rough estimates, taking only rudimentarily into account distinct consumer preferences.

Technical Services Necessary for Pay-TV—Beginning with the MSG decision, which will be discussed more extensively in section 3.3.1., the European Commission defined a product market for the wholesale provision of the technical services necessary for pay-TV.[404] So far, the European Commission has seen no need to further differentiate, and instead referred in its decision to one common market that would involve:

- The making available of a descrambling system (decoder) and, as long as there are still analogue receivers, a digital-analogue converter (integrated into a set top box or directly into satellite receivers or television sets).
- The handling of the conditional access system.
- Subscriber management.
- A transaction management system.

In the BiB decision[405] the list was extended to include:

---

[400] Rumphorst 2001, p. 3.

[401] European Commission, Newscorp/Telepiù, paragraphs 55-77.

[402] European Commission, BiB, paragraphs 28-29; European Commission, Commission decision of 2 December 1998 declaring a concentration to be compatible with the common market (Case IV/M.1327 - NC/Canal+/CDPQ/Bank America) according to Council Regulation (EEC) No. 4064/89, 14 August 1999, OJ C 233, p. 21 [hereinafter 'NC/Canal+/CDPQ/BankAmerica'], paragraph 15; European Commission, Newscorp/Telepiù, paragraph 54.

[403] European Commission, BiB, paragraph 29; European Commission, European Commission, Commission Decision 93/403/EEC of 11 June 1993 relating to a proceeding pursuant to Article 85 of the EEC Treaty (Case IV/32.150 – EBU/Eurovision System), OJ L 179, 22 July 1993, p. 23 [hereinafter 'EBU/Eurovision'], paragraphs 10, 13-20, 21-25; European Commission, Newscorp/Telepiù, paragraph 55.

[404] See also European Commission, Bertelsmann/Kirch/Premiere, paragraphs 19-21; European Commission, BIB, paragraph 30-32.

[405] See section 3.3.1.

- Electronic Programme Guides.
- Application Programme Interfaces.

This does not exclude that additional specialized product markets could be identified later if the situation so required.

Interestingly, in its BiB decision, the European Commission did not see the need to differentiate between conditional access services that control access to interactive services and conditional access systems that control access to broadcasting services. Furthermore, no distinction was made between the different infrastructures to which technical services would be connected. The European Commission saw no difference between technical conditional access services for satellite or cable transmitted services.[406] This finding is in line with the ongoing process of the convergence of transmission means that was described in Chapter 1.[407] The European Commission does not rule out the possibility that the need to make a distinction could arise in the future.[408] In contrast, it is worth noting that sector-specific law, notably Articles 6 and 8 to 13 of the Access Directive, which will be discussed in Chapter 4,[409] does strictly distinguish between technical services for digital interactive services and pay-TV.

Distinct Markets for Transmission Capacities—In its decisions, the European Commission considered that there is a separate market for cable,[410] (digital) satellite[411] and (digital) terrestrial television networks. According to the European Commission, the technical and financial conditions of operating each infrastructure differ considerably. This was explained, among other things, by the fact that each medium has a different footprint.

On the other hand, the European Commission concluded in other decisions that no distinction could be made between pay-TV services broadcast via cable, satellite or digital terrestrial means. To this extent, the European Commission argued,

406 European Commission, Commission Decision of 27 May 1998 relating to a proceeding pursuant to Council Regulation (EEC) No. 4064/89 (Case IV/M.1207 – Deutsche Telekom/Betaresearch), 27 February 1999, OJ L 53, p. 31 [hereinafter 'Deutsche Telekom/Betaresearch'], paragraph 18.

407 See section 1.4.3.

408 European Commission, BiB, paragraph 32.

409 See sections 4.4. and 4.6.

410 See European Commission, Nordic Satellite Distribution, paragraphs 61-64. The Commission argued that, from the perspective of the viewer, there were considerable differences between the possible transmission routes—terrestrial, direct-to-home satellite and cable—which would affect both the technical requirements and costs. While terrestrial transmission and satellite television only required the viewer to install an aerial or a satellite dish at his own expense, cable TV was dependent on the maintenance of a cable network, which is financed by the viewer by means of cable fees [with reference to the MSG decision of the Commission], paragraph 62.

411 See also European Commission, Nordic Satellite Distribution, paragraph 57: This decision defined a separate market for the provision of satellite TV transponder capacity and related services to broadcasters; the distribution of TV signals via satellite (transponders) was found to be a market distinct from TV distribution by terrestrial links, since considerable technical and economic differences existed between the two modes of distribution.

services were fully substitutable.[412] Again, this consideration acknowledges the impact of convergence for the pay-TV sector.

Relevant Geographical Market

According to the decisions of the European Commission, and for linguistic, cultural, licensing and copyright reasons, the pay-TV markets follow national or language borders. The same was claimed for the acquisition of programme rights,[413] and for telecommunications networks and services because of the 'national nature of the regulatory regime and of the demand of services to be delivered across the telecommunications infrastructure'.[414] On the other hand, the European Commission acknowledged that the proliferation of far-ranging transmission techniques, such as satellite networks, has the potential to extend the geographic market progressively to the entire EU territory. Obviously, such developments would be in line with the progressive realization of a single Internal market.

Still, the departure from the idea of a national character of media markets is apparently not easily realized. Even for internet services, which are genuinely considered international if not global in scope, the European Commission initially found that the different geographic product markets for internet services could essentially be considered national in nature, with the exception of certain emerging pan-European markets.[415] It took a while until the European Commission found, for example, in its AOL/Time Warner decision, that the geographical dimension of the market for online music delivery, due to the possibilities offered by digital technology, 'certainly extends beyond national borders' and could be even global in character.[416] The same was argued for the markets of technical facilities for such services—in this case player software—where the European Commission found the market to be global. It was observed that the language of the music player's file menu text could be readily adapted to support a multitude of languages. The impact of language differences was thus minimal and the localization of the player

412 European Commission, TPS, paragraph 14.

413 European Commission, Newscorp/Telepiù, paragraph 62, stated that 'nothing prevents operators from acquiring rights for more than one territory at a time, broadcasting rights are divided and sold on a mainly national basis or, at the most, by language area and the price is structured in such a way that the economic value of the contracts depends on the specific territory for which the rights are acquired'.

414 European Commission, BiB, paragraph 50.

415 European Commission, Commission decision of 24 April 2001 declaring a concentration to be compatible with the common market (Case IV/M.2222 – UGC/Liberty Media) according to Council Regulation (EEC) No. 4064/89, 16 June 2001, OJ C 172, p. 20 [hereinafter 'UGC/Liberty Media'], paragraph 16; European Commission, Commission decision of 23 July 1999 declaring a concentration to be compatible with the common market (Case IV/M.1551 - AT&T/Media One) according to Council Regulation (EEC) No. 4064/89, 30 September 1999, OJ C 277, p. 5 [hereinafter 'AT&T/Media One'], paragraph 23.

416 European Commission, AOL/Time Warner, paragraph 27.

software, meaning adapting the software to the needs of a local market, was customary and easy to accomplish.[417]

Still, the language argument is persistent. In the same decision, the European Commission argued that the geographical scope of the market for online paid-for video distribution was, due to the language requirements of the different national consumers for film and TV programming, likely to be national, similar to the case of its equivalent in the broadcasting sector, pay-TV.[418] Therefore, even though the European Commission acknowledged for online music that the language problem could be overcome thanks to technological progress, it assumes that something different applies to video services. On the other hand, the European Commission assumed that, at least for the internet, the film offering is mainly focused on US films and programmes, notably cartoons, which have an international appeal and are popular in all of the European Member States.[419] It remains to be seen whether the European Commission will apply this argument to pay-TV.

Conclusion

The analysis of the decisions of the European Commission regarding the definition of relevant markets in the pay-TV sector clearly demonstrates that market definition for the purpose of the application of general competition law is performed on a case-by-case basis, and that it can be subject to changes in the economic or technological environment. Unlike the exercise of market definition within the framework of telecommunications law, the definition of markets for a general competition law analysis is based on an actual market situation rather than on the more general assessment of the specific problem areas in a particular sector.[420] The analysis shows that market definition under European competition law is constantly 'on the move' and that the European Commission regularly reserves the right to amend market definitions where new technological or market developments so require. This also means that it is difficult to predict how the European Commission will decide in future cases. For market players in the pay-TV sector, this results in a certain degree of legal uncertainty. Because market definition is a crucial precondition for the applicarion of general competition law, general competition law is obviously not very suitable when it comes to providing consistent and predictable rules. On the other hand, the case-by-case approach is also what makes general competition law a rather flexible tool.[421]

At a more detailed level, the analysis shows that convergence plays a prominent role in the European Commission's practice of defining relevant markets.[422]

---

417 European Commission, AOL/Time Warner, paragraph 32.
418 European Commission, AOL/Time Warner, paragraph 36.
419 European Commission, AOL/Time Warner, paragraph 36.
420 Larouche 2002, p. 137.
421 See Cave/Cowie 1998, section 6. But see also the critical reflections by Larouche 2000, 347pp.
422 Instructive, European Commission, Vodafone/Vivendi/ Canal+, paragraphs 29-58. Critical Valcke 2003, p. 559, more optimistic Larouche 2002a, o. 24.

However, it can also be observed that convergence at the transport level is a more readily acknowledged fact than convergence at the service level, meaning the level at which pay-TV, interactive and other services are offered to consumers. For the service level, there is still a tendency to distinguish between information society services and broadcasting services. The analysis also shows that the Commission sometimes operates with rather general criteria and does not thoroughly investigate the different valuations that consumers may have for different content and services. Finally, it can be noted that intermediary service platforms play a distinct role in pay-TV markets, and that this fact is acknowledged in the Commission's decision practice.

3.3. Concentrative Structures and European Competition Policy

In its function as the 'watchdog' of European competition law, the European Commission has been directly involved in a number of significant merger cases that involved pay-TV markets. An examination of some of the major European Commission's decisions in this sector is interesting and relevant for this analysis because the European Commission, in its function as merger authority, did its share to shape the structures of pay-TV markets in Europe. A number of major players, or 'usual suspects', frequently re-appear in these decisions, which are either based on Article 81 and/or 82 of the EC Treaty and the relevant merger regulation. In the following paragraphs, some of the most relevant pay-TV cases are described, generally in chronological order. The description provides an overview of the dynamics of the pay-TV sector and the kinds of players involved. It also provides an overview of some of the most relevant issues for the competitiveness of this sector. The description also seeks to identify if there are any guiding principles in the European Commission's practice when dealing with such cases, and if yes, what they are. Another question is if those principles, should they exist, have evolved over the course of time. Subsequently, some of the particularly relevant aspects will be highlighted.

3.3.1. Major Decisions in Pay-TV

MSG

Beginning with the MSG decision from 1994, the European Commission demonstrated its awareness of the state of competition in pay-TV markets. It also started a trend that can be found in its later decisions, namely the recognition of the potential of pay-TV to drive digitization as well as the development of a new and innovative distribution channels and digital services. The MSG case concerned a joint venture between Deutsche Telekom (distribution infrastructure), Bertelsmann (programme rights) and Kirch (programme rights, operation of the German digital service platform Premiere) to provide pay-TV services. The goal of the planned joint venture was to offer the technical, business and administrative infrastructure for pay-TV. MSG would produce decoders, Subscriber Management System and SAS services, lease decoders to end users, provide billing services and conclude agreements with network operators and programme suppliers.

The impact of the control over the technical platform already played a major role in this European Commission's first pay-TV decision. The European Commission noted that the planned entity would be likely to establish a durable dominant position in the market for technical and administrative services in Germany, and that this could eventually result in a situation in which all newcomers would depend on access to the technical conditional access platform that is operated by the one and only leading pay-TV operator. The latter would then be in a position to influence terms, conditions and market processes in a way that was advantageous to its own services. At the same time, the new entity would also be in a position to impede or postpone new entries. It is worth noting that already then, when the future Standards Directive, which deals with the refusal to provide access to the conditional access and which is the predecessor of Article 6 of the Access Directive, was drafted, the European Commission did not focus its attention only on conditional access technology. In its MSG decision, the European Commission explicitly recognized the competitive relevance of other components of the technical platform, including the EPG and the Subscriber Management System.

In the MSG case, the European Commission found that the presence of such a powerful joint venture might deter competitors from entering the market. In particular, the European Commission attached importance to the vertical integration between the technical platform and the business of providing own access-controlled services via the same platform, thereby addressing the aspect of leverage.[423] The European Commission examined the market position that the three participants would hold in their 'native' markets and the level of vertical integration that would be the result of the joint venture. The European Commission's final decision was, indeed, motivated by the leverage argument. The prospects of the likely effects of market power that is vested in one upstream or downstream market but might be used to restrict access to a related market was a main reason why the merger was barred. The European Commission found that the joint control over several levels in the distribution chain and the resulting competitive advantages would have a) prevented possible competition between the companies involved and b) given the companies the power to determine conditions for new market entrants. This, together with the dominance of each party in their 'native' markets and vested economic interests in the new market for pay-TV services, led the European Commission to the conclusion that the situation might prejudice the parties' decisions of whether to grant access to facilities, namely, technical facilities such as conditional access, to potential competitors.

The proposed commitments (notably the implementation of a common interface in the decoder, the promotion of freely available decoders and non-restrictive

[423] See also Bertelsmann/Kirch/Premiere, paragraphs 108-112.

consumer contracts, non-discriminatory EPG representation and the commitment from Deutsche Telekom to make more digital capacity available for competitors) could not alleviate the suspicion of the European Commission. The Commission feared the potentially harmful effects of a strongly vertically integrated market structure that would not only provide incentives for discriminatory access decisions, but also offered, according to the European Commission, various possibilities of hidden discrimination with the effect that compliance with the largely behavioural undertakings was difficult to monitor. This is an argument that returns in later decisions. This is, however, also the place to remember that, as explained in Chapter 1, the idea that vertical structures would constitute as such a danger to functioning competition is by far not uncontested. In Chapter 1, it was explained that a number of influential scholars defended the view that the mere existence of vertically integrated structures is not automatically a risk to functioning competition.[424] The Commission, in its argumentation in the MSG decision did, however, little to take these concerns into account. But, as will also be shown, in later decisions the European Commission showed a tendency to abandon some of its rigid approach concerning vertically integrated structures.

Bertelsmann/Kirch/Premiere

The parties to the failed MSG merger launched a second attempt. This operation, too, ended (in two separate cases) before the European Commission and, here too, the parties did not succeed in getting the approval of the European Commission. The first case concerned the plans of Bertelsmann (in the form of CLT-UFA, which is a joint venture between Bertelsmann and Audiofina) and Kirch to jointly control Premiere (formerly controlled by Canal+ and CLT-UFA). Kirch would close down its own pay-TV platform DF1 and transfer its assets to Premiere. Premiere would subsequently use Kirch's set top box, the d-box, market programme packages, and operate its own Subscriber Management System and conditional access system. CLT-UFA would also acquire fifty per cent in BetaDigital and BetaResearch, both subsidiaries of Kirch. They would both produce pay-TV technology and offer it to third parties. The European Commission found that, after the operation, Premiere would enjoy a near-monopoly in the German pay-TV market. That this was the case, had, on the one hand, to do with the strong position of CLT-UFA in the markets for programme rights. On the other hand, it had to do with Premiere's control over the technical platform.

The European Commission concluded that a combination of Premiere and CLT's free-TV channels might induce rights suppliers to sell the rights for pay-TV and free-TV as a package to the joint venture. CLT and Premiere could then implement programme strategies for free-TV channels aimed at securing pay-TV subscribers for Premiere Digital. In so doing, the decision is based strongly on the assumption of interaction, if not competition, between free-TV and pay-TV, and that the

[424] See section 1.5.3.

prospects of pay-TV largely depend on the strength of free-TV.[425] In this constellation, Premiere and CLT-UFA would have the power to strongly limit the attractiveness of free-TV. The decision clearly articulates the wish to maintain competition between free-TV and pay-TV. On another occasion, namely the TPS case in which the European Commission showed less concern, TFI and M6, two French free-TV broadcasters, acquired a fifty per cent interest in the French pay-TV operator TPS. Here, the Commission considered that even in the event of joint control over free-TV and pay-TV channels, the competitive position of TPS regarding its rival Canal+ was so weak that there was no need to fear that TPS would even try to unduly influence competition in the free-TV market.[426]

In Bertelsmann/Kirch/Premiere, the European Commission again found the risk of a lasting monopoly in technical pay-TV services. A critical factor was, in the eyes of the European Commission, that parties would operate closely with Deutsche Telekom (the Deutsche Telekom/BetaResearch case will be discussed subsequently). Again, this is the aspect of the possible anti-competitive effects of vertical integration. Deutsche Telekom was then the dominant cable-network operator and would exclusively carry Premiere.

The European Commission was also concerned because the d-box used at the time had a proprietary API, meaning that third parties wishing to process services via the d-box would require Premiere's approval. On the other hand, the European Commission also acknowledged that the investments made by the parties in the joint venture could pave the way for other service providers who would then be able to offer access-controlled services to consumers without having to invest in an own access control infrastructure. In this respect, the proposed operation could have a positive effect and even accelerate the progress of digitization in Germany. As the European Commission explained, 'viewers' psychological barriers to additional payments are progressively lowered with the penetration of digital pay-TV'.[427]

In the end, however, the parties could not dispel the European Commission's serious doubts that, instead of paving the way for a competitive digital service market, the parties would use their position to monopolize the consumer base on a lasting basis due to technical lock-ins. The package of commitments that was suggested by the parties could not change the European Commission's opinion. The parties were willing to commit to granting access to programme material (namely to hold twenty-five per cent of the output deals of pay-TV rights open for third parties, to unbundle and refrain from acquiring free-TV rights) as well as to more openness at the technical level (compulsory licensing of the conditional access system to third

---

425 Also, European Commission, Newscorp/Telepiù, paragraph 37.

426 European Commission, TPS, paragraph 25, but see also paragraphs 28 and 29. See also European Commission, Commission decision of 10 September 1991 declaring a concentration to be compatible with the common market (Case IV/M.110 - ABC/General des Eaux/Canal+/W.H.Smith TV) according to Council Regulation (EEC) No. 4064/89, 19 September 1991, OJ C 244, p. 1 [hereinafter 'ABC'], paragraph 19.

427 European Commission, Bertelsmann/Kirch/Premiere, paragraph 121.

parties, disclosure of API information, licensing of conditional access technology to third-party decoder manufacturers, cooperation with competing cable operators). The European Commission, however, did not consider these commitments sufficiently far-reaching.

Deutsche Telekom/BetaResearch

The proposed operation between Bertelsmann, Kirch and Premiere had a second component to it, namely the plan that Deutsche Telekom and BetaTechnik would acquire joint control over BetaResearch, which is the exclusive holder of a license for the BetaResearch conditional access technology that is implemented in the d-box. BetaResearch would manufacture conditional access modules and smart cards, and provide services for third parties. Deutsche Telekom would use exclusively the BetaResearch technology when distributing pay-TV via its cable networks. Because Deutsche Telekom was by then the dominant German cable operator, the European Commission feared that the d-box technology would become the digital standard in the German-speaking area. An alternative platform could, in the best case, emerge only for a satellite transmission platform. Satellite transmission, however, was then not widely available in Germany. Second, after the transaction, Premiere, with its major programme resources, would provide digital television via the d-box technology and Deutsche Telekom's networks. This, so said the European Commission, was an additional obstacle to the development of alternative technical platforms. Third parties would only invest in an alternative technical platform if it had the corresponding opportunities for market penetration. It was unlikely, according to the European Commission, that there would be many of these opportunities given the established position of Premiere. The decision suggests that there is a close link between the success of an alternative technical platform and the market position of an established service platform with access to a vast subscriber base[428] and programme resources. This is in line with the findings of this study, which are discussed in Chapter 1. Moreover, Chapter 1 also indicates why it is critical to take this to be an automatism.[429]

Deutsche Telekom and BetaResearch also sought to dispel the doubts of the European Commission by committing to, on the one hand, cooperating with private cable operators, issuing compulsory conditional access licences and using a standardized API interface, and, on the other hand, allowing cable operators to market their own programmes and basic packages and develop their own customer relationships. However, the parties made these commitments dependent on a number of restrictions, one of which was that cable operators were not allowed to market Premiere's pay-per-view services. Cable operators would also be obliged to market Premiere's programmes unchanged and in unbundled form. Capacity restraints were not to be used as an excuse for cable operators not to transmit

[428] European Commission, Deutsche Telekom/BetaResearch, paragraph 35.

[429] See 1.5.3. See also European Commission, Deutsche Telekom/BetaResearch, paragraph 38.

Premiere's programmes. Furthermore, cable operators would make subscriber data available to Premiere. The European Commission found, correctly, that the latter was anything but common practice and that, all in all, the commitments would not be sufficient to allow cable operators to establish and operate programme platforms on their own. Interesting were the comments of the European Commission concerning the compulsory licensing of conditional access technology to decoder manufacturers and the granting of access to the technical platform for third parties. In this context, it should be mentioned that the Standards Directive had since been put in place, and that its Article 4c mandated access to the conditional access system, which was also the subject of the commitments in the Deutsche Telekom/BetaResearch case. The Deutsche Telekom/BetaResearch decision made no reference whatsoever to Article 4c of the Standards Directive. Also interesting is that the European Commission observed that the compulsory licensing of the access technology to third-party programme providers and decoder manufacturers might go 'some way to ensuring that parties are not subject to discrimination', but that it would not suffice. It would not alter the fact, so said the European Commission, that the parties to the operation still controlled the technical developments. Or, in the words of the European Commission,

> '[t]the parties are not willing to surrender their absolute control of this technology, and in particular its further development. Nor will the proposed undertakings give any alternative programme and marketing platform a realistic chance. At the same time, without the chance of an alternative programme platform, the undertakings with regard to technology, or at least to licensing for the purposes of controlling access, are rendered even more meaningless by the fact that, without a second programme platform, no alternative technological platform can be expected to be developed'.[430]

Hence, the decision expresses clear doubts whether access obligations are the optimal tool or are sufficient to create the conditions for a competitive pay-TV environment. This is a question that will be dealt with extensively in Chapter 4 (Telecommunications Law).

BiB

The next major decision in this sector was the BiB/Open decision. BiB/Open was an Article 81 EC Treaty case. It involved, among others, British Telecom and BSkyB. The plan was to form a joint venture 'BiB' to provide the necessary infrastructure for the provision of digital interactive television services and to provide such services across that infrastructure (also for use by third parties). The plan was to use the same infrastructure used by TV companies to allow the integration of interactivity in television. BiB would not only operate the technical platform, it would also offer certain services directly to consumers, such as email,

[430] European Commission, Deutsche Telekom/BetaResearch, paragraph 65.

walled garden internet access and computer game downloads. Such services were not yet provided on a significant scale in the UK. BiB would use a proprietary conditional access system and API technology, and its set top boxes would support only one proprietary EPG. Significant was the close relationship with BSkyB; BSkyB and BiB's infrastructure would be largely complementary so that consumers would have to buy only one set top box to be able to receive services from both undertakings. BiB intended to promote the proliferation of its set top box by subsidizing it. Only upon the interference of the European Commission were separate companies created for BiB's set top boxes and for the creation and operation of BiB's interactive services.[431]

The European Commission found that the modified joint venture would result in a considerable restriction of competition in the market for digital interactive television services, and that this would affect trade between Member States. Prior to the joint venture, British Telecom and BSkyB were potential competitors in the provision of interactive television services, a situation that was, so anticipated the European Commission, eliminated by the joint venture. The decision acknowledged the ongoing process of convergence, and, that as a consequence thereof, telecommunications operators would be competing with content service providers and providing content-related and e-commerce services over their networks themselves.

Nevertheless, the European Commission decided not to oppose the joint venture because it considered the conditions of Article 81 (3) of the EC Treaty to be fulfilled. Article 81 (3) of the EC Treaty allows a joint venture to be exempted from the prohibition clause in Article 81 (1) of the EC Treaty despite its restrictive effect on competition providing certain conditions are met. Those can be a contribution towards improving the production or distribution of goods or the promotion of technical or economic progress, allowing consumers to have a fair share of the resulting benefit, indispensability, and the non-elimination of competition in respect of a substantial part of the products in question. Because the joint venture would enable the provision of a new form of services that had not yet been offered, it was found to contribute to an improvement in the distribution of goods and technical and economic progress. Correspondingly, it would allow consumers to access interactive services via the TV set. To this extent, it would also benefit consumers. But the European Commission insisted on a number of conditions, one of which was that BSkyB must distribute its film and sports channels with and without interactive applications to competitors. This unbundling-condition addressed the link between BSkyB's existing operation in pay-TV and its new activities in the interactive service sector. The European Commission hereby sought to prevent an existing dominant player in a pay-TV market from exporting its market power to a new market, namely the market for interactive services. The European Commission acknowledged that pay-TV is likely to be an important driver for digital interactive

[431] Insofar, European Commission, BiB/Open, paragraph 126.

television services.[432] Other conditions imposed consisted of giving consumers the choice to acquire the set top box with or without subscribing to BSkyB's pay-TV offer and requiring that British Telecom not expand its existing cable television interest in the UK. These measures were taken to enable cable networks to compete, to create separate businesses for BiB's activities with regard to the subsidization of set top boxes, to recover profits from letting third parties use the box, and to enable the operation of interactive services. The latter was considered to ensure transparency and non-discrimination and to prevent subsidized set top boxes from being used as artificial barriers to market entry. The reason why the European Commission considered access for competitors to the (subsidized) BiB set top boxes so crucial was that it seriously doubted whether a competing infrastructure would establish itself. In the opinion of the European Commission, consumers were all the more reluctant to acquire more than one set top box because the first box would allow access to the whole range of BSkyB's programming, too.

The assumption that consumers are indeed unwilling to acquire several set top boxes (particularly if they are subsidized) is not an uncontested one. After all, consumers are willing to buy a computer in addition to a TV. Furthermore, with digitization, the set top box problem might lose much of its relevance, considering that each owner of an analogue television set will have to have a set top box irrespective of whether he or she has subscribed to a pay-TV platform.[433] Still, the hardware argument is an argument that plays an important role throughout the discussion around conditional access. And it is not limited to pay-TV markets. A very similar discussion accompanied, for example, the launch of Apple's iPod. The iPod is the music player of Apple's online music store iTunes. The iPod, which proved to be a very successful and popular hardware device among consumers, only supports one technical content protection format, meaning Apple's DRM called FairPlay. It does not support any of the standards used by competing digital music services, nor does it license its own format to rivals. Here too, the question is whether the hardware, in combination with proprietary content protection technology, could form an obstacle for competition in the online music sector. More about the iPod example can be found in the discussion further below (section 3.4.1.). At this point, it remains to be said that much will depend on the value consumers attach to a particular hardware device (for example, because it is popular, because they know how to use it or because it is clearly superior), on the consumers' expectations of the speed of technological progress, as well as on the price of the hardware.

BSkyB/Kirch

A year later, the European Commission had to decide on another important case that involved the provision of interactive television services, this time in Germany.

[432] European Commission, BSkyB/Kirch, paragraph 78.
[433] See section 1.5.2.

This was the BSkyB/Kirch merger. The plan was that BSkyB and Kirch jointly control Kirch PayTV. Because of its experience and financial resources, BSkyB would play a major role in deploying interactive television services in Germany. BSkyB said that it was not planning to enter the German pay-TV market. To this extent, and because of the strength of free-TV in Germany and Kirch's control over a proprietary conditional access infrastructure, the European Commission did not consider BSkyB a potential competitor to Kirch in Germany's pay-TV market. Consequently, a factor that could have been a strong argument against the operation was not given, at least not in the view of the European Commission. The European Commission's stance was heavily criticized by the German public broadcaster ARD in a proceeding before the European Court of First Instance against the decision in BSkyB/Kirch.[434] ARD claimed that the European Commission's approach—to assume that BSkyB would not enter the German pay-TV market in the short term, nor that Kirch would be able to maintain its dominant position alone—departed from a status quo. Instead, the European Commission should have considered the likely future evolution of the market, namely that without a merger the situation for third parties could improve considerably. Consequently, the European Commission should have regarded BSkyB as a potential competitor. The court did not follow this argumentation. It held against it that the argumentation was based on the unsubstantiated premise that financial failure by Kirch PayTV would be likely to favour access to the market by potential competitors. The two observations that seem to be implied here, namely that pay-TV markets are in their nature still national markets and that competition from foreign pay-TV operators was rather unlikely,[435] and that the absence of one dominant party does not guarantee the development of a competitive environment, will play a role in two other decisions (NewsCorp/Telepiù and Sogecable/ViaDigital) that will be discussed in more depth at a later stage.

Similar to the BiB case, one major concern about the BSkyB/Kirch merger was the proprietary conditional access technology in the d-box, which would have been used to process not only interactive services but also digital broadcasting services. According to the European Commission, this was likely to raise significant barriers to entry for potential suppliers. To this extent, Kirch, through his control over the

[434] European Court of First Instance, Judgment of the Court of First Instance of 30 September 2003, Arbeitsgemeinschaft der öffentlich-rechtlichen Rundfunkanstalten der Bundesrepublik Deutschland (ARD) v Commission of the European Communities, Case T-158/00, European Court reports 2000, p. 0 [hereinafter `ARD/European Commission'].

[435] Also, European Commission, Commission Decision of 2 August 1994 declaring a concentration to be compatible with the common market (Case IV/M.410 - Kirch/Richemont/Telepiu) according to Council Regulation (EEC) No. 4064/89, 13 August 1994, OJ C 225, p. 3 [hereinafter `Kirch/Richemont/Telepiu'], paragraph 20: ' Furthermore, the national character of the television markets seems to show that Kirch and FilmNet are not in a particularly advantageous position in order to enter the Italian market on their own with the setting-up of a new pay-TV channel as compared to other television operators in the EC. Therefore, the present transaction does not amount to an appreciable elimination of potential competition in the television market in Italy'.

technical platform, and BSkyB, because of its know-how, marketing experience and financial resources, could export their dominance from already existing markets to the new market for interactive television services in Germany. This could have had recourse to the classic leverage argument if the European Commission had not explained further: '[i]n itself entry into a new market by a dominant firm on a closely related one does not automatically lead to the creation of a dominant position'.[436] Instead, it held that the 'link between control over attractive programming and decoder technology' would raise barriers to market entry. In this respect, it seems that the European Commission had given up some of its strict approach regarding vertically integrated structures. The decisive factor in this context, said the European Commission, was the existence of first mover advantages and (indirect) network effects.[437] According to the European Commission,

> '[i]t should be recalled that the demand from "content providers" for access to an operator's digital interactive television "platform" is likely being determined by the popularity of the "platform" whilst the attractiveness of the "platform" to final consumers will be determined by the range and types of services they can find on it. It has already been noted that pay-TV is likely to be an important "driver" for digital interactive television services. As a result of its monopoly position on the pay TV market, Kirch PayTV will be the only undertaking in Germany able, in the foreseeable future, to offer pay-TV in combination with digital interactive television services. This is likely to mean that consumers will choose Kirch PayTV as it will allow them access, through the d-box, to both interactive services and pay-TV without the cost or inconveniences of having two boxes. As a result, the d-box will become the standard decoder in Germany for interactive services, as well as pay-TV'.[438]

And, unlike in its previous decisions, the European Commission held that even a de facto standard in form of the d-box technology would not necessarily amount to a dominant position for Kirch PayTV on the market for digital interactive television services if other operators were also able to supply services via the d-box.[439] In other words, the European Commission made it clear that an alternative to inter-platform competition (meaning competition between two different platforms, for example, between Canal+ and ABSat in France or competition between cable and satellite pay-TV platforms) could be intra-platform competition. Intra-platform competition, means competition between different service providers that are carried via the same pay-TV platform. But similar to the Deutsche Telekom/BetaResearch case, the European Commission was not convinced that third parties would have the possibility to compete within the platform because of the proprietary nature of the

[436] European Commission, BSkyB/Kirch, paragraph 78.
[437] See also section 1.5.2.
[438] European Commission, BSkyB/Kirch, paragraph 78.
[439] European Commission, BSkyB/Kirch, paragraph 79.

d-box technology and the fact that BetaResearch would control the technological developments.

Unlike the Deutsche Telekom/BetaResearch case, however, BSkyB/Kirch was cleared in the end; this time, the European Commission accepted the commitments that the merger parties made. Essentially, the commitments concerned the guarantee of giving third parties access to Kirch's BetaResearch d-box technology. The parties undertook to provide access to the technical platform, including the conditional access system, the subscriber authorization system, the EPG and any other technical services needed, also to providers of interactive services at fair, reasonable and non-discriminatory conditions. Again, no reference is made to the then relevant access obligations in Article 4c of the Standards Directive. Still, the European Commission pointed to the practical problems involved in granting access to the conditional access system, namely the ability of the conditional access controller to hamper third parties by means of slow proceedings or the requirement for third parties to submit their plans in advance to BetaResearch.

The parties, however, made further-reaching commitments. The d-box would be equipped with a Multimedia Home Platform (MHP) API, an open API standard, and any extensions or plug-ins developed or deployed by Kirch would maintain the openness of the API interface so that no additional licenses for developing applications to run on the API would be required from Kirch. Kirch would provide the information enabling rivals to write applications that are compatible with the d-box API and grant access to its API. Kirch undertook to provide information on d-box functionality upgrades and not to occupy the d-box's memory with applications that are not strictly necessary for the functionality of the services Kirch offers. Kirch would also be willing to licence its technology to competing platforms and would not prevent third parties from building their own conditional access into the d-box or withhold consumers from using an alternative box. Interesting from a policy point of view is the commitment to exchange consumers' analogue set top boxes for digital ones and so to promote the digital switchover. Finally, Kirch agreed to offer and develop Simulcrypt agreements for its platform and to distribute its own programmes, subject to a Simulcrypt agreement, on competing platforms.

All in all, and in anticipation of the later analysis of telecommunications law in Chapter 4, this would seem to be a very comprehensive package of commitments that go far beyond mere access obligations and that in many points clearly exceed the level of statutory obligations according to Article 4c of the Standards Directive and its successor, Article 6 (1), Annex I of the Access Directive.

Vivendi/Canal+/Seagram and Vodafone/Vivendi/Canal+

The expansion of established pay-TV providers, or more generally, players in broadcasting markets, into new and convergent service markets was subject to a number of other interesting decisions. One case concerned the merger of the media enterprises Vodafone, a major mobile telephony operator, Vivendi, a multimedia company, and the pay-TV provider Canal+. The other case concerned the merger of

Vivendi, Canal+ and Seagram, which was affiliated with the major media enterprise Universal. Vivendi, which at the time of the decision held forty-nine per cent in Canal+ and twenty-five per cent in BSkyB, would take sole control over Seagram, a Canadian enterprise involved in cinema, television and music rights, and television channels (SCI FI, 13th Street, Studio Universal).

In terms of content, the merger was expected to create the world's second largest film library, the second largest TV programming library in the European Economic Area (EEA) and move a substantial part of the theme channels' production to France, Germany, Italy and Spain. In terms of music content, the merged entity would be number one in recorded music. In terms of distribution, the parties already operated a leading pay-TV service platform in a number of Member States and were likely to become a leading player in paid-for information society service distribution via Vivendi's multi-access portal, called 'Vizzavi'. Vizzavi would provide customers with a multi-platform environment for a range of web-based interactive services that can be received across a variety of platforms, for example, on mobile phones, PCs and broadcasting set top boxes. In the case of Vodafone/Vivendi/Canal+, the parties agreed to acquire joint control of Vizzavi.

The European Commission expressed serious concerns that the combination of highly attractive content and major pay-TV, internet and mobile distribution platforms would considerably strengthen the overall market power of the merging parties and generate strong network effects and first mover advantages in their favour. It particularly stressed the competitive advantages that intermediary platforms enjoy in a multi-platform environment. Vizzavi's branded and integrated approach to the provision of services across various platforms would allow for the cross-selling and bundling of offers and thereby strengthen the position of the parties involved. This again was a factor that, in the eyes of the European Commission, favoured leveraging market power by means of bundled offers.[440] Players in traditional broadcasting markets, such as Canal+ in the pay-TV market, would be able to migrate their consumer base to other platforms, such as the internet market.[441] The fact that Vizzavi was vertically organized further added to the European Commission's concerns; consumers would be at risk of being 'walled in' and having to pay higher prices for services due to the lack of competition.[442] In addition, the Vizzavi portal would combine a potentially powerful new internet access mechanism with paid-for content and an installed customer base of pay-TV subscribers with relatively well-documented preferences who were accustomed to paying for content.[443] The European Commission, hence, had serious concerns that the parties would extend their positions of dominance into national pay-TV markets (Canal+) and their market power in national mobile markets (Vodafone) into the national markets for portals used via mobile handsets and/or set top boxes.

[440] European Commission, Vodafone/Vivendi/Canal+, paragraphs 68-70.
[441] European Commission, Vodafone/Vivendi/Canal+, paragraph 70.
[442] European Commission, Vivendi/Canal+/Segram, paragraphs 39-67.
[443] European Commission, Vodafone/Vivendi/Canal+, paragraph 75.

The European Commission, nevertheless, approved both mergers subject to the undertakings the parties made. Notably, Vivendi agreed in the Vivendi/Canal+/Seagram case to divest its interest in BSkyB, thereby cutting the link between Universal and Fox. It was further agreed not to guarantee Canal+ more than fifty per cent of the so-called 'first window rights' for Universal films. Moreover, the parties undertook to provide access to Universal music content on a non-discriminatory basis regarding the pricing and the terms and conditions, and to provide for an arbitration procedure. This would enable competitors to offer competing music services via competing portals. The decision underpins the importance of access to attractive content for the launch of alternative service platforms, as well as the advantageous position already established pay-TV players can have when they venture into new, non-broadcasting markets.

In Vodafone/Vivendi/Canal+, Vodafone and Vivendi have undertaken to open access to their mobile handsets to third-party portal providers. Interestingly, the parties did not simply commit to granting access to rivals at fair, reasonable and non-discriminatory terms, they also agreed to give subscribers to the Vodafone/Vivendi group the choice of overriding the default Vizzavi portal with the portal of their choice. Similarly, Canal+ agreed to give other portal operators access to its set top box and offer consumers the choice between rival portals. The European Commission found that 'ensuring third-party access based on consumer choice' removed the serious doubts it had regarding this merger. Most importantly, allowing for consumer choice prevents the parties from bundling their offers on a fully exclusive basis, as consumers would be allowed to choose other portals.[444] In other words, the European Commission addressed and emphasized the problem of technical and contractual consumer lock-ins.[445] The European Commission found that the key to the solution of this problem laid in the consumers' freedom to choose between competing offers.

Cegetel/AOL/Canal+/Bertelsmann

In the Cegetel/AOL/Canal+/Bertelsmann case, the European Commission showed less concern for the potential threat of dominance over the paid-for internet content market. The case dealt with a planned joint venture between Cegetel, a subsidiary of the Vivendi Group, Canal+, AOL and Bertelsmann with the aim to market, develop and provide interactive services via the internet to consumers. Interesting enough, the European Commission did not believe that the joint venture would create or strengthen a dominant position, arguing that 'the size of the market for paid-for content is currently extremely small ... given the fact there is a vast array of information providers that develop and make content available, no single provider accounts for more than a small part of the content available'.[446] In this case, the fact

[444] European Commission, Vodafone/Vivendi/Canal+, paragraphs 90-93.

[445] See section 1.5.2.

[446] European Commission, Cegetel/Canal+/AOL/Bertelsmann, paragraph 36. See also AT&T/Media One, paragraphs 25-30; European Commission, Vodafone/Vivendi/Canal+, paragraph 74.

that parties were then dominant in 'old' markets was not automatically considered by the European Commission to be an indicator that the parties would leverage their market power into new markets. Their ability to do so depended, so said the European Commission, on whether the market was already sufficiently competitive.

AOL/Time Warner

Having said that, the situation can be different where control over content comes together with control over a proprietary technology, such as an encryption technology. This was one of the findings in the European Commission's AOL/Time Warner decision. The AOL/Time Warner case concerned a combination of AOL's transmission platform (here: broadband internet access and service platform) and Time Warner's content rights (here: broadband content, in particular music). The European Commission found that the parties could move into a position to dictate the technical standards for delivering music over the internet by developing or acquiring proprietary formatting or DRM technology.[447] In combination with the breadth of the publishing rights available to the parties, the new entity would be in a position to impose its technology as an industry standard. And because of its control over the relevant content protection technology, AOL/Time Warner would be in a position to control the content business (here: downloadable music) and increase competitors' costs through excessive licence fees.[448]

The reason why the European Commission concluded in the AOL/Time Warner case that, as opposed to, for example, the MSG case, which concerned a similar constellation, the combination of Time Warner and AOL would probably not result in a dominant position in the market for broadband content was mainly based on the following argument: Time Warner would not be a unique or dominant supplier of broadband content because the parties committed themselves to breaking their links with Bertelsmann. Time Warner had also agreed not to acquire the major music publishing company EMI.[449] Similarly, one could conclude that where a dominant position in content supply is given, such a combination could create considerable gateway concerns—a situation the European Commission did not have to decide on in this case. The fact that a) Bertelsmann (which is part of a joint venture with AOL in Europe) offered to exit from AOL Europe and that b) the parties undertook not to make Bertelsmann music available online exclusively through AOL were additional reasons to clear the merger. Another undertaking that finally contributed to the clearance of the merger, and this is an important one for the given context, was that the parties agreed not to provide their broadband offering in a proprietary format that could be processed only via the AOL platform.

[447] The parallels between DRM and conditional access are pinpointed in sections 1.2. and 4.3.2.
[448] European Commission, AOL/Time Warner, paragraphs 55-56.
[449] European Commission, AOL/Time Warner, paragraphs 88-92.

NewsCorp/Telepiù

The deployment of a de facto standard for the technical platform also played a major role in the case NewsCorp/Telepiù. The NewsCorp/Telepiù case concerned the fact that NewsCorp, the entertainment company owned by Rupert Murdoch, would acquire control over Telepiù and Stream in Italy, the two major Italian pay-TV platforms. The idea was to combine the Italian operations of Telepiù and Stream under the control of NewsCorp. The European Commission found that the concentration would lead to the creation of a near-monopoly in the pay-TV market in Italy because it combined the two competing pay-TV operators and because it was unlikely that new competitors would emerge quickly from cable or digital terrestrial television.[450] The European Commission also paid particular attention to the vertical effects of the merger, namely its integration with an in-house proprietary conditional access technology. The European Commission reminded of the fact that the pay-TV sectors in Europe have so far been characterized by a strong tendency towards vertical integration between the technical and the service platform.[451] It held that the operation of a proprietary conditional access system by a monopolist was a factor that would increase barriers to entry into the related pay-TV market.[452] Summarized, the concerns were as follows: there was a chance that the new entity would grant access to its conditional access platform at unfair terms and conditions and that it would obstruct the entry of alternative pay-TV platforms that use a different conditional access technology.

For the first time, the European Commission made an explicit reference in a pay-TV decision to legal provisions that apply to conditional access. It admitted that the existence of the access rules in telecommunications law could reduce some of the concerns about the accessibility of the technical conditional access platform.[453] As regards the second concern, the chances of alternative platforms using their own conditional access technology, the European Commission referred to the Italian regulation. Italian law mandates the implementation of interoperability between different conditional access systems.[454] The European Commission pointed to interoperability solutions as a means to stimulate competition between different technical platforms. It observed, however, that difficult and lengthy negotiations between the two Italian pay-TV operators Telepiù and Stream to conclude a Simulcrypt agreement also demonstrated how difficult interoperability is to realize in practice. Eventually, at least this is the view that the European Commission expressed, it was more likely that potential competitors would seek access to the already established conditional access technology than to launch their own alternative technology.[455] This would further pinpoint the importance of access rules.

[450] European Commission, Newscorp/Telepiù, paragraph 114.
[451] European Commission, Newscorp/Telepiù, paragraph 119.
[452] European Commission, Newscorp/Telepiù, paragraph 133.
[453] European Commission, Newscorp/Telepiù, paragraph 122.
[454] European Commission, Newscorp/Telepiù, paragraph 123.
[455] European Commission, Newscorp/Telepiù, paragraph 114.

Having said that, the European Commission did not seem to have very much faith in the effectiveness of the statutory access rules either:

> 'Although the existence of access rules contained in Directive 95/47 and Directive 2002/19/EC [Access Directive] might reduce and assuage the concerns in this respect [access to the conditional access], third parties believe that access to NDS' technology can be obstructed unless NewsCorp undertakes to comply with those rules and appropriate and effective dispute settlement is put in place. Should this not be the case, long disputes on prices will arise which in the meantime will undermine the possibility to compete... The merged platform, in the absence of corrective measures, will thus be in a position to raise rival's costs by controlling third parties' access to the DTH platform services and to conditional access. Consequently, the possibility for a newcomer DTH broadcaster, which is not able to set up an alternative infrastructure, to be in a position to become operational will depend on NewsCorp's goodwill not to raise barriers when giving access to its platform services'.[456]

Another question is whether the pessimism of the European Commission regarding the benefits the competition gains from interoperability solutions is justified. One could also argue that once the Italian broadcasting market has been digitized most Italian households would have to purchase a set top box anyway. Eventually, those boxes would (also) be produced by independent hardware manufacturers who probably have an interest in producing boxes that can process a broad range of services instead of the services of one particular operator using one particular standard. Ideally, they would produce boxes that foresee some form of common interface or other interoperability solutions. In this case, one major cost factor for the deployment of an additional conditional access technology might have vanished.

In addition, the control of a vast array of attractive pay-TV content triggered, as it has in previous decisions, the concerns of the European Commission. As the European Commission observed,

> '[a]ccess to rights is even more important for pay-TV than for free-TV. In order to entice the consumer to subscribe, or to take particular productions on a pay-per-view basis, certain specific types of content are crucial'.[457]

What is remarkable about the decision in the NewsCorp/Telepiù case is that it is the first pay-TV case in which the European Commission mentions a 'fundamental right of consumers to choose'. In particular, the exercise of second window rights for programming would deprive consumers of this 'right'. In this context, second window rights are exclusive broadcasting rights to the second screening of a film.

---

456 European Commission, Newscorp/Telepiù, paragraphs 201 - 203.
457 European Commission, Newscorp/Telepiù, paragraph 183.

More precisely, the European Commission referred to the second window rights that Telepiù held. Because Telepiù also owns the first window broadcasting rights to films, Telepiù would be able to prevent third-party broadcasters from ever screening these films. According to the European Commission, this would force consumers to consume in a 'one-format-fits-all' scenario at the time, meaning during the first window, and at the price established by the pay-TV operator.[458] It would harm consumers because they could be prevented access to premium content. The reference to 'fundamental rights' in this context is a bit confusing. As was shown in Chapter 2, there is no fundamental right of individual access to particular content against broadcasters.[459] But a reference to fundamental rights under, for example, the ECHR is apparently also not what the European Commission was aiming at. This becomes obvious in the subsequent paragraph of the decision. Here, the European Commission explains that consumers should be able to choose between 'first tier' and 'second tier' pay-TV services, meaning services that are close or less close to the release of a film in movie theatres. What the European Commission probably had in mind was to prevent exclusive control over a bundle of programme rights from being used to the detriment of service differentiation and consumers' choice.

The European Commission concluded that the proposed merger would very likely eliminate any competition in the Italian pay-TV market for some time. It then considered the application of the concept of 'rescue merger' or 'failing company defence', a concept that has been applied before only rarely to respond to a situation in which the deterioration of the competitive structure as a result of concentration would occur in a similar fashion even if the concentration did not proceed.[460] The European Commission found, however, that the conditions for authorizing a rescue merger were not fulfilled.

At the same time, the European Commission feared that Stream would exit the market and that this could harm consumers. The European Commission then sought to decide which was the lesser evil—Stream's exit from the market (with the consequence that Telepiù would likely become the monopolist) or the authorization of the merger subject to conditions that could reduce at least some of the anti-competitive effects. The European Commission decided in favour of the latter.

---

458 European Commission, Newscorp/Telepiù, paragraph 192

459 See section 2.2.1.

460 European Commission, Decision of 14 December 1993 relating to a proceeding pursuant to Council Regulation (EEC) No. 4064/89, (Case IV/M.308 – Kali-Salz/MdK/Treuhand), 21 July 1994, OJ L 186, p. 38 [hereinafter 'Kali-Salz'], paragraphs 19-20. Confirmed in European Court of Justice, Judgment of the Court of 31 March 1998, French Republic and Société commerciale des potasses et de l'azote (SCPA) and Entreprise minière et chimique (EMC) v Commission of the European Communities, Joined cases C-68/94 and C-30/95, European Court reports 1998, p. I-1375 [hereinafter `Kali-Salz'], paragraphs 91, 111-120. See also European Commission, Commission Decision of 11 July 2001 declaring a concentration to be compatible with the common market and the functioning of the EEA Agreement (Case COMP/M.2314 - BASF/Eurodiol/Pantochim), 17 May 2002, p. 45, OJ L 132, p. 45 [hereinafter `BASF'], paragraphs 135-143, 144-163.

The undertakings that were finally accepted by the European Commission can be distinguished in undertakings that are related to access to programme content and undertakings that are related to access to the technical platform. To the former belong the commitments to make existing and future exclusive contracts for the licensing of film and sports rights less exclusive and to grant unbundled access to the premium content broadcast by the combined platform. As far as the second category is concerned, NewsCorp undertook to offer access to its conditional access platform and the API based on a cost-oriented non-discriminatory formula, licence its conditional access system to third parties and enter into Simulcrypt agreements. In addition, Telepiù undertook the divestiture of its digital terrestrial television business (DTT) and proposed not to enter into new DTT activities. Finally, NewsCorp proposed arbitration procedures to guarantee the effectiveness of the commitments. All in all, and taking into account:

a. That the commitments did not greatly exceed what was already provided for in telecommunications law, at least as far as access to the technical platform is concerned.[461]
b. The European Commission's previous strict position in, for example, the MSG case and the Deutsche Telekom case, where it refused to accept very similar commitments, as well as
c. The BSkyB/Kirch case, where the Commission expressed its view that access obligations alone would hardly suffice to guarantee functioning competition and demanded rather far-reaching, additional commitments,

the remark with which the European Commission accepted NewsCorp's undertakings sounded somewhat over-enthusiastic: NewsCorp went 'a long way in providing accessibility to the combined platform with a view to allowing effective competition to be achieved'.[462]

Another question is if one should follow the conclusion of the European Commission that the merger was indeed the lesser evil. One could argue that consumers would benefit from the merger because the integrated platform would offer more programmes from more service providers. Service providers would benefit from the merger because with the increasing popularity of the NewsCorp platform they would find in it an even more attractive distribution outlet.[463] On the other hand, the merger would cause further consolidation of the Italian pay-TV market, and the chances that new pay-TV platforms would establish themselves would become even smaller.[464] The European Commission had to decide between allowing a powerful and a very powerful monopoly on the Italian pay-TV market. It decided in favour of the latter. Moreover, the European Commission could not know for sure that Stream would indeed exit from the market. This decision radiates

461 See section 4.4.
462 European Commission, Newscorp/Telepiù, paragraph 257.
463 See also Evans' description of demand-side efficiencies that are the result of increasing indirect network effects in two-sided markets, using the example of dating clubs, Evans 2003, pp. 48, 59-60.
464 See Galbiati/Nicita/Nizi 2004, 17pp. (describing the situation one year after the merger took place).

a certain 'disillusion' that competition in the Italian pay-TV market is unlikely. The European Commission seemed to have abandoned its earlier strategy of promoting inter-platform competition as the best solution to promote competition and enhance consumer welfare. Apparently, it accepted that in some markets the prospects of intra-platform competition are nihil.

Whether this is true or not must remain speculation. A point of criticism is, however, that the European Commission might have been too quick in assuming that the Italian pay-TV market would not support more than one pay-TV provider. Such competition already exists in other countries, such as France or the UK.[465] Arguably, the conditions at the Italian market are different. However, the European Commission should also have considered that NewsCorp, which was partly invested in Stream, was of all possible competitors the one that was particularly well-suited to compete with Telepiù. This was, indirectly, also one of the reasons why the European Commission did not accept the failing company line of defence.[466] In the European Commission's own words:

> 'Otherwise, every merger involving an allegedly unprofitable division could be justified under merger control law by the declaration that, without the merger, the division would cease to operate'.[467]

In other words, Stream's failure to operate profitably could also have been the result of an unprofitable and unsuccessful business strategy and not of the size of the Italian pay-TV market. Indeed, this is what some third parties have argued in the NewsCorp/Telepiù case.[468] The European Commission did not pay further attention to this argument. By authorizing this merger, the European Commission may have done exactly what it wanted to avoid: authorize a merger between two potential competitors.

3.3.2. REGULATION THROUGH THE BACKDOOR?

Some scholars argued that authorizing the merger would give the European Commission at least the possibility to subject the monopolist to 'regulation' in the form of undertakings and conditions, and that the decision provided a 'regulatory blueprint' that would also be binding for the Italian telecommunications NRA.[469] If this is so, and there are indications that the European Commission is not entirely averse to this idea,[470] then the NewsCorp/Telepiù case is probably the most obvious example where the European Commission goes beyond its role as a competition

---

[465] See section 1.4.3.
[466] European Commission, Newscorp/Telepiù, paragraph 212.
[467] European Commission, Newscorp/Telepiù, paragraph 212.
[468] European Commission, Newscorp/Telepiù, paragraph 213.
[469] In this sense, for example, Caffarra/Coscelli 2003, with further arguments, p. 626; Bavasso 2003, 277pp.
[470] See also European Commission, Newscorp/Telepiù, paragraph 221.

authority and seeks to act as a 'quasi-regulator' for the pay-TV sector. Or, as one scholar phrased it, the European Commission would engage in 'regulation from the backdoor'[471] This is an issue that the following paragraphs deal with. The question is can a general policy be identified in the European Commission's decision practice, and if yes, could the decisions have an effect that goes beyond a case-by-case basis.

The previous analysis of some of the most important pay-TV decisions by the European Commission has demonstrated that the European Commission considers certain parameters essential for the contestability and functioning of competition in pay-TV markets in Europe. It repeatedly used its authority as a competition authority to preserve those parameters in the form of the authorization or non-authorization of mergers and in the form of the conditions imposed or undertakings accepted. The most important parameters that the European Commission considered crucial for competition in these decisions can be roughly divided as follows:

a. Access to programming material (in the form of transmission rights and the wholesale provision of pay-TV channels) and
b. Access to the technical platform.

Access to programme material is certainly a crucial aspect: without access to programme material neither an alternative programme nor a technical platform is likely to develop. Because this is an aspect that has been discussed already rather extensively by other scholars[472] and because there is a need to restrict the scope of the present study, the present analysis focuses most of the attention on the question of access to the integrated pay-TV platform itself. 'Access to the technical platform' for pay-TV is, in the European Commission's decision practice, a term that is understood more widely than access to the conditional access system. The European Commission acknowledged that the technical platform also comprises additional technical facilities and services, such as the API, the EPG, Subscriber Management System and Subscriber Authorisation System services as well as billing services and that they can be bottlenecks for functioning competition, too. Access to the service platform, that is the subscription and marketing platform via which services are offered to consumers, has played a lesser role so far, at least in the decisions concerning the broadcasting sector. This might have to do with the fact that the European Commission, in most of its decisions, treated the technical and the service platform as a functional unity rather than as separate components.

The European Commission's discussion of matters of access to the technical platform for pay-TV reflects to some extent the technological developments in this sector. In the MSG decision, conditional access was considered a major bottleneck to market entry. Soon additional components appeared on the stage, such as the API and the EPG. Convergence also played an important role in the decisions. It is worth noting, however, that the deployment of conditional access systems, APIs

[471] Bavasso 2003, p. 277.

[472] See the extensive discussion in the papers by Galbiati/Nicita/Nizi 2004 and Harbord/Ottaviani 2001 (discussing a number of possible remedies); Amstrong 1999; Cave/Crandall 2001 (and here in particular F18pp.); Neumann 1998, 159pp; Harbord/Szymanski 2005, 22pp.

and EPGs is still very much seen as a domain of the pay-TV operator. The possibility that conditional access, APIs and EPGs are manufactured and offered by independent manufacturers and the impact this has or could have on competition in the pay-TV market is an aspect that has not played a role so far.

Until the NewsCorp/Telepiù decision, most of the European Commission's decisions in this sector were characterized by the wish to stimulate inter-platform competition. To this extent, the European Commission put a lot of hope on competition from alternative distribution platforms, such as platforms using cable, satellite or terrestrial transmission networks. In this respect, access to programme material was considered crucial, as was the strength of a technical platform that resulted from vertical integration, exclusive control over proprietary standards, indirect network effects and first mover advantages. Within the context of inter-platform competition, questions of conditional access interoperability played a major role. In many of the decisions, it was considered the second-best solution. While interoperability solutions were considered crucial to stimulate inter-platform competition, open access to the exclusively controlled technical platform was considered necessary to create the conditions for intra-platform competition.

All in all, the European Commission's decisions in this field are characterized by the wish to keep market structures open for competition within and between pay-TV platforms. The European Commission's merger decisions are based on a process of balancing the positive and negative effects for prospective pay-TV markets. Exclusive control over the technical platform and its components, which was identified in Chapter 1 as one of the major challenges in pay-TV markets,[473] indeed played a predominant role. In this context, most of the European Commission's decisions in this field focus on finding vertical integration and the possibility of leverage. In the early decisions in particular, the European Commission showed a strong tendency to ban vertically integrated structures altogether because they provided, so said the European Commission, the conditions and incentives for leveraging market power from control of the technical platform into the pay-TV market. Over the course of time the European Commission tackled the issue of vertical integration, bottleneck control and leverage in a more differentiated manner. For example, in the BiB and BSkyB/Kirch decisions it acknowledged the positive effects vertical integration can have on innovation and the development of new services. In NewsCorp/Telepiù it concluded, perhaps too quickly, that vertically integrated structures were better than no structure. Furthermore, as later decisions such as the BSkyB/Kirch and NewsCorp/Telepiù cases demonstrated, the European Commission acknowledged that the surrounding factors play a role, too, such as the strength of a proprietary technical standard, the existence of interoperability solutions, the degree of competition in a service market, the impact of economic mechanisms such as indirect network effects and first mover advantages, as well as the business strategy pursued.

---

[473] See section 1.5.2.

The European Commission frequently used its position to shape an environment that follows stricter rules than existing legislation provides for. The analysis of the European Commission's decisions in the pay-TV sector showed that commitments often exceeded existing legal obligations, for example, with respect to access to the conditional access system. The package of commitments in the BSkyB/Kirch decision illustrated this well. Article 4c of the then relevant Standards Directive only mandated access to the conditional access system. In the BSkyB/Kirch decision, the European Commission accepted commitments from Kirch to grant access to the Subscriber Authorisation System, the EPG, the API, other technical services needed, and to technical information. This would not only apply to digital broadcasters, which was then the scope of Article 4c of the Standards Directive, but also to providers of interactive services at fair, reasonable and non-discriminatory conditions. In its decisions, the European Commission also specified what it means by 'non-discriminatory access' in pay-TV, a definition that is still missing in Article 6 of the Access Directive. According to the European Commission, discrimination-free access requires that the licenser of the decoder technology is able to conduct its business without being influenced by a programme supplier.[474]

Furthermore, on more than one occasion the European Commission expressed its view that mandated access to the conditional access system alone would probably not suffice to ensure functioning competition. Instead, it stressed the need for interoperable solutions to stimulate market contestability as a means to discipline the behaviour of single players. The European Commission also recognized that access to the technical platform alone may not be sufficient to realize intra-platform competition and that access to the service platform was of no less importance.[475] Accordingly, the European Commission also intervened in the bouquet structure.

It is worth mentioning that, in the course of time, the focus of the interoperability issue in the decisions seems to have shifted. Initially, the European Commission's decisions aimed at stimulating different standards and preventing the establishment of one dominant standard without it having been exposed to the competition test first, such as in Deutsche Telekom/BetaResearch. In its later decisions, the European Commission apparently turned its attention more towards the possibility of concluding interoperability agreements as a means to overcome the dominance of one standard.[476] Regarding access to programme material, the European Commission went to quite some length to fight the exclusivity of programme material by imposing obligations to supply wholesale material at non-discriminatory conditions and insisting on decreasing the duration of exclusive

---

[474] European Commission, Bertelsmann/Kirch/Premiere, paragraph 111. Interestingly, the Commission also noted in this context that such influence can be had if the technology is controlled by enterprises that also have interests as programme suppliers, thereby again drawing attention to the issue of vertical integration and leverage.

[475] European Commission, Newscorp/Telepiù, paragraph 257.

[476] See European Commission, BiB, Condition No. 6 of the Commissions decision; European Commission, BSkyB/Kirch, Annex I, paragraph 5 (a).

contracts or reducing the scope of output deals.[477] It also promoted different constructs of guaranteeing access for rivals to exclusive programme rights. One example was a model of 'shareware' that came up in the Bertelsmann/Kirch/Premiere decision. Here, parties proposed to open up the share of pay-TV rights they received from certain output deals with Hollywood to third parties (meaning the ability to obtain licences directly from Premiere).[478]

Finally, many of the decisions are characterized by a bottom-up approach. This is in line with a more general approach of competition law. Here, the aspect of realizing consumer welfare through functioning competition and the knowledge of the power of consumers to shape demand dominate. The decisions demonstrated how important it is that consumers are not unnecessarily impeded in their ability to choose between different services. Moreover, the decisions sought to create the conditions under which consumers could exercise choice. The European Commission repeatedly addressed the issue of technical and contractual consumer lock-ins. The question of whether consumers are willing to switch set top boxes is an important reason why the Commission repeatedly insisted on access-to-the-decoder solutions, as well as on the consumers' ability to define their own default portal. Moreover, the European Commission clearly opposed a situation in which providers made the subscription to basic packages a condition for the subscription to film or sports channels. Companies would have to provide unbundled access for consumers to programme channels.[479] The European Commission also emphasized that it was not desirable to make the purchase of a BiB-subsidized set top box and digital satellite dish conditional on the subscription to BSkyB's services.[480] Information problems were even discussed in the decisions, for example, when the Commission insisted on a neutral and non-discriminatory style of presentation within the EPG. Aspects of the consumers' choice also played a role in the way exclusive programme rights were licensed. In this context, the European Commission, in its NewsCorp/Telepiù decision even coined the notion of a 'right of consumers to choose' as a protection-worthy aspect when assessing the competitive effects of a merger on the pay-TV sector.

Formulated somewhat poignantly, one could conclude that the European Commission in its role as competition authority had and still has its own ideas of what effective remedies are and are not. The European Commission regularly pushed parties to go beyond existing regulation, and the fact that in most of the cases it did not even refer to the current legal situation—to the extent that it was relevant—speaks for itself. In other instances, the European Commission was more explicit, such as in the NewsCorp/Telepiù case where it expressed little faith in the

477 See also Harbord/Ottaviani 2001, 18pp. and Harbord/Szymanski 2005, 22pp. suggesting a number of similar remedies.
478 European Commission, Bertelsmann/Kirch/Premiere, paragraph 124.
479 European Commission, Bertelsmann/Kirch/Premiere, paragraphs 125, 134.
480 European Commission, BiB, paragraph 129 and condition No. 2.

effectiveness of the existing legal framework.[481] Statutory access obligations would not be sufficient to prevent NewsCorp from abusing its position, for example by exercising control over the technical developments. This interpretation would be in line with the findings of the European Commission in its earlier decisions.

In its function as a competition authority, the European Commission has its own toolbox from which it can draw to create the conditions for a competitive environment. An instrument that is important in shaping the development of pay-TV markets in merger cases is, in addition to the authorization or non-authorization of a merger, the acceptance or rejection of the parties' commitments. The aim of commitments is to reduce the market power of the merging parties and to restore the conditions for effective competition.[482] Often, a major distinction is made between behavioural and structural commitments. Structural commitments solve structural problems. Telepiù's suggestion to divest its terrestrial network is an example of structural commitment. A commitment to grant access to the conditional access system is an example of behavioural commitment. The European Commission claimed in its earlier decisions to be very critical of behavioural commitments, in particular if they consisted of a promise not to abuse certain aspects of the dominant position.[483] In its Notice on Remedies, the European Commission clearly explained that

> '[w]here the competition problem is created by control over key technology, a divestiture of such technology is the preferable remedy as it eliminates a lasting relationship between the merged entity and its competitors'.[484]

However, the distinction between structural and behavioural commitments is losing much of its significance in this sector, certainly after the decision of the European Court of First Instance following ARD's attack on the BSkyB/Kirch decision.[485] Citing earlier jurisdiction, the European Court of First Instance said that the decisive question was whether behavioural commitments, and the court explicitly mentioned in this context the granting of access to essential facilities, were equally capable of preventing the emergence or strengthening of a dominant position.[486]

---

[481] European Commission, Newscorp/Telepiù, paragraphs 201 - 203.

[482] European Commission), Commission Notice on remedies acceptable under Council Regulation (EEC) No. 4064/89 and under Commission Regulation (EC) No. 447/98, 2 March 2001, OJ C 68, p. 3 [hereinafter 'Notice on Remedies'], paragraph 2.

[483] See the European Commission decisions in the cases MSG, Deutsche Telekom/BetaResearch, Vivendi/Canal+/Seagram, to name but some.

[484] European Commission, Commission Notice on remedies, paragraph 29.

[485] European Court of First Instance, ARD/European Commission, paragraphs 186, 193. See, more generally, the discussion in Bavasso 2003, pp. 336-339.

[486] European Court of First Instance, ARD/European Commission, paragraph 193; Judgment of the Court of First Instance, Gencor Ltd v Commission of the European Communities, 25 March 1999, Case T-102/96, European Court reports 1997, p. II-879 [hereinafter 'Gencor'], paragraph 319.

Commitments (also referred to as 'modifications' or 'remedies') are offered by the parties themselves, and it is up to the European Commission to decide whether they suffice to remove its concerns about possible negative effects of the merger on the competitive structure. They are self-obligations made by the parties and accepted by the European Commission on a case-by-case basis.[487] They are no regulations. According to the European Court of Justice,

> 'Commitments, on the other hand, impose detailed obligations to be met within short periods of time, compliance with which is ensured by an effective, binding arbitration procedure which reverses the burden of proof.[488]

One consequence of commitments is, however, their effect on a switch in the burden of proof. The monitoring of abuses, according to Article 82 of the EC Treaty, requires proof of a dominant position and an abuse of this position. In contrast, a commitment transfers the burden of proof of compliance to the undertakings concerned by the operation. And because commitments impose detailed obligations coupled with a binding arbitration procedure they offer a greater legal certainty to the market players than Article 82 of the EC Treaty.[489] Having said that, the only way the European Commission can enforce commitments is by making the authorization of a merger subject to compliance with them. Where undertakings are in breach of the obligations from their commitments, the European Commission is entitled to revoke the clearance decision.[490]

This begs the question whether the European Commission uses its position as competition authority to create (if not law than at least) an alternative 'pay-TV' policy. Because competition law decisions are case-by-case decisions, it is not adequate to distil more general rules from previous decisions or predict how future cases will be decided. Nevertheless, some parties have tried to use precedents in earlier European Commission decisions to enforce a particular decision practice in newer decisions. This was what ARD did in front of the European Court of First Instance in the BSkyB/Kirch case. ARD claimed that, following its practice in earlier decisions in Bertelsmann/Kirch/Premiere and Telekom/BetaResearch, the European Commission should not have authorized the BSkyB/Kirch merger.[491] The European Court of First Instance was not very impressed by this argument. It observed that the fact that the European Commission rejected similar commitments in previous cases was of no relevance for this case as it was not comparable to the earlier cases. In addition, specific commitments could not be considered in isolation

---

487 European Commission, Commission Notice on Remedies, paragraph 9.

488 In this sense also European Court of First Instance, ARD/European Commission, paragraph 203, see also paragraph 267.

489 European Court of First Instance, ARD/European Commission, paragraphs 202-203.

490 European Council, Merger Regulation, Articles 6 (2), 8 (2). The European Commission can also impose fines and penalties, according to the Council's Merger Regulation No. 139/2004, Articles 14 (2) and 15 (2).

491 See European Court of First Instance, ARD/European Commission, paragraph 232.

and the European Commission would have to assess the proposed commitments in each case as a whole.[492] In the Sogecable/ViaDigital case, the ruling of the European Court of First Instance followed a similar line.[493] This case dealt with the complaint of a number of Spanish cable operators against a decision of the European Commission in a case concerning a merger between Sogecable and Via Digital, two major pay-TV operators in Spain. The European Commission acknowledged that the merger would probably lead to a dominant position of the joint entity in the Spanish pay-TV market. But at the request of the Spanish government according to Article 9 (3)a of the Merger Regulation No. 4064/89, the European Commission referred the case to the Spanish authorities for decision. The Spanish authority subsequently cleared the merger. The applicants in the case that went before the European Court of First Instance complained that the European Commission should have decided rather than refer the case to the Spanish authorities. One of the applicants' arguments was that the European Commission breached the spirit of Article 9 of the Merger Regulation No. 4064/89 and its own best practice in pay-TV merger cases. According to the applicants, the Spanish market was in this case a substantial part of the common market and the European Commission had before systematically prohibited concentrations with a Community dimension where they had the effect of excluding competitors from the market. The applicants' argument was that:

> 'it was necessary for the European Commission to examine the present concentration to ensure that the pay-TV market in Spain remained accessible to competitors. In that way the European Commission could have ensured that similar concentrations were treated in the same way in all the Member States. Furthermore, the applicants observe that the European Commission aims to liberalise the telecommunications sector. According to them, the European Commission is best placed to ensure that concentrations do not jeopardise the achievement of the aims of the Community telecommunications policy in a substantial part of the common market such as Spain'.[494]

The parties further observed that the application of national law would create a risk to the uniformity of the policy currently implemented by the European Commission in the markets concerned.[495]

The court did not follow the parties' argumentation. First, it stipulated that the European Commission had broad discretion in assessing whether to refer a concentration or not. Second, reasons of consistency or uniformity with earlier

---

[492] European Court of First Instance, ARD/European Commission, paragraphs 244-245.

[493] European Court of First Instance, Judgment of the Court of First Instance of 30 September 2003, Cableuropa SA and Others v Commission of the European Communities, Joined cases T-346/02 and T-347/02, European Court reports 2003, p. 0 [hereinafter 'Cableuropa/European Commission'].

[494] European Court of First Instance, Cableuropa/European Commission, paragraph 162.

[495] European Court of First Instance, Cableuropa/European Commission, paragraph 169.

decisions were not reasons that should influence the decision of the European Commission. The court made very clear that:

> 'the fact that, in a given sector, the European Commission has decided itself to examine the concentration and has prohibited certain concentrations in the past can in no way prejudice the referral and/or examination of a later concentration because the European Commission is required to carry out an individual appraisal of each notified concentration according to the circumstances of each case, without being bound by previous decisions concerning other undertakings, other product and service markets or other geographical markets at different times. For the same reasons, previous decisions of the European Commission relating to concentrations in a specific sector cannot prejudge the decisions to be taken by the European Commission on a request for referral to the national authorities of a concentration taking place in the same sector'.[496]

Conclusion

In conclusion, and in response to the research question asked at the beginning of section 3.3.2., it can be said that the decisions of the European Commission have no binding effect beyond a concrete case. The European Commission is not a second telecommunications NRA, even if it does sometimes tend to evoke the impression. The case-by-case character of its decisions and the legal uncertainty that results from the non-prejudicial character of its decisions are a further powerful argument why one should not overestimate the 'policy-making' ambitions or abilities of the European Commission in its function of competition authority. Assuming any 'rule making' power would not only contradict the intended nature of decision-making in merger cases, it would also mean a circumvention of formal rule making procedures based on democratic and Community principles. This is not desirable. Finally, 'rule making' in the pay-TV sector is not only a question of economic considerations; there are also public information policy and other public policy interests to be considered, such as the promotion of pluralism, diversity, fair access opportunities, European integration and the Internal Market, etc.—questions on which only little time is spent in merger decisions.

Having said that, the analysis of the European Commission's decisions also showed that a competition authority could be a powerful and well-informed body to prevent the emergence of potentially anti-competitive structures from the very beginning. To this extent, the European Commission contributed significantly to shaping some of the existing pay-TV markets in Europe. One advantage of the case-by-case character of its decision-making process is that it is in this respect rather flexible and open for new technical or economic developments. A review of its decisions gives some insight into the market dynamics and the competition

[496] European Court of First Instance, Cableuropa/European Commission, paragraph 191.

problems involved in the pay-TV market. And in many instances, the cases generated a range of interesting, sometimes innovative solutions to potential competition problems; solutions that have been elaborated in close cooperation with the market parties themselves.

3.4. Access Control and Antitrust

Section 3.3. showed how the European Commission applies merger law to achieve open and competitive structures in European pay-TV markets. In section 3.4., European antitrust law and in particular Articles 82 and 81 of the EC Treaty (abuse of market power) are central, although the focus of the analysis lies on Article 82 of the EC Treaty.[497] The most interesting question in this context is to what extent European antitrust law provides an effective instrument to address and stop anti-competitive behaviour. More specifically, section 3.4. focuses on the forms of anti-competitive behaviour that were described in Chapter 1 as particularly relevant to this context, namely the refusal of access (section 3.4.1.), discriminatory behaviour (section 3.4.2.) and bundling strategies (section 3.4.3.).[498] So far, neither the European Commission nor the European Court of Justice have had to decide on such cases in the pay-TV sector. For this reason, all reflections on the possible outcome of cases remain somewhat speculative.

3.4.1. Refusal of Access

In the pay-TV sector, refusals of access can take the form of direct refusals to let competitors use a conditional access system, an API or programme resources that are under the exclusive control of the dominant pay-TV provider. There are also other forms of market behaviour that can have a similar effect, such as delayed access, the withdrawal of access from existing clients,[499] unfavourable pricing and contract conditions, and tying agreements.[500] Some situations have been addressed by sector-specific law, notably by Articles 5 (1)b and 6 of the Access Directive, provisions that will be discussed in Chapter 4. In addition to existing sector-specific

---

[497] For Article 82 of the EC Treaty to apply, a number of general conditions must be fulfilled, which will not be discussed here, but which have to be identified on a case-by-case basis. Such conditions are the definition of a relevant product market, the finding of market power and that abuse must have an effect on trade between Member States and take place in a substantial part of the European Internal Market.

[498] See also sections 1.5.2. and 1.5.3.

[499] European Commission, Access Notice, paragraphs 99-100.

[500] See also European Commission, Access Notice, paragraph 94: 'Three important elements relating to access which could be manipulated by the access provider in order, in effect, to refuse to provide access are timing, technical configuration and price'.

regulation[501] and cases in which no such regulation exists general competition law can apply.[502]

Article 82 (a) of the EC Treaty contains a broad general principle that stipulates that companies in dominant positions must not refuse to supply their goods or services if refusal to supply would significantly impact competition. Accordingly, one could consider whether the refusal to provide, for example, conditional access services to rivals qualifies as a refusal to supply and whether Article 82 (a) of the EC Treaty would apply. One argument why access claims probably exceed the scope of the concept of refusal to supply is that the granting of access to a facility goes beyond the mere duty to supply. The duty to provide access amounts to the use of the operator's own resources, which are possibly resources that are used internally and have not even been offered to third parties.[503] The duty to provide access can be, hence, a more far-reaching duty than the supply of services to third parties. The obligation to share one's own assets with competitors can result in considerable conflicts with commercial interests and economic freedoms, including the right to property and the freedom not to be forced to promote competitors at one's own cost.[504] In addition, the sharing of one's resources could trigger considerable security risks for the resource operator, as well as capacity problems and financial losses. All these are reasons why the European Court of Justice and scholars have argued that the obligation to share one's resources should remain subject to stricter conditions than Article 82 (a) of the EC Treaty provides for and should be reserved to exceptional circumstances.[505] Arguably, mandating access on a formal legal basis and by way of a constitutional law-making process is the preferable route to strike the needed balance.

---

[501] European Commission, Access Notice, paragraph 58.

[502] See sections 4.4. and 4.5.

[503] See European Court of Justice, Judgment of the Court of 26 November 1998, Oscar Bronner GmbH & Co. KG v Mediaprint Zeitungs- und Zeitschriftenverlag GmbH & Co. KG, Mediaprint Zeitungsvertriebsgesellschaft mbH & Co. KG and Mediaprint Anzeigengesellschaft mbH & Co. KG, Reference for a preliminary ruling: Oberlandesgericht Wien – Austria, Case C-7/97, European Court reports 1998, p. I-7791 [hereinafter 'Bronner'], Opinion General Advocate Jacobs, delivered on 28 May 1998, Paragraph 56: ' ...it is apparent that the right to choose one's trading partners and freely to dispose of one's property are generally recognized principles in the laws of the Member States, in some cases with constitutional status. Incursions on those rights require careful justification'.

[504] See European Commission, Commission Decision of 29 July 1987 relating to a proceeding under Article 86 of the EEC Treaty (Case IV/32.279 – BBI/Boosey & Hawkes), 9 October 1987, OJ L 286, p. 36 [hereinafter 'BBI/Boosey & Hawkes'], paragraph 19: 'A dominant undertaking may always take reasonable steps to protect its commercial interests, but such measures must be fair and proportionate to the threat. ... There is no obligation placed on a dominant producer to subsidize competition to itself'. The judgements of the European Court of Justice in, for example, the cases Bronner and Magill support this view, European Court of Justice, Judgment of the Court of 6 April 1995, Radio Telefis Eireann (RTE) and Independent Television Publications Ltd (ITP) v Commission of the European Communities, Joined cases C-241/91 P and C-242/91 P, European Court reports 1995, p. I-743 [hereinafter 'Magill'].

[505] See European Court of Justice, Bronner, paragraph 26; European Court of Justice, Magill, paragraph 50. Areeda 1990, pp. 852-853. Temple Lang 1994, pp. 478-483; McGowan 1996, pp. 805-806.

The European Commission's past decision practice in Essential Facilities Cases might suggest otherwise. The problem of access refusals to bottleneck facilities, albeit never decided at the European level in a pay-TV case, is not new and has led to the application and development of the Essential Facilities Doctrine in European competition law. The Essential Facilities Doctrine specifies the conditions under which a refusal to supply or grant access can be considered abusive behaviour in the sense of Article 82 of the EC Treaty. Instead of providing yet another detailed discussion of the Essential Facilities Doctrine cases that were decided by the European Court of Justice and the European Commission, this study will refer to the excellent treatises in which the doctrine has already been discussed at length[506] and proceed to the main conclusions. Both the European Commission and the European Court of Justice concluded that there could be circumstances in which the overall public interest in open access to resources or facilities can outweigh the economic interests and freedoms of a single operator to maintain exclusive control over a facility. In their competition law analysis, economic reasoning prevails; public information policy considerations about open access, such as public broadcasters' access and pluralism, do not play a role. Important economic considerations in this context are the existence of vertically integrated structures and leverage in the form of access refusals.

The Essential Facilities Doctrine says that any dominant company that controls a so-called 'essential facility'[507] and that refuses access to competitors without objective justification[508] or that grants access only on terms less favourable than those that it offers its own associates, acts in breach of Article 82 (a) of the EC Treaty providing:

a. The party requesting access is dependant on the facility for supplying customers and where constructing an alternative facility is not an economically viable option, in other words: a facility is 'essential'.
b. The capacity of the facility is adequate to carry the additional traffic taking the owner's own long-term plans into account.
c. The potential entrant's traffic meets all relevant technical standards.
d. The entrant is willing to pay adequate compensation for using the system, and
e. The access request is reasonable and there is no objective justification for refusing access.

For the telecommunications sector, and therefore also for the example of conditional access, the European Commission specifies in its Access Notice the conditions under which the Essential Facilities Doctrine applies. For example, it

[506] See the papers by Temple Lang 1994; Furse 1995; Scherer 1999; Larouche 2000, pp. 179-217; Jones/Sufrin 201, pp. 387-418; Llorens-Maluquer 1998, p. 213, Deselaers 1995, to name but some.

[507] An essential facility in the sense of the Essential Facilities Doctrine may be a product, a service, content, infrastructure, technical facilities or access to a physical thing such as a harbour or an airport. See also Opinion General Advocate Jacob in the case Bronner, paragraph 50.

[508] As to relevant justifications in this context according to the European Commission, see European Commission, Access Notice, paragraph 91.

lists possible relevant justifications, such as an overriding difficulty of providing access to the requesting company or the need for a facility owner to have sufficient time and opportunity to use the facility in order to recoup its investments.[509] If the conditions are fulfilled, a competition authority can impose the obligation to provide access to the facility in question.

Perhaps the most difficult question when applying the Essential Facilities Doctrine is the question of when a facility is 'essential'. The Access Notice explains that

> '[i]t will not be sufficient that the position of the company requesting access would be more advantageous if access were granted - but refusal of access must lead to the proposed activities being made either impossible or seriously and unavoidably uneconomic'. [510]

And according to the European Court of Justice, an essential facility must be 'indispensable for carrying on a particular business'.[511]

The application of this definition to facilities in the pay-TV sector is subject to controversial and incoherent discussions among scholars; this all the more as no case law exists for this sector—at least not at the European level. Some commentators seem to support the view that at least the conditional access is an essential facility.[512] Cave and Cowie argued that, should a programme provider wish to have access to an existing subscriber base, it must first secure access to the dominant conditional access system because establishing an alternative decoder base able to bypass the dominant gatekeeper is prohibitively expensive.[513] Similarly, Enßlin argued that, at least for the German market, no third undertaking possessed the technical and financial power to build a new, alternative pay-TV distribution structure alongside the systems of Kirch and Deutsche Telekom.[514] At the same time, the same commentators, Cave and Cowie, claimed that, given the wide variety of organizations active in the business of billing customers and the ease of acquiring subscriber management technology, it is unlikely that a subscriber management system was an essential facility.[515] Moreover, without further explanation, Enßlin claimed that the EPG was an essential facility (however, referring exclusively to the Basic Navigator), while the service platform was not.

At the heart of the controversy is the question whether viable alternatives are available. As the European Commission stated: '[u]nder existing case law, a

[509] European Commission, Access Notice, paragraphs 91, 93.

[510] European Commission, Access Notice, paragraph 91(a).

[511] European Court of Justice, Judgement of the Court of 29 April 2004, IMS Health GmbH & Co. OHG v NDC Health GmbH & Co. KG, Case C-418/01, European Court reports 813 [hereinafter 'IMS Health'], paragraph 38; European Court of Justice, Bronner, paragraph 41; European Court of Justice, Magill, paragraph 53.

[512] Cave/Cowie 1996, p. 135; Enβlin 2000, pp. 117 – 136; Bavasso 2003, p. 273.

[513] Cave/Cowie 1996, p. 135.

[514] Enßlin 2000, p. 133.

[515] Enßlin 2000, p. 133.

product or service cannot be considered "necessary" or "essential" unless there is no real or potential substitute'.[516] The European Court of Justice argued along the same lines and specified that the denial of access amounts to an abuse only if it is apparent that there are serious technical, legal or economic obstacles that make it 'impossible, or even unreasonably difficult' for any other market player to duplicate the facility.[517] In the Bronner case, the European Court of Justice made it clear that the Essential Facilities Doctrine does not apply if alternative distribution channels exist, even if they are possibly less optimal.[518] The burden of proof lies with the party requesting access. The court continued and held that for a facility to be regarded as indispensable 'it would be necessary at the very least to establish ...that it is not economically viable to create a second home-delivery scheme'.[519] This also means that it is not enough that it is extremely difficult, expensive and time-consuming for a particular rival to duplicate the facility. The prospects would in effect have to 'deter(s) any prudent undertaking from entering the market' (stress by the author).[520] In the following paragraphs, three examples of facilities are given that may or may not qualify as an essential facility. The purpose of the examples is to elaborate arguments of how to assess when the Essential Facilities Doctrine might be applicable in pay-TV cases. The examples deal with the issues of market definition, essentiality and scope of the Essential Facilities Doctrine.

One important aspect in the process of identifying whether there are viable alternatives to a facility is the way in which the relevant market is defined. This is why the first example—conditional access—demonstrates some of the problems a competition authority can encounter when defining the relevant market.

Example 1 - Conditional Access

A conditional access system could qualify in a specific market as an essential facility if rivals had no viable alternatives to distribute access-controlled services to

---

516 European Commission, Commission Guidelines on Market Analysis and the Assessment of Significant Market Power, paragraph 81. As the Commission further correctly observes in this context, the fact that a given facility is not essential for an economic activity in a distinct market does not mean that the owner of this facility might not be in a dominant position. The Commission gives the example of a network operator who can be in a dominant position despite the existence of alternative competing networks, if the size or importance of its network affords him the possibility to behave independently from other network operators.

517 European Court of Justice, Bronner, paragraph 44; European Court of First Instance, Judgment of the Court of First Instance (Second Chamber) of 15 September 1998, European Night Services Ltd (ENS), Eurostar (UK) Ltd, formerly European Passenger Services Ltd (EPS), Union internationale des chemins de fer (UIC), NV Nederlandse Spoorwegen (NS) and Société nationale des chemins de fer français (SNCF) v Commission of the European Communities, Joined cases T-374/94, T-375/94, T-384/94 and T-388/94, European Court reports 1998, p. II-03141 [hereinafter 'European Night Services'], paragraph 208.

518 European Court of Justice, Bronner, paragraph. 43.

519 European Court of Justice, Bronner, paragraph 46.

520 European Court of Justice, Bronner, Opinion Advocate General Jacobs, paragraph 66.

subscribers in that market and therefore depended on access to the established conditional access system.

When adopting a rather broad market definition—the market for technical facilities for pay-TV—one might find that in a given market several conditional access systems are available. There are at least two or more competing providers of access-controlled services operating alongside each other and using different conditional access solutions in a number of Member States.[521] In addition, in Europe alone there is a considerable range of independent conditional access providers, many of which also operate internationally and offer their services to third-party broadcasters.[522] The Essential Facilities Doctrine does not apply to markets in which alternative systems are available, or in which market precedents suggest that it is not entirely unreasonable to launch an alternative system.

One could also adopt a narrow market definition. One could argue that the relevant market is not the market for technical facilities for pay-TV, but for one particular conditional access system. Subscribers who have purchased one set top box could be extremely unwilling to switch to an alternative system, in particular if the existing conditional access solution is not interoperable and/or the programme of the first platform is particularly attractive. This might justify the conclusion that each individual established conditional access system is a market in itself and that no alternative systems are available in the market. In fact, this would mean that each proprietary conditional access system is an essential facility, irrespective of whether alternative systems are available and irrespective of whether the operator of a particular system is dominant in the total conditional access market.

The outcome in the second alternative—narrow market definition—results in an unlimited—and unreasonable—extension of the Essential Facilities Doctrine to virtually all assets of property. The criterion of dominance was replaced by 'essentiality',[523] which is not in line with the basic principles of competition law.[524] The doors were wide open for political instead of economic arguments in the competition law analysis. Political arguments as a motive for mandating access, however, are better reserved for the sector-specific legislator rather than the competition authority.[525]

Having said that, a conditional access solution might not be essential for market entry. Still, there can be competition or public information policy reasons why foreclosing access to a particular technical platform is not desirable. This, however, is an aspect that probably falls outside the scope of the Essential Facilities Doctrine and inside the scope of sector-specific regulation.[526]

---

521 See section 1.4.3.
522 See section 1.4.3.
523 Larouche 2000, pp. 207, 209.
524 Larouche 2000, p. 212.
525 In this sense also Larouche 2000, p. 216.
526 See sections 4.2.2. and 4.4.

Example 2 – Apple's FairPlay
The second example is not a pay-TV example, but originates from the online music market. It deals with Apples' DRM solution FairPlay. Nevertheless, there are a number of interesting parallels with the pay-TV case. Moreover, Apple's Fair Play was probably the first case, at least to the knowledge of the author, in which a competition authority in Europe had to decide if an electronic content management system qualifies as an essential facility.

Apple's online music store iTunes sells music for download via the internet. Apple itself does not produce music, but it has concluded licensing agreements with the major record labels and many leading independent labels.[527] Apple sells individual songs or whole tracks via its portal, iTunes. The songs can be played on Apple's own portable music player, the iPod. The music is protected using Apple's proprietary DRM solution FairPlay, but what is important to note is that Apple's iPod only supports the FairPlay standard. The iPod does not support any of the rivalling standards used by competing download services, and Apple does not license FairPlay to rivals. Users of the Apple iPod can download unprotected MP3 files to their iPod, but if they wish to download DRM protected music they can only go to the iTunes music store.

One will see similarities and differences with the pay-TV case. The differences are that iTunes sells music, and that the music is sold via the internet. Apple does not use a conditional access solution, but a DRM solution, and is therefore able to control access and the way content is used. Furthermore, iTunes did not begin with a subscription model, but specialized in individual downloads and billing. However, these differences are less consequential than would seem at first sight. As mentioned in Chapter 1, conditional access systems can also be combined with elements of usage control in order to protect intellectual property rights in the distributed content.[528] In addition, pay-TV platforms offer individual download services (on demand), and pay-TV services can be offered via the internet, too. In both cases, technical content control solutions support the selling of digital content services via intermediary platforms.

Some of the concerns that were voiced against the use of technical content control and (vertically integrated) intermediary platforms are very similar to the pay-TV case. A prominent example is RealNetworks' complaint that the iPod does not process any other technical protection systems, including RealNetworks' DRM solution, Harmony. By doing this, so said RealNetworks, iTunes monopolized all of the consumers who bought an iPod. In the end, however, it was another enterprise that challenged iTunes in Europe. The French entertainment company VerginMega filed in 2004 a complaint against Apple Computers France with the French

[527] For more information see <www.itunes.com> (last visited on 20 March 2005).
[528] See section 1.2.

competition authority, the Conseil de la concurrence.[529] VirginMega offers its own music download service and uses for this purpose a different DRM solution, namely Microsoft's DRM. Because of the proprietary nature of the iPod, consumers who buy digital music files from VirginMedia cannot, so is the argument of VirginMega, transfer these files to their iPod. VirginMega requested a licence from Apple for FairPlay so that it could encode its music files in the FairPlay format. Apple refused. VirginMedia claimed that the refusal to grant access to the FairPlay DRM constitutes an abuse of a dominant position according to French competition law and Article 82 of the EC Treaty.

The French competition authority found that Apple was probably dominant in all three relevant French markets, namely the DRM market, the market for portable music players and the music download market. It then recalled the jurisdiction of French courts and the European Court of Justice in Essential Facilities Doctrine cases. It identified three aspects that were in its opinion relevant when deciding whether FairPlay is an essential facility:

a. The actual usage habits of consumers regarding music download,
b. Possibilities to circumvent the problem of lacking interoperability, and
c. The developments in the market for portable music players.

The Conseil de la concurrence concluded that FairPlay was not an essential facility for the following reasons: First, the competition authority found that only a minor share of the market would listen to music from a portable device, the majority would listen to music via the computer or burn songs onto a CD. Second, it thoroughly described a way consumers could get around the existing lack of interoperability and download music from VirginMega onto their iPod. Third, the French competition authority found that the market for portable music players was sufficiently competitive and offered several products in addition to the iPod. In other words, there were alternatives available that could process VirginMega's music. In conclusion, the French competition authority did not consider FairPlay an essential facility because consumers had a choice: the iPod was not necessary to listen to VirginMega's music, alternatives were available, and iPod owners were not excluded from access to VirginMega's music. In addition to its doubts whether the FairPlay DRM was an essential facility, the French competition authority also questioned the causality between VirginMega's lesser economic success and the access refusal. It argued that Apple probably had the more successful business strategy and was for this reason market leader, thereby raising the free-rider issue. It furthermore found that the market for online music was actually competitive as there were at least two major operators active in that market.[530]

The French case provides a number of arguments that can also be useful in assessing whether it is adequate to apply the Essential Facilities Doctrine to the

[529] Conseil de la concurrence, Décision du 9 novembre 2004 relative à des pratiques mises en œuvre par la société Apple Computer, Inc. Dans les secteurs du téléchargment de musique sur Internet et des baladeurs numériques, Case No. 04-D-54 [hereinafter 'Apple Computer'].

[530] Conseil de la concurrence, Apple Computer, paragraphs 96-103.

conditional access case in the first example above. To begin with, it is worth noting that the French competition authority acknowledged that a technology that implements a proprietary standard could constitute an essential facility. In other words, it is not the facility itself but the standard that is embedded in the facility that can make it essential for market entry for others. The European Commission reached a similar conclusion in the Microsoft Europe case. In this case, Sun Microsystems Inc, a competing manufacturer of computer operating systems, complained that Microsoft refused to disclose the technology necessary to allow the interoperability of its Sun server operating system[531] with the Microsoft Windows Client operating system. This would prevent Sun Microsystems from competing in the workgroup server operating system market. The European Commission established that Microsoft was abusing its dominant position by refusing to supply interoperability information to Sun, that is the specifications for the protocols used by Windows workgroup servers in order to provide file, print, group and user administration services.[532] The Commission noted that 'interoperability with the client PC operating system is of significant competitive importance'.[533] As the Commission also upheld, due to the lack of interoperability an increasing number of consumers were locked into a homogeneous Windows solution at the level of workgroup server operating systems.[534] This impaired the consumers' ability to benefit from alternative, innovative products. Likewise, it limited the competitors' prospects to enter the market and/or develop new products.[535] Similar to Apple's defence before the French competition authority, Microsoft claimed in an Interim procedure before the President of the European Court of First Instance against the Commission's decision that the Essential Facilities Doctrine did not apply, among others, because there were alternative ways of getting around the lack of interoperability meaning that alternative products were available.[536] The President of

---

[531] The European Court of First Instance referred to 'work group operating systems' as operating systems designed and marketed to deliver services such as the sharing of files stored on services, sharing printers and the administration of the manner in which authorized users can access network services. The work of the work group operating system depends on interoperability with the client PCs and their operating systems. European Court of First Instance, Order of the President of the Court of First Instance of 26 July 2004, Microsoft Corp. v Commission of the European Communities, Case T-201/04, European Court reports 2004 p 0 [hereinafter 'Microsoft Europe'], paragraphs 10-11.

[532] European Commission, Commission decision of 24 March 2004 relating to a proceeding under Article 82 of the EC Treaty (Case COMP/C-3/37.792 Microsoft), 21 April 2004, C(2004)900 final [hereinafter 'Microsoft Europe'], paragraphs 546-692.

[533] European Commission, Microsoft Europe, paragraph 586, see also the explanation in paragraphs 587-637.

[534] European Commission, Microsoft Europe, paragraphs 642-646, but also claiming that 'customers will thus have a tendency to underestimate the importance of interoperability in their purchasing decisions', paragraph 643. It also should be mentioned that Microsoft criticized the survey as not being representative, see paragraphs 654-655.

[535] European Commission, Microsoft Europe, paragraphs 694 and 695-701.

[536] European Court of First Instance, Microsoft Europe, paragraphs 100-104.

the European Court of First Instance reserved the final decision to the Court dealing with the substance of the case.[537]

Back to the conditional access example, the French competition authority followed the standing legislation of the European Court of Justice, namely that the Essential Facilities Doctrine must remain reserved to exceptional circumstances, notably those where alternative technologies and distribution channels are not available.[538] To this extent, it is probably the third argument of the French competition authority that is the most relevant when applied to an access-to-a-conditional-access case. As far as the first argument is concerned—alternative reception devices—much will depend on how consumers will access pay-TV services. As was explained in Chapter 1, it is possible to receive access-controlled broadcasting services also via, for example, a PC, providing the PC is equipped with a TV card and media player software, and is connected to a broadcasting network. Pure software-based conditional access solutions would not even require substantial investments in additional hardware. At the time of writing, the reception of broadcasting services via the PC is still the exception. The majority of consumers tend to prefer to receive such services via the TV. The second argument—consumers being able to get around the lack of interoperability—is interesting because it takes both the supply side and the demand side into account when identifying the essentiality of a resource. Another question is, however, if consumers would really follow this route. Even though the procedure that the competition authority described was rather simple, its point of reference was a fairly technically experienced user. In terms of the third argument—competition in the technical-facilities market, and here more specifically the market for the consumer reception device, the set top box—an independent market for set top boxes already exists in some countries. Furthermore, once digitization has been completed, most households will probably need set top boxes or will have to be otherwise equipped to receive digital television. Arguably, this would lower some of the entry barriers for newcomers in the pay-TV service market. Newcomers could also stimulate demand for new or additional conditional access technology. Already now, the prices for set top boxes are falling. Likewise, the internet as a platform for the distribution of pay-TV can bring a new dynamic to the sector. The digitization of broadcasting platforms will also make it more attractive for new entrants from non-broadcasting sectors, such as infrastructure operators and entertainment industry representatives, to enter the pay-TV sector with the consequence of increasing demand for alternative systems.[539] One important aspect in this context is interoperability. Arguably, one pressing obstacle to the development of an alternative conditional access system is the lack of adequate interoperability

[537] European Court of First Instance, Microsoft Europe, paragraphs 198 -225.

[538] See European Court of Justice, Bronner, paragraph 26; European Court of Justice, Magill, paragraph 50; Europe Court of Justice, IMS Health, paragraph 49.

[539] See also Galbiati/Nicita/Nizi 2004, p. 21, highlightening the possibility to sustain competition models in the pay-TV sector that are not based on 'competition for the market'.

solutions. In areas in which interoperability solutions are available, this might be a further argument to hold that the dependency upon one particular conditional access system and one particular set top box is less pressing with the consequence that it is not an essential facility. Interestingly, the French competition authority made its point clear that a competitive market must not mean that consumers have access to all content services through one and the same device.

Example 3 – Access to the EPG

It is widely acknowledged that, in a multi-channel environment, access to a popular EPG, meaning an EPG that is used by a large share of the audience, can be important to reach consumers, especially for smaller, yet unknown niche channels. So far, no cases are known in which the European Commission had to define a distinct market for EPGs. Unlike search engines and browsers on the internet, EPGs are commonly used for internal purposes, namely by the entity that also operates the pay-TV platform. Here, the role of the EPG is to guide the consumer to the programmes and services that are offered via the platform. The example of the EPG illustrates another difficulty with the application of the Essential Facilities Doctrine: what should be done in cases in which a market has not yet been defined and in which a facility is used entirely for internal purposes? In the decision practice of the European Court of Justice,

> 'it is sufficient that a potential market or even hypothetical market can be identified. Such is the case where the products or services are indispensable in order to carry on a particular business and where there is an actual demand for them on the part of undertakings which seek to carry on the business for which they are indispensable'.[540]

Still, forcing an enterprise to share a facility that it has not offered publicly is a particularly far-going intrusion on the property rights of the EPG controller.[541] This would be an additional argument to apply the Essential Facilities Doctrine only in exceptional cases. Indeed, this was also the stance of the European Court of Justice in three of the leading Essential Facilities Cases. All three cases, Magill, Bronner and IMS Health, concerned a similar constellation in which a facility was originally used for internal purposes only. In Magill, it concerned programme information, in Bronner it was a home-delivery system for newspapers and in IMS Health it was a particular brick structure for the presentation of regional sales. It was in these decisions that the European Court of Justice substantially narrowed down the applicability of the Essential Facilities Doctrine. One additional criterion—in addition to the fact that a facility must be indispensable, not duplicable and that refusal of access to that facility must have a negative impact on competition in a secondary market—that the court developed in these cases and which has not yet

540 European Court of Justice, IMS Health, paragraph 44.
541 Where a facility has been offered to third parties already, non-discrimination principles might apply.

been discussed here, is the condition that the refusal of access must prevent the emergence of a new product for which there is a potential consumer demand.[542] Transferred to the EPG example, this means that access to the EPG is first and foremost needed to provide already existing services to consumers, namely pay-TV services and other services that are carried via a particular digital platform. The lack of a new product, the provision of which is only possible if access to the EPG is given, could be an argument why the EPG is not an essential facility.

This argument was developed in Magill and later in the case of IMS Health.[543] The latter concerned a competitor's request to obtain a licence to use a brick structure for the presentation of regional sales, which is developed by the defendant, IMS Health, and on which IMS Health holds intellectual property rights. The competitor wished to provide the same services in the same national market. The court found that refusal of access to a facility that is essential for operating in a secondary market may be regarded as abusive only where the access requester does not intend to just duplicate the goods or services already offered in that secondary market, but also intends to produce new goods or services not yet offered and for which there is a potential demand.[544] Correspondingly, a refusal of access to the EPG would only be in conflict with competition law principles if access to the EPG was needed to provide services that do not exist yet. In contrast, access to programme data in order to produce and operate an independent, comprehensive EPG could be a case for the Essential Facilities Doctrine providing such a service does not yet exist in a market.[545]

The EPG example is also instructive for another reason. Providing the Essential Facilities Doctrine is applicable, would it be useful? Access to the EPG is still no guarantee that the audience will take notice of a particular programme. Often, the decisive factor will be the position on the EPG list. Programmes that are placed at position 135 have a lower chance of getting consumers' attention than programmes at position five. The question whether the Essential Facilities Doctrine can be used to not only enforce access to a particular facility, but also to enforce certain modalities of access has not yet been clarified. This is to say, it is still unclear if the doctrine can be used to enforce a particular position in an EPG—providing an EPG were an essential facility. The Essential Facilities Doctrine does not provide any guidelines as to the scope of the actual access obligation. In a complex multi-layered multi-standard service environment, remedies may not be as straightforward as mandating open access. Accessibility and compatibility are conditions that depend on complex procedures, including the disclosure of information on the technical interface, the implementation of common standards and interfaces, etc. The example of the EPG clearly illustrates this. To provide another example, access

---

[542] European Court of Justice, IMS Health, paragraph 38; European Court of Justice, Magill, paragraph 54; European Court of Justice, Bronner, paragraph 40.

[543] European Court of Justice, IMS Health, paragraph 32.

[544] European Court of Justice, IMS Health, paragraph 49.

[545] European Court of Justice, Magill, paragraph 54.

to an API, meaning the permission to operate via a third-party operator's API, is not sufficient if not accompanied with the technical information that enables third parties to write compatible applications.

The situation is rendered more difficult by the fact that, as far as access to the EPG is concerned, editorial and public information policy considerations will come to the fore. Arguably, the EPG operator also enjoys a certain editorial freedom, similar to that of the editor of an off-line programme guide. The EPG operator may have content-related reasons not to grant a rival access to its EPG, for example, because the content of the rival's programme does not fit with its programme offering, because it is not attractive enough to attract sufficient viewers, or because it simply does not want to tolerate the opinion presented in the rival's programme. On the other hand, there can be public information policy interests why certain programmes, such as public service programmes, should not only be accessible but also have a prominent position. This is another reason why the Essential Facilities Doctrine is probably of limited value in particular in the broadcasting sector where content-related considerations play a prominent role.

Conclusion

'A competitive advantage is not the same as an essential facility'.[546] It is likely that most facilities in digital pay-TV will not easily qualify as an 'essential facility'. It is also unclear to which extent the Essential Facilities Doctrine concept is able to provide satisfactory answers for access conflicts in this sector. Because of the many insecurities and difficulties in defining whether its conditions are given, the Essential Facilities Doctrine does not provide potential market players with the legal security that is necessary when launching a new business. The remedy, obliging an undertaking to share its facility, is probably not in all cases the appropriate and sufficient solution.[547] One may doubt, therefore, whether the Essential Facilities Doctrine is indeed such a 'powerful tool to pry open markets'[548] or whether its potential is genuinely overestimated.[549]

In most of the cases, the question of whether a conditional access system, a programme bundle, and an API or EPG constitute an essential facility or not, can probably be left open. A dominant company that selectively discriminates against a particular competitor—for example, to discourage it from competition by denying access to an important facility on the same terms granted to other companies—is likely to be acting unlawfully, even if the facility is not necessarily essential. As the European Commission explicitly states in its Access Notice:

[546] Temple Lang 1997, p. 73.

[547] Yoo 2002, 246-247; McGowan 1996, p. 806; Bittlingmayer/Hazlett 2002, 299pp and 308pp., differentiating Waterman, 528pp. See also section 4.8.6.

[548] Furse 1995, p. 473.

[549] In this sense, also Scherer 1999, pp. 315, 318.

> '... it is clear that a refusal to supply a new customer in circumstances where a dominant facilities owner is already supplying one or more customers operating in the same downstream market would constitute discriminatory treatment which, if it would restrict competition on that downstream market, would be an abuse. ... In the absence of any objective justifications, a refusal would usually be an abuse of the dominant position on the access market'.[550]

This means that the Essential Facilities Doctrine is actually only relevant in cases where an operator does not offer this facility to third parties, but uses it exclusively internally. In such cases, it is already very questionable whether the obligation to grant access is too substantial an infringement on the facility operator's property rights, and the conditions of the Essential Facilities Doctrine too vague as that it would be desirable to mandate access merely on the grounds of general competition law without specific codification in material law.[551]

3.4.2. Discriminatory Treatment

Once a provider of a conditional access system or any other bottleneck facility has opened its facility to third parties, unjustified refusal of access or the provision of access under discriminatory conditions, is very likely to be held in breach with Article 82 (c) of the EC Treaty. The preconditions for the application of Article 82 (c) of the EC Treaty are that the operator opened its facilities or provided access to third parties under substantially dissimilar conditions without reasonable commercial or comparable justification for the different treatment. This applies irrespective of whether the facility is essential or not. Another precondition is that the further requirements of Article 82 are fulfilled. As the European Commission explains in its Access Notice,

> 'the dominant company's duty is to provide access in such a way that the goods and services offered to downstream companies are available on terms no less favourable than those given to other parties, including its own corresponding downstream operations'.[552]

Incompatible with Article 82 (c) of the EC Treaty could be a case of unjustified and discriminatory refusal of access to, for example, the EPG or a conditional access system. Discriminatory could also be the unfavourable presentation of a programme in an EPG if it is not done according to objective criteria. A discriminatory act may be the charging of different access prices or the imposing of different access conditions except where 'such discrimination would be objectively

---

550 European Commission, Access Notice, paragraph 85.

551 European Court of Justice, Bronner, opinion of Mr Advocate General Jacobs, paragraphs 56, 57. In this sense also Larouche 2002, p. 216, and more general Larouche's critique of the EFD in pp. 211–217.

552 European Commission, Access Notice, paragraph 86.

justified, for example on the basis of cost or technical considerations or the fact that the users are operating at different levels'.[553] One important case of the application of the principle of non-discriminatory treatment is price discrimination, such as the charging of different prices for different customers.[554] It could apply, for example, where an operator of a conditional access system charges different prices for different operators for the use of the system. As Larouche pointed out, however, in more diversified markets it can be difficult to determine whether the price charged to competitors is discriminatory or not.[555] When, for example, can two providers of a niche service be considered equal, and when is it justified to charge one a higher price for access to the conditional access infrastructure than the other? Are online music services comparable to online newspapers? Under which conditions is it justified to give one broadcaster a more prominent position in an EPG than another broadcaster? Obviously, comparability is a problem that does not only arise in context with pricing issues.

Knowing when a certain allegedly discriminatory practice is justified is another problem. For example, pay-TV providers, who at the same time produce the set top box technology, can find it economically efficient to offer set top boxes to consumers who subscribe to their platform at no or minimal costs and sell the same set top box to non-subscribed consumers at a high price. Here, the provision of the set top box can be used as an additional incentive for consumers to subscribe to a particular platform and thereby make this platform more attractive for potential subscribers and providers of access-controlled services. Although a discriminatory practice (consumers who are subscribed receive the set top box for free, while others have to pay), this strategy might be an acceptable strategy as it is intended to get the two sides, consumers and providers of access-controlled services, on board, and thereby make the pay-TV platform more attractive for both sides.

Geographical Discrimination

One form of discriminatory behaviour, which is particularly relevant within the context of European competition law, is geographical discrimination. Chapter 1 illustrated how electronic access control is used to re-enforce national borders in an environment of borderless transmission channels.[556] From a European law perspective, geographical discrimination is an issue of Internal Market principles and the European policies concerning the free flow of services. The latter was dealt

---

[553] European Commission, Access Notice, paragraph 120.

[554] Another question is whether prices are unfair or excessive. In this case, they could (also) fall under Article 82(a) of the EC Treaty. Prices are excessive when they have no reasonable relation to the economic value of the product supplied. See e.g. European Court of Justice, Judgment of the Court of 14 February 1978, United Brands Company and United Brands Continentaal BV v Commission of the European Communities, Chiquita Bananas, Case 27/76, European Court reports 1978, p. 207 [hereinafter 'United Brands'], paragraph 235.

[555] Larouche 2002, p. 227.

[556] See section 1.5.2.

with in Chapter 2.[557] However, it can also be an issue of European competition law in so far as Articles 81 or 82 of the EC Treaty apply. For the time being, no cases are known that would concern forms of geographical discrimination in pay-TV markets. The following considerations must hence remain speculative.

An example of geographical discrimination is an agreement between two service platform operators in different countries not to make access-controlled services available across the border and, thereby, refrain from competing with each other in non-native markets. This constellation is comparable to the situation in cases concerning export restrictions in other sectors, which has been repeatedly criticized by the European Commission and the European Court of Justice for its incompatibility with European competition law. To name but two examples, in the Volkswagen case, the European Commission found that agreements that make it difficult for nationals of one European country to purchase products from another European country could infringe Article 85(1) of the EC Treaty (now: Article 81 of the EC Treaty).[558] The case concerned agreements between Volkswagen and its subsidiary, an exclusive Italian importer of Volkswagen, Autogerma. Autogerma forced Italian Volkswagen dealers to sell vehicles only to Italian customers by threatening to terminate their dealer contracts. German customers who wanted to buy Volkswagen vehicles in Italy complained about this practice. The European Commission concluded that the restrictions on the Italian dealers, performed through Autogerma, constituted an infringement of Article 85(1) of the EC Treaty since it implemented a market-partitioning policy. The European Court of First Instance confirmed the view of the European Commission in so far as it found that practices restricting competition and reinforcing the compartmentalization of markets on a national basis would hold up the economic interpenetration that the EC Treaty intended to bring about.[559] In the second case, JCB Service, the European Court of First Instance decided that the effect of exclusive distribution agreements on the Internal Market and on the free movement of services is a consideration that can play a role when assessing the amount of fines to impose for anti-competitive behaviour.[560] The case concerned a UK company that entered into exclusive agreements with its national distributors to establish exclusive territories outside

---

[557] See section 2.2.4.

[558] European Commission, Decision of 28 January 1998 relating to a proceeding under Article 85 of the EC Treaty (Case IV/35.733 — VW), 25 April 1998, OJ 1998 L 124, p. 60 [hereinafter 'VW'], paragraphs 151-153.

[559] European Court of First Instance, Judgment of the Court of First Instance (Fourth Chamber) of 6 July 2000, Volkswagen AG v Commission of the European Communities, Case T-62/98, European Court reports 2000, p. II-02707 [hereinafter `Volkswagen'], paragraph 179.

[560] European Court of First Instance, Judgment of the Court of First Instance (First Chamber) of 13 January 2004, JCB Service v Commission of the European Communities, Case T-67/01, European Court reports 2004, p. 0 [hereinafter `JCB'], paragraph 182.

which authorized distributors were prevented from selling JBC's products.[561] The European Court of First Instance confirmed that agreements to partition the Internal Market can be in conflict with Article 81 (1) of the EC Treaty.[562]

Parallels can be found in agreements between pay-TV platform operators and third-party providers of access-controlled services in which the latter agree that the services they provide via this platform cannot be received by consumers outside a specific geographical territory. The question is if such behaviour constitutes a concerted action and whether it can be proved that the objective is to distort competition within the Internal Market. For example, it could also be a result of the practice of licensing programme rights on a national basis. However, another question is how viable this argument is in a time that technological content control allows to track usage irrespective of the country or residence of the user.[563]

One could also think of a limitation of the market to the prejudice of consumers according to Articles 82 (b) and (c) of the EC Treaty. Access-controlled service platforms are essentially marketing platforms. Restricting the access of consumers from particular countries on grounds of nationality means that consumers from other Member States are excluded from receiving services that are offered via this particular platform. For example, consumers in Luxembourg, which does not have its own pay-TV platform, could be excluded from digital access-controlled services that are offered in Germany—even if transborder reception were theoretically possible. This conduct might limit markets to the disadvantage of consumers and could amount to unjustified geographical discrimination. Again, no relevant case law is known so far for the pay-TV sector. The European Court of Justice, however, found in other cases that certain practices that distinguish between customers on grounds of their nationality or their location could amount to abusive behaviour in the sense of Articles 82 (b) and (c) of the EC Treaty. Cases concerned, for example, export rebates and the pricing of industrial sugar that were not based on the supply and demand of a Member State's industrial sugar market, but on the customers' location;[564] the pricing of bananas at a selling price that differed according to the Member State in which the customer was established;[565] and different prices charged for milk-packaging machinery and cartons in different Member States.[566] Still,

---

[561] See also European Commission, decision 2002/190/EC of 21 December 2000 relating to a proceeding under Article 81 of the EC Treaty (Case COMP.F.1/35.918 - JCB), 12 March 2002, OJ L 69, p. 1 [hereinafter ' JCB'].

[562] European Court of First Instance, JCB, paragraph 103.

[563] See also European Court of Justice, Coditel II, paragraph 19; European Commission, Report on the Application of the Cable and Satellite Directive, p. 7-8, see also section 2.2.4.

[564] European Court of First Instance, Judgment of the Court of First Instance (Third Chamber) of 7 October 1999, Irish Sugar plc v Commission of the European Communities, Case T-228/97, European Court reports 1999, p. II-02969 [hereinafter `Irish Sugar'], paragraphs 140-141.

[565] European Court of Justice, 14 February 1978, United Brands, paragraphs 183, 233.

[566] European Court of First Instance, Judgment of the Court of First Instance of 6 October 1994, Tetra Pak International SA v Commission of the European Communities, Case T-83/91, European Court reports 1994, page II-755 [hereinafter `Tetra Pak II'], paragraph 170.

treating customers from different countries differently could be justified on the grounds of different market conditions in the different national markets.[567] In the case of pay-TV, again, the way programme rights are licensed on a national basis might be an argument to justify geographical restrictions.

3.4.3. TYING

Other, and within the context of this study, very relevant forms of anti-competitive behaviour are tying practices. Tying means that an undertaking makes the purchase or licensing of one product (the tied product) conditional on the purchase or licence of another product (the tying product).[568] Outside the competition law analysis, the notion of 'bundling' is often used interchangeably. It should be noted, however, that from a legal point of view a distinction is made between 'tying' and 'bundling' (see below). Chapter 1 explained that pay-TV involves a variety of possibilities for bundling, meaning tying, each of which could be a potential case of anti-competitive tying depending on the circumstances.[569] From the range of possible examples, this section will focus, after a more general discussion of tying and the applicability of competition law, on two exemplary cases that were introduced in Chapter 1. These are the tying of the subscription to a pay-TV channel to the purchase of a set top box from the same operator, and channel bundling.

In competition law, tying refers to the practice of linking the supply of services to the acceptance of additional conditions or services that stand in no direct relation to the original object of the contract, the so-called natural link and that are not justified by other objective reasons, including considerations of commercial usage.[570] Tying, thus, can also be seen as a form of qualified refusal of supply, namely when the requester of a service/product is not willing to accept the tied service/product.[571] The challenge for competition authorities and courts is to assess when the joint selling of two products is an anti-competitive strategy to leverage market power from the market for the tying good into the market for the tied good. In order to assess the compatibility of tying strategies with antitrust law, the effects for both the competition and consumers should be considered.[572]

As mentioned in Chapter 1, the effects of tying strategies can be divided into welfare effects for consumers and effects for competition.[573] Among the possible effects for consumers are, on the one hand, reduced transaction costs, benefits in

---

567 European Court of Justice, United Brands, paragraph 184.
568 Rubinfeld 1998, p. 24.
569 See section 1.4.3.
570 See European Court of Justice, Tetra Pak II, paragraph 82. As the Court further stressed, even where tied sales of two products are in accordance with commercial usage or where there is a natural link between the two products in question, such sales may still constitute abuse within the meaning of Article 86 unless they are objectively justified.
571 Van Geffen/Nooij/Theeuwes 2002, 109pp.
572 See section 1.5.3.
573 See section 1.5.3.

terms of price, value and the compatibility of a particular service.[574] On the other hand, tying can result in contractual lock-in situations in the form of long-term subscription contracts or subscriptions to large bundles. Tying strategies can also lead to technical lock-in situations, for example, in the form of an arrangement that ties the provision of a specific, often subsidized, set top box to the subscription to one particular proprietary pay-TV platform. In this case, the choice for a particular pay-TV platform would also determine in the short term the range of information services consumers are technically able to receive.

As far as the effects of tying or bundling for competition are concerned, they are often difficult to measure precisely and can range from 'little impact on the ability of rivals to compete' to 'total exclusion from competition'.[575] A variety of factors must be considered, such as the height of the production costs for the manufacturer of competing products in the market for the tied good, and if competitors have the ability to offer added value to consumers and thereby make the additional purchase of their competing product attractive. The conditions can be different in one-sided and multi-sided markets. In the latter, competitors have more choices to differentiate their products and offer added value at one or the other side of the market.

Compatibility with Antitrust Law

Tying can be in conflict with Article 81 (1)e and/or 82 (d) of the EC Treaty. If tying is practiced by a dominant enterprise, Article 82 of the EC Treaty applies.[576] Preconditions for the application of Article 82 (d) of the EC Treaty, on which the following analysis will focus, are the finding of a dominant position in the market for the tying product, the finding of two distinct products or services, the finding that these two products or services are offered in a way that leaves customers or consumers no choice to obtain the products/services separately, and, finally, the finding that an anti-competitive effect results from that practice.

Two landmark cases are the Hilti and the Tetra Pak II case. The Hilti case concerned a manufacturer of a range of products used for fastening materials into place, including its nail gun or 'power-actuated fastening systems' ('PAF systems') and nails that were compatible with the system. Hilti pursued a policy of selling PAFs to end users or distributors only when they were purchased with the necessary complement of nails. Alternatively, Hilti would reduce the discounts when end users or distributors bought nails from competitors. Hilti's competitors, Eurofix and Bauco, who also produced nails, complained that this behaviour was intended to

---

[574] For an in-depth discussion see the papers by Nalebuff 1999; Bakos/Brynjolfsson 1999 and 2002a; Katz/Shapiro 1998; Van Geffen/Nooij/Theeuwes 2002, p. 95pp., with further references; Bishop/Walker 1999, p. 209pp. and 292pp.

[575] See the paper by Leveque 2000.

[576] For an overview on the legal discussion, see Van Geffen/Nooij/Theeuwes 2002, p. 95pp.; Garzaniti 2000, p. 180; Katz/Shapiro 1998, p. 47pp.: 'The legal treatment of tying remains confused'. See also European Commission (1998), Access Notice, paragraph 103.

restrict their sale of nails for Hilti PAFs. The European Commission decided that Hilti-compatible PAFs and Hilti-compatible nails[577] were two separate markets and not, as Hilti claimed, one market. It thereby established that the first condition for the finding of anti-competitive tying, the existence of two separate product markets, was given. The European Commission then found that Hilti left customers with no choice for the source of their nails and, in so doing, abused its dominance in the market for nail guns. Making the sale of Hilti's nail guns (the tying product) conditional on buying a corresponding complement of nails (the tied product) constituted tying.[578] The European Commission was herein confirmed by the European Court of First Instance[579] and the European Court of Justice.[580]

Similarly, in the Tetra Pak II case, the European Commission decided that Tetra Pak's refusal to sell machines for the packaging of liquid and semi-liquid milk products separately from the aseptic packaging cartons that Tetra Pak also produced and sold, was a form of anti-competitive tying. In its decision, the European Commission further elaborated on when products are considered distinct, and when the tying of such products is not justified by objective reasons. Again, the decision was confirmed by the European Court of First Instance[581] and the European Court of Justice.[582] The European Commission concluded that the cartons and the packaging machines were distinct products and belonged to different product markets.[583] The effect for competitors was that Tetra Pak's position in integrated distribution systems was further strengthened and competitors would find themselves in an 'extremely uncomfortable position'.[584] The European Commission did not follow Tetra Pak's argumentation that synergistic effects at the level of research and development and after-sales services would be sufficient to constitute a 'natural link' between the machine and the type of packaging it uses. Neither would the European Commission accept that, as Tetra Pak claimed, the system of tied sales was justified because it benefited users by enabling Tetra Pak to offer consumers a

[577] European Commission, Commission Decision of 22 December 1987 relating to a proceeding under Article 86 of the EEC Treaty (Case IV/30.787 and 31.488 - Eurofix-Bauco v. Hilti), 11 March 1988, OJ L 65, p. 19 [hereinafter 'Hilti'], paragraph 55.

[578] European Commission, Hilti, paragraphs 74-75.

[579] European Court of First Instance, Judgment of the Court of First Instance of 12 December 1991, Hilti AG v Commission of the European Communities, Case T-30/89, European Court reports 1991, p. II-1439 [hereinafter `Hilti'].

[580] Judgment of the Court of 2 March 1994, Hilti AG v Commission of the European Communities, Case C-53/92 P, European Court reports 1994, p. I-667.

[581] European Court of First Instance, Judgment of the Court of First Instance of 6 October 1994, Tetra Pak International SA v Commission of the European Communities, Case T-83/91, European Court reports 1994, p. II-755 [hereinafter 'Tetra Pak II'].

[582] European Court of Justice, Judgment of the Court of 14 November 1996, Tetra Pak International SA v Commission of the European Communities, Case C-333/94, European Court reports 1996, p. I-5951.

[583] European Commission, Commission Decision of 24 July 1991 relating to a proceeding pursuant to Article 86 of the EEC Treaty (IV/31043 - Tetra Pak II), 18 March 1992, OJ L 72, p. 1 [hereinafer ' Tetra Pak II], paragraph 11.

[584] European Commission, Tetra Pak II, paragraph 117.

comprehensive performance guarantee.[585] The European Court of Justice followed the tendency of the European Commission to interpret Article 82 (d) strictly:

> 'Consequently, even where tied sales of two products are in accordance with commercial usage or there is a natural link between the two products in question, such sales may still constitute abuse within the meaning of Article 86 unless they are objectively justified'.[586]

Closer to the subject matter of this study is the Microsoft Europe case of the European Commission. The case concerned, among other things, Microsoft's practice of making the availability of its Windows operating system conditional on the simultaneous acquisition of Microsoft's Windows Media Player, much to the concern of competing producers of media players.[587] The European Commission found that the conditions for a finding of anti-competitive tying according to Article 82 (d) of the EC Treaty were fulfilled: Microsoft was dominant in the market for operating systems (the tying product), a media player (tied product) is a product that is distinct from an operating system, Microsoft did not offer consumers the possibility to buy Windows without the Windows Media Player and this practice negatively affected competition in the market for media players.[588] To remedy what the European Commission found to be an anti-competitive tying practice, it ordered Microsoft to unbundle its products and offer a version of the Windows operating system that did not include the Windows Media Player.[589] The remedy applies to Windows systems that are licensed directly to end users (home users via retail and corporate customers) and licensed to intermediaries.

All three decisions made it quite clear that, in their decision practice, the European Commission and the European Court of Justice/First Instance are quick to consider the joint selling of products or services tying in the sense of Article 82 (d) of the EC Treaty unless it is objectively justified. In the Hilti and Tetra Pak II decisions, the European Commission and the European Court of Justice/First Instance further made it clear that the threshold for the objective justification of a tying practice is very high—commercial practice, compatibility, synergies, safety aspects,[590] and the so-called natural link do not necessarily provide sufficient justification. The rule of proportionality applies, and enterprises must take, where available, recourse to less infringing practices than tying.[591] Moreover, the fact that

585 European Commission, Tetra Pak II, paragraph 118.

586 European Court of Justice, Tetra Pak II, paragraph 37.

587 'A media player is a software product that is essentially able to play back such audio and video content, that is to say, to understand that digital content and translate it into instructions for the hardware (for example, loudspeakers or a display)', European Commission, Microsoft Europe, paragraph 60.

588 European Commission, Microsoft Europe, paragraphs 799-977.

589 European Commission, Microsoft Europe, paragraph 1011.

590 European Court of First Instance, Hilti, paragraph 118. European Commission, Tetra Pak II, paragraph 119.

591 European Commission, Tetra Pak II, paragraph 119.

two products are closely associated does not, on its own, constitute a 'natural link' in the sense of Article 82 (d). As long as alternative manufacturers provide the product or service, an enterprise principally has no right to treat two services or products as integrated services.[592] In the Microsoft Europe case, the European Commission underlined that the distinctness of products for the purpose of an Article 82 of the EC Treaty analysis has to be assessed with a view to consumer demand. If there was no independent consumer demand for the allegedly tied product, then the products at issue were not distinct and a tying charge would be to no avail.[593] Hence, the fact that the market provides a product or service separately can be evidence for separate consumer demand.[594]

Example 1—Joint Selling of Set Top Boxes and Pay-TV subscription

If one transfers these principles to the case of the bundling of set top boxes and pay-TV subscriptions, this could mean that making the subscription of a pay-TV service dependent on the purchase of a particular set top box could constitute anti-competitive tying. For Article 82 (d) of the EC Treaty to apply, the pay-TV provider would have to have a dominant position in the pay-TV market, pay-TV subscriptions and set top boxes would have to be two distinct services/products, they are not offered separately and the joint offer affects competition. Providing dominance is given, and set top boxes and subscriptions are sold jointly, the question is whether the selling of set top boxes and of pay-TV subscriptions are two distinct services. One could try to argue that the set top box and the subscription platform are necessary components of a single wider system or are supplied together by nature of custom. The argument could go along the lines that consumers would not buy a set top box if they didn't have a subscription to a pay-TV service.[595] Whether this is a valid claim or not depends on several factors. For example, where more than one pay-TV service is available, there could be demand for set top boxes that is independent of the pay-TV subscription. Moreover, set top boxes are not only used to decrypt content; they are also instruments that transform digital signals into a format that is readable for analogue TV sets. In addition, with increasing sophistication, set top boxes offer more and more functionality than just the decryption of pay-TV programmes, such as the processing of interactive applications, billing services, etc. It follows that any generalization and basic assumption or denial of a 'natural link' between two set top boxes is difficult to make, and that it is the circumstances of the concrete case that matter.

The question is then whether there is an objective justification to be found for tying the two services/products together. The fact that the set top box is often necessary to decrypt pay-TV services, that purchasing the set top box and the pay-

592 European Commission, Tetra Pak II, paragraph 119.
593 European Commission, Microsoft Europe, paragraph 803.
594 European Commission, Microsoft Europe, paragraph 804.
595 See, for example, the argument of Microsoft in the Microsoft case, European Commission, Microsoft Europe, paragraphs 801, 809.

TV subscription from the same operator can guarantee service compatibility or that only then the security of the encryption system is safeguarded, are not factors that—judging from previous decisions—are likely to convince the European Commission or the European Court of Justice.

Selling a subscription and set top box jointly could also affect competition in the set top box market, because making the provision of a set top box dependent on the subscription to the pay-TV platform limits the chances for competing set top box manufacturers to sell set top boxes to the same consumer base. The impact of the tying strategy on competition will depend on how much room is left for competition. There could be room for competition in a situation in which independent set top box manufacturers produce boxes that can process a range of additional services—providing such services exist—and so offer consumers some added value as compared to the set top box of the first platform operator. An interesting question is what the situation would be if pay-TV providers gave set top boxes away for free in order to attract additional subscribers. One could argue, as Microsoft did in the Microsoft Europe case, that there is no anti-competitive effect from tying in such situations.[596] The pay-TV provider would not even compete in the set top box market because it was not trying to make a profit in this market. The European Commission held against this that even if a (tied) product were given away for free, the effect could be anti-competitive, notably as far as complementary products to the tying product were concerned. The European Commission found that tying could have 'spill-over effects' on the competition of related products, such as DRM solutions, set top boxes or the online delivery of content, as a result of indirect network effects.[597] This finding was one of the aspects that Microsoft contested in the Interim proceedings before the European Court of First Instance. Microsoft claimed that the European Commission's spill-over argument had no basis in economic theory and would fail to reflect market realities.[598] The European Court of First Instance was ready to agree that Microsoft's arguments raised important questions, but felt that the interim measure proceedings were not the place to deal with them.[599] If one were to use the argument of the European Commission in the pay-TV example, one 'spill-over effect' could concern the contestability of the pay-TV market. Tying the subscription to a pay-TV service and the purchase of a set top box can discourage the entry of new pay-TV providers.[600] Whether this is true would also depend on whether the technology used in the set top boxes is a proprietary technology that the first operator refuses to licence to other operators or set top box manufacturers, or whether it can be made compatible. One consequence of this, very broad, interpretation by the European Commission

[596] European Commission, Microsoft Europe, paragraph 840. Another question is if there is a case of predatory picing, see below.
[597] European Commission, Microsoft Europe, paragraphs 842, 897.
[598] European Court of First Instance, Microsoft Europe, paragraphs 329.
[599] European Court of First Instance, Microsoft Europe, paragraphs 394, 404.
[600] See section 1.5.3.

was that the rules on anti-competitive tying could not only be used with regard to competition in the tied product market but, more generally, with regard to the overall competitive structure.

Example 2—Channel Bundling
The second example relates to channel bundling. For example, consumers must subscribe to one niche channel before they can receive a premium channel, so-called buy-through contracts. The question if these services have a natural link might not always be easy to answer. Providing there are separate markets for the provision of premium services (sports, films) and other content, this would be an argument in favour of two distinct services. Both types of services, and most certainly premium services, can probably be offered independently and on a à la carte basis.

Tying channels together can have positive and negative effects on competition and consumer welfare.[601] Instructive is the argumentation of the former British regulatory authority for broadcasting, the Independent Television Commission (ITC), in a tying case concerning the UK pay-TV operator BSkyB. The ITC recognized that there are valid reasons for offering basic and premium channels in one package, such as the ability to guarantee income to programme providers, pass off the risk of investment to the platform operators and in this way stimulate investment in new services (which would again increase consumer choice), and to realize economies of scope. However, as the ITC also noticed, buy-through obligations could (and already did) easily result in a situation where consumers were forced to subscribe to channels they did not want in the first place. According to the ITC, the question was if a wider range of channels would result in increased consumer welfare. And if it did not, were there alternative, better risk-bearing options?

According to the ITC, similar considerations would also apply to basic channel bundling or rather: tying. Having said that, the question whether basic channel tying, meaning the packaging of different basic channels in a larger programme bouquet, would also be considered tying will also depend on the way markets are defined and whether separate demand for the individual channels is given. If the basic channels in question were found to be offered in one and the same market, the typical tying constellation would not be given.[602] Still, it could be a case of 'bundling.

---

601 See section 1.5.3.

602 Compare, European Court of Justice, Tetra Pak II, paragraph 36; European Court of First Instance, Hilti, paragraphs 66 and 67; European Commission, Microsoft Europe, paragraph 803: '[t]he distinctness of products for the purposes of an analysis under Article 82 therefore has to be assessed with a view to consumer demand. If there is no independent demand for an allegedly tied product, then the products at issue are not distinct and a tying charge will be to no avail'.

Bundling

Tying is often distinguished from bundling[603] although the notions of bundling and tying are closely related and easy to confuse. The difference between tying and (pure) bundling is that the tied product is available on a stand-alone basis under tying, but not under (pure) bundling. This suggests that at least in the case of pure bundling, the finding of anti-competitive bundling practices seems to be less dependent on the evidence that the components are distinct and offered in different markets.[604] With regard to mixed bundling, the tied product can be offered separately, but the (monopoly) product is sold together with a complementary product at a single price that is less than the sum of the products sold individually.[605] One example of mixed bundling is discounts that are given for the purchase of additional services or products.

Mixed bundling was the subject of another investigation against BSkyB, this time performed by the UK Office of Fair Trade (OFT). A number of BSkyB's competitors had complained to the Office of Fair Trade that BSkyB was abusing its dominant position with respect to its channel pricing. A subscriber to BSkyB's premium channels (for example a film channel) could get discounts if he or she also subscribed to BSkyB's premium sports channels. The OFT found that BSkyB's action bordered on anti-competitive action but that there was no evidence of abuse.[606] The OFT acknowledged, however, that 'bundling poses a dilemma'. It recognized that where the fixed costs were high, for example, because of the costs of programme rights, it was 'neither unnatural nor undesirable for suppliers to offer discounts to consumers taking additional products as the incremental cost of supplying those extra products to consumers is relatively low'.[607] Accordingly, in OFT's investigation, it was not so much the fact that BSkyB would tie together two services that did not have a 'natural link' that played a role—the decisive criterion in the theoretical tying constellation.[608] The fact that concerned the OFT was that a subscriber who subscribed jointly to two BSkyB services would have to pay less than a subscriber who subscribed to these or similar services from different

603 Larouche 2000, p. 263.

604 See also Larouche 2000, p. 263.

605 Rubinfeld 1998, pp. 25-26.

606 Office of Fair Trade (OFT), `BSkyB: The outcome of the OFT's Competition Act investigation`, London, 2002, available at < www.oft.gov.uk> (last visited on 20 March 2005) [hereinafter 'Office of Fair Trade 2002'], paragraphs 4.18-4.24. Similar to the UK, in France the National Regulatory Authority (CSA) also decided to pay closer attention to anti-competitive bundling strategies in pay-TV; Rapport sur la mission confiée par le Ministre de l'Économie, des Finances et de l'Industrie à M. Jérôme Gallot, Directeur Général de la Concurrence, de la Consommation et de la Répression des Fraudes, La Television Numerique Terrestre: Enjeux et modalités de mise en œuvre au regard des règles de concurrence, notamment au travers de la question de la distribution February 2002, available at <www.minefi.gouv.fr> (last visisted on March 14), pp. 16-17.

607 Office of Fair Trade 2002, paragraph 4.20.

608 Article 82(d) of the EC Treaty reads: 'making the conclusion of contracts subject to acceptance by the other parties of supplementary obligations which, by their nature or according to commercial usage, have no connection with the subject of such contracts'.

operators separately. As a consequence, a rival supplier of a premium sports channel might be locked out of the market because subscribers to other BSkyB premium channels (for example, premium film channels) got a discount if they also subscribed to, for example, Sky Sports 1.[609]

The application of the concept of anti-competitive tying to situations in which the distinct-product criterion plays no or a lesser role, considerably extends this concept. In order to prevent the extension from becoming unreasonable, additional criteria are needed to distinguish anti-competitive bundling from acceptable bundling strategies. One interesting suggestion was made by Larouche, who suggested requiring the demonstration that the product is an 'essential facility'[610] in accordance with the criteria developed in Essential Facilities Doctrine cases. According to Larouche, unbundling is a means to 'enable competitors of the dominant firm to have access to the broken-down components of services offered by the dominant firm'.[611] Applicants would have to prove that access to this product is absolutely necessary for rivals to compete. The OFT went a different way and took recourse to the concept of predatory pricing. The OFT asked whether the incremental price of an additional channel in a bundle of channels would be more or less than the incremental cost of supplying the product. Where the incremental price was below the incremental cost, the pricing would be a strong indication of anti-competitive behaviour, especially in the presence of evidence of foreclosure of competitors. In the BSkyB case, however, the OFT had only limited evidence of foreclosure of rival channels resulting from mixed bundling, which is why the OFT did not declare BSkyB's practice anti-competitive.[612]

Another, similarly difficult question is how to effectively remedy anti-competitive tying and bundling. Unbundling, which is the obligation to offer services separately and at an adequate price, is probably the most obvious answer. However, as the above analysis shows, it is not always the act of tying products or services together that puts functioning competition and/or consumer welfare at risk. Tying or bundling can also be the trigger for anti-competitive pricing strategies, such as in the OFT case, or it is a qualified form of refusal to supply, such as in the Hilti or the Tetra Pak II case. Here, other remedies might be better and lastingly fit to end anti-competitive behaviour, such as imposing access obligations or monitoring the pricing behaviour, as well as initiatives to stimulate competition in the market for the tying product so as to eliminate dominance in this market. For example, in the channel bundling cases, a more effective remedy than unbundling might be the close monitoring of exclusive programme licensing deals so that competitors get a chance to offer attractive premium channels themselves. In the set top box/pay-TV subscription case, possible remedies could be standardization initiatives to stimulate the independent set top box market and/or the production of

609 Office of Fair Trade 2002, paragraphs 4.18-4.24
610 Larouche 2000, 267pp.
611 Larouche 2000, p. 263.
612 Office of Fair Trade 2002, paragraphs 4.21-4.23.

set top boxes that are able to process different formats or conditional access technologies. Here, mandating the licensing of proprietary conditional access solutions to competitors and independent set top box manufacturers, or encouraging the implementation of common interfaces to make set top boxes compatible with several systems might be a more promising route to take than simply requiring unbundling.

Conclusion

Tying and bundling practices have been examined by competition authorities and courts because of their potentially harmful effects on competition and the consumers' freedom to choose from a range of different offers. The bundling of one or more products or services is a widespread business practice in pay-TV. Technical and contractual lock-ins or lock-outs of the kind described in Chapter 1 can be the consequence. Section 3.4.3. found that the widespread practice of tying the subscription to a pay-TV platform to the purchase of a set top box as well as the practice of channel bundling can, under certain circumstances, qualify as abuse of market power according to Article 82 (d) of the EC Treaty. Under particular circumstances, they can also conflict with Article 81 (1)e of the EC Treaty. The European Commission and the European Court of Justice have shown a tendency to treat the tying cases they dealt with rather strictly and to ban them unless there is clear objective justification. Difficulties with which competition authorities or judges will be confronted are the need to identify whether a product or service is distinct from the product or service it is tied to, whether the practice must be considered anti-competitive, and whether there are efficiency or similar considerations that may determine that such behaviour is desirable and should not be banned. A final difficulty will consist of finding the appropriate remedy. Unbundling obligations might sometimes be inadequate or too weak to ensure functioning competition. Other remedies, such as mandatory access, price control, standardization or other measures to stimulate competition in the market for the tying product should also be considered as possible remedies.

3.5. Competition Law and Consumer Interests

'In times of change there can be only one answer to the question how best to protect the consumer: **protect competition**!'[613] (original emphasis). According to the former European Commissioner for Competition Policy, Mario Monti, consumer welfare is at the heart of EC competition policy.

When applying general European competition law to pay-TV cases, the European Commission perceives electronic access control primarily as an alternative means of financing electronic services.[614] In this function, electronic

613 M. Monti, Keynote speech at the European Competition Day in Madrid, February 26, 2002.
614 See section 3.2.2.

access control can contribute to improving consumer welfare. This became evident, for example, in the European Commission's BiB decision. When applying Article 81 (3) of the EC Treaty to the proposed operation, the European Commission found that, as almost all households in the UK possess a TV set, the purchase of a BiB/BSkyB digital set top box would give consumers access to interactive services via television screens. The introduction of a new service of this type was therefore beneficial to consumers.[615] The fact that the services of BiB/BSkyB were encrypted and that consumers first had to purchase a decoder and take out a subscription is not an issue under competition law. Competition law protects competition and not individual consumers' access to certain products or services. Considerations such as whether prices for access might, from a social welfare or public interest point of view, be too high, conditions unfair or single consumers excluded from access to content of particular importance, lie principally outside the scope of general competition law assessment. They might be an issue under media law.[616] As far as competition law and policy are concerned, consumer interests are best served in markets that are competitive and generate choice from a range of quality services—this is a view that the European Commission repeatedly expressed in its decisions. Protection-worthy interests of consumers might be at stake where consumers are not offered this choice. In the NewsCorp/Telepiù case, the European Commission evoked a 'fundamental right of consumers to choose'.[617] European competition law focuses on the broader scope of consumer welfare in general. Consumer welfare is the goal, and functioning competition the way to get there. European competition law focuses on the behaviour of enterprises in the market place in order to ensure the goal is reached. Although Articles 81 and 82 of the EC Treaty only apply to enterprises,[618] consumer interests and the reduction of consumer surplus are factors a competition authority considers when judging the compatibility of market behaviour with competition law.

Consumer interests and expectations enter the competition law analysis at different stages. They are taken into account when defining a market in the form of the aspect of demand-side substitutability. For example, in the NewsCorp/Telepiù decision, while in the process of defining the relevant markets, the European Commission asked the Italian consumer associations to assess the relationship between free-TV and pay-TV, and to determine why consumers view both services differently.[619] Harming consumer interests can trigger the application of competition law, such as in Article 82 (b) of the EC Treaty. The positive effects of market behaviour on consumer welfare can justify or exempt such behaviour, such as in Article 81 (3) of the EC Treaty. Finally, consumers are an important index when monitoring the compatibility of economic activities and competition law and when

---

[615] European Commission, BiB, paragraph 163.
[616] See sections 2.2.1. to 2.2.3.
[617] European Commission, Newscorp/Telepiù, paragraph 192
[618] Wyatt & Dashwood 2000, p. 544.
[619] European Commission, Newscorp/Telepiù, paragraphs 32-33.

detecting cases of non-compliance. In the European Commission's Notice on the Handling of Complaints by the European Commission under Articles 81 and 82 of the EC Treaty, the European Commission explicitly encourages citizens to report suspected infringements of competition rules and has even created a specific instance for this purpose.[620] The Consumer Liaison Office in the Directorate General for Competition is responsible for receiving information and requests concerning competition problems faced by consumers. Its' tasks include alerting consumer groups to competition cases when their input might be useful, and advising them on the way they can provide input and express their views, establishing contacts with National Competition Authorities regarding consumer protection matters and to intensify contacts between the Competition and other Directorates Generals.[621]

Still ongoing is the question to which extent consumers can be active parties in competition law procedures. According to Article 7(2) of Council Regulation on the Implementation of the Rules on Competition,[622] 'natural [...] persons who claim a legitimate interest' can file an application with the European Commission to examine whether there is an infringement of Articles 81 or 82 of the EC Treaty. In its Notice on the Handling of Complaints by the European Commission under Articles 81 and 82 of the EC Treaty, the European Commission stated that consumer associations could lodge complaints with the European Commission. The European Commission also said that eventually 'individual consumers whose economic interests are directly and adversely affected in so far as they are the buyers of goods or services that are the object of an infringement can be in a position to show a legitimate interest'.[623] In practice, the European Commission occasionally accepted[624] and even actively encourages complaints that are lodged by individual consumers.[625] Still, the European Commission is entitled to prioritize its selection of cases and concentrate on cases that have a particular political, economic or legal significance for the Community.

---

[620] European Commission, Notice on the handling of complaints by the Commission under Articles 81 and 82 of the EC Treaty, 27 April 2004, OJ C 101, p. 65 [hereinafter 'Notice on the Handling of Complaints by the European Commission under Articles 81 and 82 of the EC Treaty'].

[621] For more information visit <europa.eu.int/comm/competition/publications/competition_policy_and_the_citizen/consumer_liaison> (last visited on 20 March 2004).

[622] European Council, Council Regulation (EC) No. 1/2003, of 16 December 2002 on the implementation of the rules on competition laid down in Articles 81 and 82 of the Treaty, 4 January 2003, OJ L 1, p. 1 [hereinafter 'Regulation on the Implementation of the Rules on Competition'].

[623] European Commission, Notice on the Handling of Complaints by the European Commission under Articles 81 and 82 of the EC Treaty, paragraph 37.

[624] European Commission, Commission decision of 9 December 1998 relating to a proceeding pursuant to Article 85 of the EC Treaty (Case IV/34466 – Greek Ferries), 27 April 1999, OJ L 109, p. 24 [hereinafter 'Greek Ferries'], paragraph 1.

[625] European Commission, EU Competition policy and the consumer, Booklet, Brussels, 2004, available at <europa.eu.int/comm/competition/publications/consumer_en.pdf> (last visited on 20 March 2005).

3.6. Competition Law and Non-Economic Interests

It would exceed the scope of this study to investigate closer the very controversial discussion of the extent to which general public information policy interests of the kind that were described in Chapter 2 play or can play a role in competition law procedures.[626] This is a discussion that can fill books and on which excellent treatises have already been written.[627] Although the opinions are divided, there seems to be at least some agreement that competition and public information policy may, at times, pursue similar goals. The opinions part when it comes to the question whether it is desirable to instrumentalize general competition law as a tool to realize goals such as freedom of expression, pluralism, diversity, the broad accessibility of information and the availability of information of general public interest.

Goals that play a role both in competition and information policy are the contestability of the information market and functioning economic competition in order to generate a broad and diverse choice of content for consumers at fair and reasonable conditions. Accordingly, it is often argued that general competition law is perfectly sufficient to also realize what can be said to be some of the most important public information policy objective, namely pluralism and diversity. However, a comparison between the long-term goals of the two regimes already shows disparities. Successful competition policy is almost inevitably characterized by a 'survival of the fittest' attitude. In contrast, public information policy is determined by the idea of guaranteeing the survival of politically or socially desirable content providers (such as public broadcasters) even if or just because they would most probably not survive the free market reality. This already indicates that information policy often pursues very different, if not contradictory, outcomes/goals of competition policy.[628] The much criticized financing of public broadcasting is just one example, and the granting of favourable access for public broadcasters to pay-TV platforms and EPGs another. The listing of areas in which general competition law and public information policy collide can be easily continued. Internal growth or vertical concentrations can be the result of a functioning competitive selection process or sound economic thinking without necessarily leading to anti-competitive behaviour. Consequently, general competition law does not sanction internal growth. On the other hand, the presence of dominance in a media market and preventing that such dominance is abused to influence journalistic competition is a recurring issue in media regulation. Similarly, it is also possible that even where market behaviour is found to conflict

[626] See section 2.2.

[627] See instead the study by Valcke, with further references. See also Diesbach 1998, p. 126-133; Van de Gronden/Mortelmans 2003, 16pp.; critical Van den Beukel/Nieuwenhuis 2000, p. 123: '... laat zien dat het concept van pluriformiteit en het concept van volledige mededinging een one-night-stand verhouding hebben: op het eerste gezicht lijkt het wel aardig maar daarna blijkt dat de karakters van beide begrippen te veel verschillen'. Urek 1991, p. 33pp.

[628] Herdzina 1999, p. 36; Schmidt 1981, pp. 54-69.

with the objectives of competition law, public policy considerations can be a reason not to ban such practices. For example, in Germany, the US company Liberty Media sought to buy the German cable network from Deutsche Telekom. Although the merger raised considerable concerns about the possible creation of a dominant position in the German cable market, it was nevertheless encouraged by the German authorities in the hope that Liberty Media would invest in Germany's broadband roll-out.

The previous examples suggest that general competition law could be helpful in realizing public information policy goals in some situations. This could be true at the Member State level.[629] In other situations, it is unlikely that the application of general competition law principles will lead to a satisfactory outcome.[630] As far as the European level is concerned, the European Commission probably does not even have the authority to pursue public information policy goals within the framework of competition law procedures. The crucial and more fundamental question in this context is whether it is actually desirable to use general competition law purposefully as a means to realize further-reaching public policy objectives. The result could be an even stronger politicization of an instrument that is primarily designed to protect competition. Making general competition law a tool for the realization of public information policy interests risks exposing it to rent-seeking, conflicts of authority, and questions that must be solved by parliaments rather than competition law authorities.

3.7. Conclusion

European competition law has its own response to bottleneck control, technical and contractual lock-ins or the information problem. The set of tools that competition law provides can be roughly divided into ex ante merger control and ex post antitrust control.

The European Commission's merger and concentration policy is one important instrument for shaping markets for access-controlled services and keeping them contestable. Over the course of time, the European Commission has gained valuable experience in handling mergers in the pay-TV sector and in influencing market structures, enabling it to prevent situations in which dominant parties can use their control over technical or content resources to impede market entry and competition. Arguably, merger control can be particularly helpful in the initial phase, meaning at a time in which powerful operators that are already established in older media

[629] European Council, Merger Regulation No. 4064/89, Article 21 (3). Wyatt/Dashwood 2000, p. 673. In some countries, however, the application of general competition law in order to also enforce information political interests might even conflict with the constitutional order, such as in Germany, where competition law falls under the authority of the Federal State, while broadcasting law is a matter of the Länder, KEK 2000, p. 50.

[630] See also Larouche 2000, p. 339, with further references.

markets seek to position themselves in these new markets and ensure strong market footholds. On the other hand, and in particular in new markets, it can be difficult to predict the course developments will take.

The European Commission's merger decisions in this area reach quite far and often even beyond existing sector-specific regulation to minimize anti-competitive influence and monopoly control over conditional access, the EPG and the API, or more generally, control over the consumer base. Within this context, it became clear that the finding of bottleneck situations depends on the economic and technical circumstances of a particular case, and that it is difficult to generalize. The power of the European Commission as a competition authority is not restricted to allowing or banning a certain proposal. In addition, it can influence the behaviour of the parties toward a proposed merger far beyond the point at which the merger is concluded by means of undertakings and conditions. Occasionally, the Commission seems to find it difficult to draw the line between its authority as a competition law authority and possible ambitions as a parallel regulatory authority. Having said that, the European Commission takes its decisions on a case-by-case basis, which is why they do not have a binding effect in other cases. Nevertheless, an analysis of the Commission's decisions in this sector can provide insight into the dynamics of pay-TV markets and a level playing field for the development and testing of possible solutions to keep those markets competitive.

It is particularly interesting to note that the potential of mandated access, the traditional response to bottleneck problems, for example, in telecommunications law, is critically and realistically viewed by the European Commission. The sole existence of access obligations would not prevent parties from abusing dominant market power in one way or another. Eventually, alternative initiatives directed at stimulating competition and the emergence of inter-platform competition would be the more reliable and lasting safeguards for functioning competition. Within this context, interoperability solutions played a prominent role in the European Commission's pay-TV decisions as did the restrictive licensing practices of programme content and the freedom of consumers to switch between services. To this extent, the decisions are characterized by a certain bottom-up approach. They repeatedly illustrate that the existence of competition ultimately depends on whether consumers are not unnecessarily impeded in their choices between different services. The European Commission, accordingly, did not fail to address the freedom of consumers to choose between services, free from technical, contractual or other lock-ins.

The conclusions regarding the potential of antitrust law to ban ex post anti-competitive practices to monopolize the consumer base are mixed. One finding of Chapter 3 is that the potential of the Essential Facilities Doctrine, the classic competition law instrument used to address bottleneck situations, should not be overestimated. The conditional access system, the API, the EPG, a particular marketing platform or programme rights will probably qualify as 'essential facilities' only in very exceptional circumstances. Moreover, the potential of general competition law to realize further-reaching public policy interests in the accessibility and availability of (selected) content is rather limited. Obliging an enterprise to share its resources with a competitor is a rather far-reaching interference with its economic freedoms. Mandated access should be reserved to cases in which access to a facility is absolutely indispensable for competitors to offer a service. The degree of market power of the controller of that facility, the existence of alternative solutions and the question whether mandated access is the optimal remedy are important aspects in this context.

At least as far as the economic aspects are concerned, the principle of non-discriminatory treatment may offer the greater practical value. Non-discrimination is also the keyword when assessing certain practices that seek to exclude consumers in one country from access to services that are offered in another country.

Contractual or technical lock-ins could fall under the prohibitions on anti-competitive tying. The finding of anti-competitive tying should depend, among others, on whether there is a distinct market for a service or product. No less difficult will be to identify when a practice is anti-competitive and when it is the result of sound economic thinking and innovation, and as such principally desirable and beneficial. Finally, the example of tying also demonstrated how difficult it can be to find effective remedies. Unbundling may only be the optimal remedy to some cases. In other cases, stimulating competition in the market for the tying product might be the preferable route to follow. A second, more theoretical consideration is to which extent 'old' measures can or should be applied to the 'new economy' where dominance sometimes seems to be almost the logical and inherent course of development as a result of network effects and first mover advantages.

More generally, the analysis has shown that some of the well-known drawbacks of general competition law also apply to in pay-TV sector. Examples are the uncertainties of market definition in a period of development, the formation of new markets and business models, or the difficulties of defining dominant market power and anti-competitive behaviour. In addition, the responsible authorities will face considerable information deficits, not only because of the lack of experience in and of information about new sectors, but also because of the difficulties in gathering the information necessary to define markets and judge individual market behaviour and its effect on competition. The lack of predictability and legal certainty can be a further reason why relying on general competition law might not be appropriate, especially not where freedoms or resources of fundamental importance are concerned, such as the protection of the freedom to property and the freedom of expression.

There is also a time factor. Lengthy proceedings can work in favour of the established party while artificially extending the critical period of market entry. This time aspect can make legal proceedings a virtual race against time. This is particularly true in markets such as those for access-controlled services and conditional access, where a lot has to do with being the first to establish a dominant standard. This is not least a consequence of network effects and lock-in effects that give many modern information markets their own dynamics. The comment made by Tom Miller, the Attorney General from Iowa on the US Microsoft Judgement, pinpointed the issue: 'Much of the Microsoft announcement deals with the browser—but the browser war is over. Microsoft has won'.[631] By the time a court comes to assess anti-competitive behaviour, business, services and market places are likely to have changed dramatically.

In conclusion, competition law can provide useful tools to address some anti-competitive practices and strategies and to remedy occurrences of anti-competitive bottleneck control or the monopolization of the consumer base. Competition law also goes to some lengths to protect consumer interests. Due to its shortcomings in terms of predictability, scope and (sometimes) efficiency, however, it is not likely that general competition law will replace sector-specific regulation in pay-TV in the short or medium term. Still, general competition law is a fall-back option to address questions sector-specific regulation falls short of addressing in this sector.

631 Cited in Krim/Cha, Washington Post from 12 July 2001, p. A01

Chapter 4

Conditional Access and Telecommunications Law

4.1. Introduction

Control over technical bottlenecks has become a pressing issue in today's competition and public information policy. With the increasing sophistication and diversification of the distribution patterns for digital content, entering service markets is a matter of access to a growing number of different technical facilities and competing standards. These technologies are often owned and controlled by one or several market players who have their own vested economic interests in delivering electronic services through a specific infrastructure. This situation is characteristic of the pay-TV sector.[632] In Chapter 1 we saw that monopolistic control over the technical pay-TV platform or elements thereof, such as the conditional access system, the EPG or the API, opens possibilities to impede the activities of potential and actual rival pay-TV service providers, particularly when the control is exercised by powerful vertically or horizontally integrated operators. The goal of this chapter is to provide a critical overview of the existing sector-specific solutions that deal with technical bottlenecks in pay-TV.

The regulator's task is difficult: regulatory intervention should not discourage the proliferation of new facilities, such as conditional access, if they are to be seen as promising drivers of the future 'knowledge-based economy'. Furthermore, regulatory intervention must leave sufficient room for market mechanisms to develop an efficient and functioning service environment, and for service operators to launch viable and profitable business models. On the other hand, the purpose of intervention is to prevent the abusive use of such facilities when they pose a threat to market contestability. Protecting and stimulating market contestability in this context means allowing for new market entries, disciplining the market behaviour of dominant players and stimulating functioning competition. Stimulating competition is, however, not the only task. Regulators must seek to strike a balance between relying on market mechanisms and realizing consumer rights and interests and non-economic goals, such as pluralistic and diverse service offerings. The overall goal is to stimulate the development of a rich service offering and optimal conditions for competitors and consumers.

---

[632] See section 1.4.3.

At the transport level, the Access Directive shapes the framework for the regulation of access issues in Europe. The Access Directive is the outcome of a revision of the former European telecommunications law. It replaced, among others, the Standards Directive, which was the first European directive to address questions of access to conditional access systems. The Access Directive is part of the revised European Communications Framework from 2002. The overriding purpose of the revised Framework is to adopt a horizontal technology-independent approach to the regulation of access issues related to the telecommunications infrastructure.[633] This should foster the convergence of broadcasting, telecommunications and information technology, and the gradual removal of sectoral provisions in favour of the application of general competition law. Accordingly, the Access Directive brings conditional access issues that were formerly regulated in separate directives under the umbrella of one common framework that regulates access to electronic telecommunications networks and associated facilities.

This chapter starts with an analysis of how Europe approaches the regulation of control over conditional access and other facilities of the technical pay-TV platform, notably the API and the EPG. The analysis will show that, despite its commitment to a technology-independent approach, the Access Directive still maintains two divergent access regimes for bottleneck situations: one regime applies to selected bottlenecks in pay-TV and is laid down in Articles 5 (1)b and 6 of the Access Directive. The other applies to all other telecommunications facilities and services, and is laid down in Articles 8 to 13 of the Access Directive. The fact that the Access Directive maintains two different access regimes can, as will be shown, lead to a number of frictions and problems in the practical application of the rules it covers. Second, the chapter demonstrates that the Communications Framework is strongly based on the concept of strict separation between the transport level and facilities that are affiliated with the transport of signals on the one hand, and the service level and content-related questions on the other. This chapter also explains why this approach fails to satisfactorily respond to the realities of digital pay-TV markets. It, furthermore, demonstrates that questions about individual consumer's access to the pay-TV platform have been neglected.

The analysis starts with a brief introduction to the history of bottleneck regulation in pay-TV (section 4.2.) and to the goal and scope of the Access Directive (section 4.3.). This is followed by a description of Article 6 of the Access Directive (section 4.4.) and of Article 5 (1)b of the Access Directive, which deals with the regulation of access to other bottlenecks in pay-TV (section 4.5.). Next, the flexible approach to access regulation in telecommunications law, under the directive's Articles 8 to 13 (section 4.6.) is described. The chapter continues with reflections on the likely impact of the different regulatory approaches on future market developments (section 4.7.), after which both access concepts are compared (section 4.8.) and conclusions drawn (section 4.9.). It is not the intention of this

[633] European Commission, 1999 Communications Review, p. iii.

chapter to present an exhaustive analysis of whether the Access Directive, in its present form, is the best solution to regulate the future telecommunications market. It is also not the intention of this chapter to comment in detail on matters such as the criteria of market definition, the identification of significant market power, the 'toolbox approach', etc. This chapter will bring forward some more fundamental points of criticism of the way the regulation of pay-TV services is approached under the Access Directive.

4.2. Regulating Access to Conditional Access—A Brief History

4.2.1. STANDARDS DIRECTIVE

The first European initiative to regulate pay-TV issues dates back to 1990/1991. The story of conditional access regulation began with a failed attempt at standardization. Then, the European Commission was actively promoting the HD-MAC standard for analogue satellite television, and along with it, a unique conditional access system called Eurocrypt as the sole satellite encryption standard. Eurocrypt was a joint initiative of Philips, Thomson, Bosch and the DG XIII (Information Society) of the European Commission. The plan was to incorporate the Eurocrypt standard in the HD-MAC Directive of 1991.[634] The plan failed, among other things, due to the resistance of competing market forces among which were BSkyB and other enterprises. Ironically, BSkyB countered the Eurocrypt proposal with the suggestion to promote the use of 'different decoders' instead of one common standard.[635] Subsequently, BSkyB was among the first to establish a proprietary conditional access standard.

In response to the HD-MAC debacle, the European Commission performed a volte-face and left the question of conditional access standards entirely up to the industry. Yet, the idea was to promote a standardized conditional access. In its Resolution on the Development of Technology and Standards in the Field of Advanced Television Services,[636] the European Council emphasized the need for a 'European non-proprietary encryption/conditional access system serving a number of competing service providers'. Meanwhile, Europe experienced the fragmentation

[634] Council Directive 92/38/EEC of 11 May 1992 on the adoption of standards for satellite broadcasting of television signals, 20 May 1992, OJ L 137, p. 17 [hereinafter `HD-MAC Directive'], Article 6: 'In the case of all services using the D2-MAC standard, which are encrypted and employ a conditional access system, Member States shall take all the necessary measures to ensure that only a conditional access system that is fully compatible with D2-MAC, and standardized as such by a European standardization organization by 1 July 1993, is used'.

[635] 'Down to the Wire', *Cable & Satellite Europe*, July 1991, p. 16.

[636] European Council, Council Resolution of 22 July 1993 on the development of technology and standards in the field of advanced television services, 3 August 1993, OJ C 209, p. 1 [hereinafter 'Resolution on the Development of Technology and Standards in the Field of Advanced Television Services'], paragraph 1 (iv).

of the Internal pay-TV market into several national territories with incompatible conditional access systems and set top boxes.

The European Council entrusted the Digital Video Broadcasting (DVB) Consortium, an industry-led consortium of broadcasters, manufacturers, network operators, software developers, regulatory bodies, etc., with a mandate to resolve the controversial conditional access discussion and initialize an industry-led agreement to adhere to one common single conditional access system. Since it's founding in 1991, the DVB consortium has developed into a pan-European platform for major European media interest groups, consumer electronics manufacturers, common carriers and regulators. Its goal was, and still is, to oversee the development of digital television in Europe and promote this development through, among other things, its standardization activities.[637] The DVB consortium develops specifications for digital television systems that are turned into standards by international standards bodies such as the European Telecommunications Standards Institute (ETSI) or the European Committee for Electrotechnical Standardization (CENELEC). The DVB consortium, however, failed to reach an agreement due to internal frictions and diverging interests between the DVB consortium members. The major operators in the early European markets for access-controlled services and conditional access systems objected heavily to the common interface solution that was proposed by the opposing interest groups within the DVB consortium (notably the European Broadcasting Union (EBU)).[638]

The European Commission, still under the impact of the HD-MAC debacle and afraid of discouraging potential major investors from investing in the digital television infrastructure,[639] was reluctant to enforce any standard that was not fully backed by the industry. Consequently, the Commission did not seize the chance to introduce the issue of conditional access interoperability into the drafting process for the successor to the HD-MAC Directive, the Standards Directive, despite the Council's call in its 1993 resolution for an open common interface. Equally, the European Parliament did not take the opportunity to promote a non-proprietary conditional access system and thereby affirm its commitment for content diversity and open access.[640] The first drafts of the Standards Directive would even completely ignore the issue of conditional access regulation.[641]

When it became apparent that the DVB consortium would fail to reach an agreement, the European Council indicated that

> 'it is willing, however, to introduce regulatory measures, if required, under the conditions that: (i) adequate and timely consensus among economic agents, including broadcasting organizations, to ensure the harmonious evolution of

---

637 For more information see <www.dvb.org> (last visited on 20 March 2005).

638 See overview in Kaitatzi-Whitlock 1997, pp. 101-104.

639 Kaitatzi-Whitlock 1997, p. 109.

640 Kaitatzi-Whitlock 1997, p. 107.

641 Kaitatzi-Whitlock goes even so far as to suspect that the 'development of a proprietary system may have constituted part of a hidden agenda', Kaitatzi-Whitlock 1997, p. 106.

> the market is lacking; and/or (ii) the requirements of fair and open competition, consumer protection or other significant public interest so demand, in order to facilitate the achievement of this objective and the protection of those interests'.[642]

Equally, public broadcasters and other industry players voiced their fears that control over conditional access facilities might give major pay-TV operators a means to exclude other broadcasters from accessing consumers. It was argued that the vertical organization of pay-TV enterprises (which was already characteristic then) demanded the provision of additional legal safeguards to discipline pay-TV operators. Aspects of consumer protection were also put forward whereby, irrespective of the technical platform (encrypted or free-TV television), consumers must be offered a variety of content while being protected from incompatible equipment that would hinder the reception of other competing services.

As a result, Article 4c of the Standards Directive was introduced. The aim of Article 4c of the Standards Directive was to ensure open entry into the pay-TV market to third-party broadcasters by mandating open access to conditional access facilities on 'fair, reasonable and non-discriminatory terms'. It was left to the Member States to further specify the access obligation.

Since then, EU Member States have implemented the Standards Directive. Article 4c of the Standards Directive, therefore, broadly influenced existing national regulations on access to the conditional access system.[643] The majority of Member States adopted the provisions almost identically. Only a few countries adopted further-reaching regulations, such as the UK, Germany and Italy.

4.2.2. Access Directive

In the course of the regulatory reform of the telecommunications sector, Article 4c of the Standards Directive was transformed into Article 6 and Annex I of the Access Directive. The Access Directive reflected the Commission's view that the digital broadcasting sector still required sector-specific access regulation because markets were not yet sufficiently competitive. Despite a number of further-reaching proposals by the European Parliament, Article 6 and Annex 1 of the Access Directive took Article 4c of the Standards Directive over almost word for word. At the same time, the Access Directive revises the former framework that regulated issues of access to telecommunications infrastructures and facilities. Originally, the regulation of access to and the interconnection of other selected

---

[642] European Council, Council Resolution EEC/47/95 of 27 June 1994 on a framework for Community policy on digital video broadcasting, 23 November 1995, OJ L 281 [hereinafter 'Council Resolution of Community Policy on Digital Video Broadcasting'], p. 51, paragraph 3.2.

[643] See European Commission, The Development of the Market for Digital Television in the European Union, Report in context of Directive 95/47/EC of the European Parliament and of the Council of 24th October 1995 on the use of standards for the transmission of television signals, 9 November 1999, COM(1999)540 final [hereinafter 'Standards Directive Implementation Report'], pp. 2-4.

telecommunications facilities fell under the former Open Network Provisions (ONP) Framework. The Communications Framework replaced the ONP Directives, and covers, together with the Access Directive, in far more general terms, access to all technical bottleneck facilities in the telecommunications sector (the so-called horizontal approach). It was argued that the distinction between access to networks and access to digital gateways was questionable since the two would raise similar problems and require comparable solutions. Moreover, it was said that separate regulatory frameworks for different telecommunications infrastructures and associated services would lead to inconsistencies and could potentially distort competition.[644] Instead, there was a need for technology-independent regulation, the so-called horizontal approach.[645] Consequently, part of the reform was to merge all of the existing initiatives on access issues into a single directive.[646]

Correspondingly, the Commission observed that both technology and the digital broadcasting sector had evolved beyond the scope of the former Standards Directive. New services would take digital television beyond the scope of traditional broadcasting and into convergent markets, and service providers would extend their offerings to include non-broadcasting services such as interactive and online services. Correspondingly, new bottlenecks and gatekeeper functions that were not covered by the Standards Directive would develop. Examples given were the EPG and API functions, access to cable television networks or multiplexes.[647]

4.2.3. UNIVERSAL SERVICE DIRECTIVE

Some of the provisions that are or that, as will be shown in this chapter, could be relevant for the pay-TV sector were implemented in the Universal Service Directive. The Universal Service Directive deals with questions of consumer[648] access in the telecommunications sector. The goal of the Universal Service

---

644 European Commission, 1999 Communications Review, p. vi.

645 European Commission 1999 Communications Review, p. 13. The Commission explained 'technological neutrality' in the sense that legislation should define the objectives to be achieved, and should neither impose, nor discriminate in favour of, the use of a particular type of technology to achieve those objectives.

646 European Commission, 1999 Communications Review, p. 28.

647 European Commission, Standards Directive Implementation Report, p. 27.

648 The Communications Framework handles different notions that all can refer to 'consumers'. While 'consumer' is defined as any natural person who uses or requests a publicly available electronic communications service for purposes outside his or her trade, business or profession (Article 2 (i) of the Framework Directive), consumers, which are party to a contract with the provider of publicly available electronic telecommunications services, are also referred to as 'subscribers' (Article 2 (k) of the Framework Directive). Then there is the 'end user' according to Article 2 (n) of the Framework Directive, which can be either a consumer or a provider of broadcasting or information society services, as long as he or she is not providing public telecommunications networks or publicly available electronic telecommunications services. Occasionally, users are also referred to as 'citizens' and 'users' (Recital 12 of the Universal Service Directive). The use of the different notions throughout the Communications Framework seems to be random (see, for example, Recitals 7, 8, 12 of the Universal Service Directive). This chapter will continue to use the notion of consumer.

Directive is twofold. First, its goal is to protect the position of consumers. The rules of the Universal Service Directive are meant to combat exclusion and to create the conditions for the public availability of affordable and good quality services through effective competition and choice.[649] Second, the directive deals with circumstances in which the needs of end users are not satisfactorily met by the market.[650] The overall rationale behind the Universal Service Directive is to protect consumers and foster social inclusion in the knowledge-based society across Europe.[651]

One underlying idea of the directive is that consumers drive competition in telecommunications markets. Enterprises with significant market power that charge excessive or predatory prices to consumers, apply unreasonable bundling strategies or show undue preferences to certain consumers, not only inhibit the realization of general public interests, they also inhibit entry or distort competition.[652] Apart from strengthening the position of consumers in order to stimulate competition, there is also a need to strike the right balance between relying as much as possible on market mechanisms and competition to achieve a high level of choice and quality, and ensuring regulatory intervention to uphold a minimum number of consumers' rights throughout the European Union.[653]

One part of the Universal Service Directive deals with the protection of consumer rights and interests in access to publicly available telecommunications services. It includes detailed provisions on the information that should be provided in subscription and similar contracts between service providers and consumers.[654] It foresees a right to withdraw in the event of modifications to the original contract.[655] It provides for comparative pricing information,[656] quality controls[657] and the public availability of directory services[658] It authorizes NRAs[659] to remedy predatory pricing and unbundling strategies.[660] Furthermore, it encourages Member States to

[649] See Universal Service Directive, Recitals 26, 33, 49 and Article 1 (1).

[650] Universal Service Directive, Article 1 (1). European Commission, 1999 Communications Review, pp. 14-15. See also the overview of the development of the universal-service concept in the US and Europe in Bavasso 2003, pp. 382-387.

[651] European Commission, 1999 Communications Review, p. 25.

[652] Universal Service Directive, Recital 26.

[653] Proposal for a Directive of the European Parliament and of the Council on universal service and users' rights relating to electronic telecommunications networks and services, 19 December 2000, OJ C 365, p. 238 [hereinafter 'Proposal for a Universal Service Directive'], Explanatory Memorandum, section III.

[654] Universal Service Directive, Articles 20 (1)-(3).

[655] Universal Service Directive, Article 20 (4).

[656] Universal Service Directive, Article 21.

[657] Universal Service Directive, Articles 22, 23.

[658] Universal Service Directive, Articles 25, 29.

[659] In this Chapter, Chapter 4, if not said otherwise NRA refers to communications NRA, meaning the NRA responsible for the supervision of the communications sector.

[660] Universal Service Directive, Article 17 (2).

ensure the accessibility and affordability of specific services where it is deemed necessary and desirable for public interest reasons.[661]

Another part of the Universal Service Directive deals with the so-called universal service obligations. Universal service obligations serve the goal of imposing obligations on designated operators to ensure that a defined minimum set of services of a specified quality are available to all consumers, including consumers with special needs, at an affordable price. This is the notion that certain services or facilities play such a fundamental role for society and competition that they must be available to everyone in good quality and at an affordable price.[662] This is to avoid a situation in which parts of the population are excluded for practical, geographical or financial reasons from accessing telecommunications services whose universal accessibility is in the public interest.[663] Providers on whom a universal service obligation is imposed have to provide upon request access to all end users at a constant quality and affordable price.[664] So far, universal service obligations apply to fixed telephony networks, directory enquiry services and public pay phones. Internet access also falls under the universal service obligation.[665]

As far as the broadcasting sector is concerned, the Universal Service Directive provides for a broadcasting-specific variation of the universal service obligation in Article 31. This is the must-carry concept that was discussed in detail in Chapter 2.[666] As Chapter 2 also mentioned, however, the must-carry rules do not deal with consumer access to a technical broadcast distribution platform and they do not apply to pay-TV platforms.[667]

It will be shown that the Universal Service Directive responds to a number of issues identified in Chapters 1 and 2,[668] namely the need for market transparency, the adequacy and fairness of contractual conditions, consumer choice, technical and contractual lock-ins, the information problem, etc. Another question is whether the aforementioned provisions of the Universal Service Directive obligate conditional access platform operators, notably pay-TV operators. Where relevant, this chapter will take a closer look at this question.[669] One provision in the Universal Service Directive, however, that is directly relevant for competition in pay-TV markets, is the provision on the interoperability of digital television consumer equipment in Article 24 of the Universal Service Directive (see section 4.4.2.).

---

661 Universal Service Directive, Recitals 26, 29, 31, 33, 34, 49, Articles 19, 24, 25, 29 and 32, to name but some.
662 Hills (1993), 65pp. Also the paper by Xavier 1997.
663 Universal Service Directive, Recital 25.
664 Universal Service Directive, Articles 1 (2). Instructive Larouche 2000, pp. 375 – 378.
665 Universal Service Directive, Recital 8.
666 See section 2.3.3.
667 See section 2.3.3.
668 See sections 1.5.2., 1.5.3. and 2.2.
669 See sections 4.4.4. and 4.6.6.

4.3. Goal and Scope of the Access Directive

The goal of the Access Directive is ambitious: 'this Directive harmonises the way in which Member States regulate access to, and interconnection of, electronic telecommunications networks and associated facilities' (Article 1 (1) of the Access Directive). In other words, the Access Directive seeks to establish a uniform, harmonized approach towards the treatment of technical bottleneck issues at the transport level.[670]

4.3.1. DEFINITION OF ACCESS IN THE ACCESS DIRECTIVE

The term 'access' is understood in the widest possible sense as

> 'the making available of facilities and/or services, to another enterprise, under defined conditions, on either an exclusive or a non-exclusive basis, for the purpose of providing electronic telecommunications services' (Article 1 (1) of the Access Directive).

'Access' in the sense of the Access Directive refers to the relationship between the provider of electronic telecommunications networks, services and associated facilities, and the user of such technical resources.[671] Article 1 (1) of the Access Directive reads:

> 'The aim is to establish a regulatory framework, in accordance with internal market principles, for the relationships between suppliers of networks and services that will result in sustainable competition, interoperability of electronic communications services and consumer benefits'.

It regulates, for example, the relationship between the controller of a conditional access system and a broadcaster seeking access. It does not regulate the relationship between the controller of a conditional access system and a subscriber.[672] In other words, the Access Directive does not deal with cases in which a pay-TV operator refuses access to a particular consumer or offers access on unfair, unreasonable or discriminatory terms.[673] According to the Access Directive, the interests of consumers in access to diverse and numerous broadcasting services are safeguarded by the imposition of the obligation for conditional access operators to provide conditional access on fair, reasonable and non-discriminatory terms.[674] In this way,

[670] European Commission, 1999 Communications Review, pp. 25-28.
[671] Article 2 (h) of the Framework Directive defines user as 'a legal entity or natural person using or requesting a publicly available electronic communications service'.
[672] In this sense also Bavasso 2003, p. 72
[673] See also section 2.2.1.
[674] Access Directive, Recital 6; Article 5 (1) b) of the Access Directive. In this sense also Bavasso 2003, p. 79.

the directive ensures that a wide variety of programming and services are available.[675]

Having said so much about the consumer side, it is time to take a closer look at how the Access Directive regulates competitor access to technical bottlenecks. The Access Directive[676] distinguishes between two kinds of bottlenecks: Articles 8 to 13 of the Access Directive regulate questions of access to electronic telecommunications networks, services and associated facilities (section 4.6.), whereas the regulation of conditional access is treated as an exception under Article 6 of the Access Directive (section 4.4.).

4.3.2. ELECTRONIC TELECOMMUNICATIONS NETWORKS, SERVICES AND ASSOCIATED FACILITIES—DEFINITIONS

'Electronic communications networks, services and associated facilities' basically comprise all of the facilities at the transport level that can be involved in the process of transmitting signals. The term 'communications networks' means all of the resources at the network level that enable the transmission of signals by wire, radio, optical or other electromagnetic means, including satellite networks, fixed and mobile terrestrial networks, the local loop, the internet, networks used for radio and television broadcasting and cable TV networks.[677] The Access Directive also regulates access to physical infrastructures including buildings, ducts and masts.

'Associated facilities' refers to the enhanced services at the upper levels of the technical distribution chain that support the provision of telecommunications services[678] via networks.[679] In other words, associated facilities are facilities that support the transport function (as opposed to content-related services that are associated with the service level). Associated facilities in the sense of the framework can be operational support systems, number translation systems or roaming and switching services as well as the conditional access system.

In practice, the notion of associated facilities still leaves many questions unanswered. The example of the EPG may illustrate that clear-cut distinctions can be difficult to make in an environment in which technical facilities are closely integrated into the process of making content accessible for consumers.[680] According to the Access Directive, the EPG is considered an associated facility in the sense of

[675] Access Directive, Recital 10.
[676] Article numbers without further reference are articles of the Access Directive.
[677] Framework Directive, Article 2 (a).
[678] The Framework Directive defines 'electronic communications services' as a service normally provided for remuneration which consists wholly or mainly in the conveyance of signals on electronic communications networks, including telecommunications services and transmission services in networks used for broadcasting, but excluding services providing, or exercising journalistic control over, content transmitted using electronic communications networks and services; Article 2 (c) of the Framework Directive.
[679] Framework Directive, Article 2 (e).
[680] See sections 1.2. and 1.4.3.

the Communications Framework. It is true that the EPG (as any other electronic information agent) also has a purely facilitative transport function by leading consumers to the content they wish to access. This function, however, is subordinate to its real task, namely to provide content, meaning information about content services. Having said that, Recital 2 of the Access Directive states, '[s]ervices providing content [...] are not covered by the common regulatory framework for electronic communications networks and services'. Following this interpretation, one would probably not classify EPGs as an associated facility in the sense of the Communications Framework. The Framework Directive, however, listed the EPG explicitly as an example of an associated facility. It is apparent that facilities, or rather services, such as the EPG that operate at the interface between the transport and the service level can combine technical and content-related aspects. A strict distinction between content and technical facilities not only leads to considerable legal uncertainty, it is also difficult to practice. The Access Directive responds with a kind of 'ostrich' strategy: it simply excludes all content-related aspects of the EPG from the scope of the Communications Framework.[681]

The EPG is not the only borderline case. How about web browsers, programme lists and search engines that fulfil very similar functions? Are they 'associated facilities' even if their main function is to provide consumers with content? DRM solutions are also interesting. DRM protection schemes are designed to provide technical end-to-end protection for content at all levels of the process from the point of initial distribution to the point at which end users view and/or listen to the content. So far, however, the main function of DRM solutions is typically associated with the protection of content once it has been delivered to the consumer. This is a reason not to consider the DRM as an element of the transport level, but to categorize it as part of the service level, and hence as a facility that does not fall under the Communications Framework. However, DRM solutions can have functions and effects very similar to those of conditional access systems. More than providing pure content protection, they involve elements of identification, authorization and enforcement similar to a conditional access system. Often, DRM solutions will even integrate conditional access technologies or elements thereof.[682] Does this mean that DRM systems must be qualified as 'associated facilities' in the sense of the Communications Framework? If DRM systems can be considered associated facilities in the sense of telecommunications law—and there are good reasons to argue in this direction—DRM system operators would face obligations and provisions regarding access, interoperability and consumer protection that apply to conditional access operators.

[681] See, for example, Article 6 (4) of the Access Directive, see also below, section 4.5.

[682] European Commission, Report from the Commission to the Council, the European Parliament and the European Economic and Social Committee on the implementation of Directive 98/84/EC of the European Parliament and of the Council of 20 November 1998 on the legal protection of services based on, and consisting of, conditional access, Brussels 24 April 2003, COM(2003) 198 final [hereinafter 'Report on the Implementation of the Conditional Access Directive'], p. 23.

4.3.3. CONDITIONAL ACCESS (IN THE SENSE OF THE ACCESS DIRECTIVE)—DEFINITIONS

Contrary to the broad, technology-independent definition of telecommunications networks and associated facilities, the definition of conditional access for the purpose of the Access Directive is rather restricted and transmission-medium dependent. It refers to

> 'any technical measures and/or arrangements whereby access to a protected radio or television broadcasting service in intelligible form is made conditional upon subscription or other forms of prior individual authorisation'.[683]

It is interesting to compare this definition to the definition in the Conditional Access Directive that was drafted to protect conditional access against unauthorized circumvention.[684] The Conditional Access Directive acknowledges the fact that conditional access solutions are also used within the context of the delivery of information society services, which is why it defines conditional access as 'any technical measure and/or arrangement whereby access to the protected broadcasting or information society services in an intelligible form is made conditional upon prior individual authorization'.[685] The definition in the Framework Directive does not reflect this.

683 Framework Directive, Article 2 (f).
684 See also sections 1.1 and 2.2.
685 Article 2 (b) Conditional Access Directive.

Not included in the Communications Framework:

Service level
Pay-TV platforms, broadcasting, paid-for content services, interactive applications

Included in the Communications Framework:

Teleservices
Subscriber management services, conditional access, API, EPG, operational support systems

Network and carrier services
Routing, transcontrol, Internet backbone, multiplex

Spectrum, physical infrastructure
Wire and wireless telecommunications network, local loop, cable, satellite, terrestrial and broadband networks

Figure 9—Scope of the Communications Framework. Figure 9 provides an overview of the services and facilities that are covered by the Communications Framework. Services carrying content do not fall under European telecommunications law. At the same time, the technical aspects of the transmission of such services, meaning to the extent that service providers use teleservices, network and carrier services and the spectrum or physical network, are covered by the Communications Framework.

4.4. The Exception: Access to Conditional Access—An Absolute Approach

4.4.1. SCOPE OF ARTICLE 6 OF THE ACCESS DIRECTIVE

Article 6 of the Access Directive gives a clear and exhaustive definition of the subject of its regulation; it exclusively refers to conditional access services for digital television and radio[686] broadcasting services, anticipating the end of analogue broadcasting.[687] The distinction between access to conditional access systems for broadcasters and for non-broadcasters could have far-reaching practical consequences. As will be shown, the regulatory regimes for broadcaster access to

[686] The former Article 4 c of the Standards Directive did not apply to digital radio services.
[687] Access Directive, Article 6 and Annex I; Framework Directive, Article 2 (f).

the conditional access solution under Article 6 and access to the remaining technical bottlenecks under Articles 8 to 13 differ considerably on some points. It is also likely that the divergent regulations will generate very different market outcomes (section 4.7.).

One consequence of the narrow understanding of conditional access in the Access Directive is that it explicitly addresses only selected aspects of conditional access.[688] It does not apply to the other elements of the technical pay-TV platform such as the operating system, the return channel, the set top box memory, the billing and subscriber authorization infrastructure, the EPG or other technical information. The European Parliament suggested in one of its earlier proposals for the directive to include a more general clause that covers, in addition to conditional access, all of the facilities associated with digital television[689] even if they are as far-related functions as the return channel and the storage capacity of decoders. But the proposal was subsequently rejected. Access to APIs and EPGs, however, is regulated in Article 5 (1)b and Annex I, Part 2 of the Access Directive (section 4.5.).

Article 6 of the Access Directive does not apply to conditional access devices that control access to non-broadcasting services, meaning IP-based (webcasting) or individualized telecommunications services that do not fall under the traditional definition of broadcasting[690] even if the signals are transmitted together with the broadcasting signal and are received via the same consumer equipment device. Correspondingly, providers of webcasting, interactive services, e-commerce and similar services that do not fall under the definition of broadcasting do not benefit from access rights, with the effect that a broad range of potential competitors are excluded from the scope of the Access Directive. Here too, the European Parliament suggested to extend Article 6 to include at least interactive services that are an integral part of the TV services delivered to viewers. This proposal did not find its way into the final version of the directive either.[691] Conditional access systems that control access to information society services services will fall, if at all, under another access regime than Article 6 of the Access Directive. However, it remains unclear what this regulatory regime will be. An option that this study will look into in more detail further on, is the applicability of Articles 8 to 13 of the Access Directive (section 4.6.).

The European approach to bottleneck regulation for digital television services highlights the difficulty of the theoretical distinction that is made between

---

[688] See insofar sections 1.2. and 4.3.3.

[689] European Parliament, European Parliament, European Parliament, Recommendation for second reading on the Council common position for adopting a European Parliament and Council directive on access to, and interconnection of, electronic communications networks and associated facilities (Access Directive), 20 November 2001, A5-0434/2001 [hereinafter 'Recommendation for Second Reading Access Directive'], Amendments 9, 10, 26.

[690] As to the discussion of the definition of broadcasting under the influence of new technical developments, see Helberger 1999, pp. 7-8 and 10-13. See also section 4.3.3.

[691] European Parliament, Recommendation for Second Reading Access Directive, Amendment 14.

conditional access solutions for broadcasting services and conditional access solutions for non-broadcasting services. This distinction makes it very difficult to classify services based on new transmission technologies or converging media. At the moment, the development of the service market is greatly affected by the phenomenon of convergence. The associated development of conditional access systems suggests that the future lies in advanced set top boxes that are capable of controlling access to broadcasting as well as to a wide range of interactive service applications.[692] In response, some of the Member States have already moved towards a less technology-dependent approach.[693]

4.4.2. ACCESS OBLIGATION

Article 6, Annex 1, Part 1 (b) of the Access Directive mandates an absolute, unconditioned ex ante access obligation:

> 'All operators of conditional access services… are to offer to all broadcasters, on a fair, reasonable and non-discriminatory basis… technical services enabling broadcasters' digitally transmitted services to be received by viewers or listeners authorised by means of decoders administered by the service operator'.

Article 6 is a behavioural rule that addresses the individual conditional access operator. As a rule, conditional access solution providers are not in a position to freely determine their contracting partners or the terms of access. The mere fact of having control over a conditional access facility triggers an unconditional access obligation—unconditional in the sense that Article 6 does not specify any reasons to legitimately deny access. Contrary to other existing concepts of access to facilities—notably in telecommunications and general competition law[694]—the access obligation in Article 6 of the Access Directive is absolute. The application of Article 6 does not depend on a particular market structure, be it the existence of significant market power, entry obstacles or the level of vertical integration. All conditional access solution operators are obliged to grant access, providing they are not using the conditional access facility exclusively for internal purposes. Article 6 (1) and the Annex are based on the assumption that each conditional access is a potential obstacle to market entry.

Among other things, Chapter 1 illustrated why, from an economic point of view, this is a highly simplified assumption and that certain market conditions must be fulfilled before an enterprise can and will find it profitable to use monopoly control over a conditional access to hinder entry into the pay-TV market.[695] Moreover, certain welfare and general interest arguments could justify a certain behaviour as

[692] See sections 1.2., 1.4.1. and 1.4.3.
[693] For example the UK: Oftel (now: Ofcom) 1997.
[694] See 3.4.1.
[695] See section 1.5.3.

well as monopoly control over a conditional access facility. Article 6 of the Access Directive, however, does not leave room for NRAs to consider such aspects.

It should be noted, however, that unlike the former version of Article 4c of the Standards Directive, Article 6 (3) of the Access Directive entitles Member States to withdraw or amend access obligations for operators that lack significant market power. In the event that a Member State chooses to permit its NRA to impose access obligations only on enterprises with significant market power, the NRA has to perform a market analysis according to Article 15 and 16 of the Framework Directive to define the relevant conditional access market. So far, the Commission, in its recommendation on relevant product and service markets, has only identified a

> 'market for wholesale ancillary technical broadcasting services across all relevant transmission platforms, unless specific national situations in respect of switching costs and available transmission platforms justify a narrower market definition'.[696]

The Commission has not yet defined a wholesale market for conditional access or associated facilities, because, as the Commission argues, Articles 5 and 6 leave it for Member States to determine whether to place access obligations only on conditional access operators with significant market power. Only in the event that Member States decide to restrict access obligations to significant-market-power operators, must NRAs perform a market analysis. [697]

In the Explanatory Memorandum to the Recommendation on Relevant Markets, the European Commission points out a number of aspects that can be taken into account when performing the market analysis. For example, one aspect is that enterprises seeking access to ancillary technical broadcasting services may be interested in delivering or negotiating access to a sufficiently high number of end users to sustain a viable business rather than accessing all delivery platforms or all possible end users.[698] Hence, the relevant market does not have to consist of the whole sector for broadcasting or subscription services, but could consist of one pay-TV platform and its installed consumer base. The Commission draws attention to the fact that, in a perfectly convergent environment, consumers are, in principle, able to switch between different platforms and that in this case, the market would have to be defined with the corresponding broadness. On the other hand, where the switching costs are high, the conditional access that is associated with one particular platform would eventually have to be regarded as a separate market.[699] For

---

696 European Commission, Recommendation on Relevant Product and Service Markets, Explanatory Memorandum, p. 38.

697 See also Oftel (now: Ofcom) 2002, paragraph 45.

698 European Commission, Recommendation on Relevant Product and Service Markets, Explanatory Memorandum, p. 38.

699 European Commission, Recommendation on Relevant Product and Service Markets, Explanatory Memorandum, p. 38.

example, in Member States in which several pay-TV platforms compete with each other there are two possible scenarios. First, switching costs for consumers could be low, for example, because consumers can use the same set top box for all services and because consumers can choose according to an à la carte model rather than from a small number of large bundles.[700] In this case, competing broadcasters could find substitute conditional access solutions on other platforms. The market would have to be defined with the appropriate broadness with the consequence that perhaps no pay-TV platform would have a monopoly position as far as conditional access is concerned. In areas in which the switching costs are high, however, the different conditional access solutions for the different platforms could not be considered substitutes. In this case, the conditional access system of one particular platform could constitute a market on its own, and the operator of that conditional access would be the operator with significant market power. This discussion is comparable to the discussion about the definition of mobile telephony markets: does the relevant market consist of one particular operator's mobile telephony network and all of its subscribers, or of the network of all mobile telephony operators together?[701]

Even if NRAs performed a market analysis and found that a particular conditional access operator did a) not have significant market power and that therefore, b) no access obligation should apply, NRAs would still have to observe certain conditions before they could withdraw or amend an access obligation. An access obligation can only be amended or withdrawn if doing so does not have a negative effect on the end users' ability to access broadcasting services, or on the prospect of effective competition for retail broadcasting services and/or conditional access or other associated services.[702] Parties affected by such an amendment would be given an appropriate period of notice so that they can bring forward and discuss potential concerns with the NRA. Taking a closer look, it seems that the requirement to not affect the end users' accessibility might be rather difficult to apply in practice. Obviously, if a third-party broadcaster does not have the right to access a particular conditional access system, subscribers to the platform using that system would not be able to watch the broadcaster's programme; at least not via the pay-TV platform they are subscribed to. The rival broadcaster would still have the option of operating its own system, using the system of another pay-TV platform in that market or delivering the services in free-TV format,[703] Of course, it is possible that the rival decides that it is not attractive or viable to deliver in free-TV format or to install its own conditional access system. In this case, consumers would not be able to access the rival's services. The end users' accessibility would be affected,

---

[700] See sections 1.4.3., 1.5.2. and 1.5.3.

[701] European Commission, Commission Guidelines on Market Analysis and the Assessment of Significant Market Power, paragraphs 68-69. See also section 3.4.1.

[702] Articles 6 (3)a and b.

[703] Were there no alternative system, the first platform would be the only pay-TV platform and would have a monopoly position, meaning that Article 6 (3) of the Access Directive would not apply.

with the consequence that the access obligation could not be withdrawn according to Article 6 (3) of the Access Directive. This would mean that the permissibility of withdrawing an access obligation would depend on circumstances within the sphere of a competitor, namely its willingness or ability to operate its own conditional access system. The same would be true if there were several conditional access platform operators, none of which have significant market power, but all of which refuse the broadcaster access to their conditional access systems. In this case, no platform would benefit from a relaxation of the access obligations because of the uncooperative behaviour of the others.

Moreover, withdrawing the access obligation would change the form of competition between pay-TV operators. Two operators who distribute their programmes through the same pay-TV platform would be competing on one platform, the so-called intra-platform competition. Two operators who distribute their services through different platforms would be competing between platforms, meaning compete in inter-platform competition. It is very difficult to say what the effects of both forms of competition would have on the conditional access system or the pay-TV sector, and whether they would have a positive or negative effect on competition and consumer welfare.[704] For example, a broadcaster's inability to access a particular platform could be an incentive to develop its own platform. In other words, withdrawing an access obligation could have a negative impact on the prospects of competition within one platform, but it could stimulate competition between different pay-TV platforms, technical innovation and investment, as well as the development of new and more attractive service offerings. In conclusion, it is questionable if, and if yes, how, Article 6 (3) of the Access Directive can be applied in practice.

One improvement that Article 6 of the Access Directive brings compared to the former situation under the Standards Directive is that Article 6 acknowledges that it is not adequate to impose access obligations where alternative facilities are in principle available, or where it could be expected that competitors undertake adequate efforts to develop an alternative themselves. This was demonstrated in Chapter 3 when discussing the applicability of the Essential Facilities Doctrine.[705] Compared to the former Article 4c of the Standards Directive, the revised version of the access obligation in Article 6, Annex I of the Access Directive is narrower in scope. It stipulates that the access obligations only concern conditional access systems upon 'whose access services broadcasters depend on to reach any group of potential viewers'.[706] This addendum approximates the concept of sector-specific conditional access regulation and the concept of access obligations under general competition law. For conditional access operators, this could mean that once alternative conditional access systems are offered, they could deny access with the

[704] See section 3.3.1.
[705] See section 3.4.1.
[706] Access Directive, Article 6, Annex I Part I 1 (b).

argument that broadcasters can switch to another system. The exact scope of the provision, however, remains unclear. The provision recalls the jurisdiction of the European Court of Justice in essential facilities cases.[707] Interpreted within the framework of the Court's judgement in the Bronner case, Article 6, Annex 1 of the Access Directive could be understood in a sense that broadcasters would have to accept less favourable solutions or even undertake adequate efforts to establish alternative solutions themselves.[708] Within the framework of the Bronner judgement, a conditional access operator would not be subject to the access obligation unless the broadcaster succeeded to show that it was not economically viable for him to install an alternative distribution scheme. Moreover, the individual incapacity to install such a system would not justify the imposition of an access obligation as long it was not impossible or unreasonable for any other broadcaster to implement its own conditional access system.

4.4.3. INTEROPERABILITY

Another issue that arises in addition to the question of open access to a third party's conditional access facility, is that of interoperability. Interoperability is one of the issues that clearly demonstrate that the Communications Framework combines matters of competition and consumer protection policy while promoting general public information policy objectives, such as the free flow of information and media pluralism:

> 'Interoperability of digital interactive television services and enhanced digital television equipment, at the level of the consumer, should be encouraged in order to ensure the free flow of information, media pluralism and cultural diversity. It is desirable for consumers to have the capability of receiving, regardless of the transmission mode, all digital interactive television services, having regard to technological neutrality, future technological progress, the need to promote the take-up of digital television, and the state of competition in the markets for digital television services'.[709]

The Communications Framework does not define the notion of interoperability. It only defines interconnection in the sense of the physical and logical linking of public telecommunications networks.[710]

'Interoperability' is commonly used to refer more broadly to the

[707] See section 3.4.1.

[708] See European Court of Justice, Bronner, saying that for the Essential Facilities Doctrine to apply neither would it be sufficient that it is not economically viable for one particular operator to install an alternative distribution scheme himself, nor would individual incapacity generate an access obligation, as long as it is not impossible or unreasonable for any other broadcaster to establish alternative facilities, paragraphs 44-46.

[709] Framework Directive, Recital 31; Article 18 (1). Also: Access Directive, Recital 9.

[710] See Access Directive, Article 2 (b).

'[c]apability to provide successful communications between end-users across a mixed environment of different domains, networks, facilities, equipment, etc. from different manufacturers and (or) providers. In this context the communication is meant between end-users or between an end-user and a service provider'.[711]

Making services or different pay-TV platforms interoperable means creating a 'harmonized' environment in which different components are able to interoperate in a way that consumers are technically able to receive services from different providers using different technical facilities and technical standards.[712] For pay-TV, interoperability primarily means that consumers are able to receive, with their reception equipment, services from different operators, irrespective of the software they use, be it the encryption software, be it the software with which applications are written and transmitted. In this respect, it is less the aspect of physical access or the linking of different networks, but rather questions of standardization of technical facilities at the higher level of the technical transport chain that are paramount. More specifically, this is first and foremost the aspect of standardization of such technical elements that are implemented in the consumers' equipment, notably the set top box.

Standardization can be achieved in different ways: by mandating one particular standard (the HD-MAC or the Eurocrypt standard, for example), by stimulating the adoption of open standards (for example, open source software and middleware), by mandating a common interface (for example, MHP for the API) or by ordering enterprises to make their services or facilities compatible with each other (for example, interconnection obligations for public telephone networks).[713] The public promotion of open standards[714] is also an attempt to stimulate interoperability. Different examples of these different ways of achieving standardization and, more generally, the way standardization is approached in the pay-TV sector, will be provided in the following paragraphs.

In the next paragraphs, three different aspects of interoperability as regards conditional access platforms will be looked at more closely: the first aspect is interoperability between two different conditional access systems so that the consumer can be reached through one and the same set top box. The second aspect of interoperability concerns the interoperability of a set top box's middleware, or

[711] ETSI User Group STF228 Progress Report, Sophia-Antipolis, 9 February 2003, available at <portal.etsi.org> (last visited on 14 March 2005). Closely related is the notion of 'compatibility', which is often used to refer to the compliance of a facility with common technological specifications. Scheuer/Knopp 2003, p. 3.

[712] See Access Directive, Recital 9.

[713] For an in-depth discussion of policy options to realize interoperability, see the study prepared for the European Commission by Oxera 2003; also the study by Contest Consultancy 2003.

[714] In the context of this study, open standard is understood as a standard for which the specifications are available to third parties to use, irrespective of whether this is done for free or for a licence fee, or whether the standard itself is proprietary or non-proprietary.

more specifically, the API. Here, interoperability means that a set top box's API can process its own applications as well as those provided by unaffiliated third-party service providers. This second aspect has received by far the most attention during the revision of the conditional access provisions. API interoperability is considered to be particularly important in conjunction with the processing of more advanced, interactive digital services.[715] The third and last aspect of standardization, the standardization of consumer equipment for the reception of digital television signals in general (for example, the TV), is not dealt with in the Access Directive, but in the Universal Service Directive.

The Standardization of Consumer Equipment

Article 24, Annex VI of the Universal Service Directive deals with the standardization of consumer equipment for the reception of digital television signals in general. According to Article 24 of the Universal Service Directive, 'any digital television set' intended for the reception of digital television signals is to be fitted with at least one open interface socket to permit the connection of, for example, smart cards from pay-TV operators. Digital television sets should be able to pass all the elements of a digital television signal, including information relating to interactive and access-controlled services.[716] As a result of Article 24, Annex VI of the Universal Service Directive, it should be possible to connect set top boxes to any digital television set in Europe.[717] Note that the provision seeks to introduce a common interface—a solution that has also been discussed in context with the conditional access—for digital television sets so that set top boxes and other devices can be attached. It does not impose any obligation on set top box manufacturers to implement a common interface so that several different conditional access systems can be linked and made interoperable. Moreover, it does not require set top box manufacturers to provide for a common API standard.

Interoperability Between two Different Conditional Access Systems

Interoperability between competing conditional access platforms might be of even more practical importance than the 'access to the decoder' question. This is due to the fact that the majority of access-controlled service providers that are active on the European markets are affiliated with relatively large commercial content service providers who use a conditional access system to market their own services.[718] Where an access-controlled service provider intends to operate its own conditional access platform, one main obstacle to market entry is interoperability with the established system. The more popular the established standard and the stronger the indirect network effects, the more difficult can it be for newcomers to get a foothold

---

715 See also section 1.4.1.

716 Universal Service Directive, Annex VI, paragraph 2.

717 Universal Service Directive, Recitals 32 and 33.

718 See section 1.4.3.

in the market and convince consumers and third-party (content) service providers to switch to a new, non-interoperable conditional access platform.[719]

Over the past years, a number of solutions to make rivalling conditional access systems interoperable in a way that all service providers could access consumers via one set top box were discussed, including:

- Mandating one common standard for all conditional access services.
- Ensuring that all operators can use the same encryption system, or
- Designing sufficiently open boxes to allow users of other conditional access technologies to dock.[720]

After the failure of Eurocrypt,[721] another serious attempt was never made at the European level to promote one common conditional access standard. It is likely that a common standard would have soon been outdated due to the developments in the conditional access technology sector. Within the context of the Communications Framework, standardization is considered a process that should remain a primarily market-driven process. The Communications Framework refrains from promoting explicitly one particular conditional access standard. Taking a closer look at Article 6 and Annex I, Part 1, however, one could also argue that the Access Directive indirectly continues to support the establishment of one common conditional access standard. Article 6 encourages all broadcasters to use the established conditional access system. Consequently, for broadcasters using the existing conditional access system there is no need to establish a second one. As a practical result, Article 6 promotes a kind of common standard for the conditional access system itself, namely, the standard of the first mover on the market that succeeds in establishing its conditional access system. Once an operator succeeds in establishing a conditional access system that operates with economic efficiency, it is likely that this system will evolve into a de facto standard.[722]

The DVB consortium proposed the Simulcrypt and Multicrypt solutions as possible standards the industry could agree on. Both the Simulcrypt and Multicrypt solutions seek to make one and the same set top box fit to receive a choice of services using different conditional access technologies. The DVB consortium defines Simulcrypt as an

719 See the in-depth discussion in 1.5.2. and 1.5.3.

720 See also European Commission, Commission Staff Working Document on Barriers to widespread access to new services and applications of the information society through platforms in digital television and third generation mobile communications, 00.00.00, COM (0000)000, Brussels, [hereinafter 'Commission Staff Working Document on Barriers to Widespread Access'], available at <europa.eu.int/information_society/topics/telecoms/regulatory/publiconsult/documents/211_29_en.pdf> (last visited 20 March 2005), p. 9: '[O]peness (…) and interoperability can be achieved either by choice, by design or by law'.

721 See section 4.2.1.

722 See sections 1.5.2. and 1.5.3.

'architecture that allows a service to be transmitted with the entitlement messages for multiple CA systems. A decoder supporting a particular CA system can extract the relevant entitlement messages and ignore the others'.[723]

Simulcrypt solutions require an agreement between different conditional access operators to use one particular conditional access system. The operator of the selected system agrees to process signals that are protected by another conditional access solution. In contrast, the Multicrypt solution is a built-in solution, meaning that the set top box is fitted with a common interface so that consumers can insert a number of security modules that belong to different conditional access solutions.[724] So far, however, industry-driven interoperability solutions have not been very successful.[725] Only a small number of systems have a common interface that makes systems interoperable. To the extent that interoperability solutions exist at all, they are predominantly Simulcrypt solutions.[726]

The only provision concerning the interoperability between two different conditional access systems in the Access Directive can be found in Annex I, Part 1 (c) of the directive. Here, the Access Directive stipulates that the holders of intellectual property rights to conditional access products and systems must ensure that licences are granted to consumer equipment manufacturers on fair, reasonable and non-discriminatory terms. More importantly, holders of such rights may not subject the granting of licences to conditions that prohibit, deter or discourage the inclusion in the same product of a common interface allowing the connection with several other conditional access systems. Consumer equipment manufactures that intend to implement a Multicrypt solution should be able to do so. Whether they choose to do so is entirely up to them. Moreover, contracts covering the licensing of conditional access technology may not prohibit, deter or discourage the implementation of means that are specific to another access system under the precondition that the licensee complies with relevant and reasonable conditions concerning the security of transactions of conditional access system operators.

When one of the Member States, Spain, sought to go beyond the European Framework and mandate a particular interoperability solution, the European Court of Justice found this behaviour incompatible with Internal Market principles. The Spanish government had introduced a compulsory licensing regime for conditional

---

723 DVB, The DVB Glossary, available at <www.dvb.org/documents/dvb_glossary.pdf> (last visited on 20 March 2005).

724 DVB, The DVB Glossary, available at http://www.dvb.org/documents/dvb_glossary.pdf (last visited on 20 March).

725 See, for example, Article 24, Annex VI of the Universal Service Directive. See also section 1.4.3.

726 See European Commission, The Development of the Market for Digital Television in the European Union, Report in context of Directive 95/47/EC of the European Parliament and of the Council of 24th October 1995 on the use of standards for the transmission of television signals, 9 November 1999, COM(1999)540 final [hereinafter 'Standards Directive Implementation Report'], pp. 18-20 and country reports.

access operators.[727] License registration required that the conditional access system complied with particular technical specifications and the Multicrypt standard. The Court ruled that national registration requirements would be a restriction to the freedoms guaranteed under Articles 28 and 49 of the EC Treaty. According to the court, those restrictions were neither necessary nor justified in attaining the objective, which is, according to the Spanish government, mainly to increase transparency and transpose the Standards Directive.[728] Unfortunately, the court did not consider whether mandating a particular standard could have justified the regulation for public interest reasons.[729] The Court simply stated that it is 'incompatible in principle with the freedom to provide services to make a provider subject to restrictions for safeguarding the public interest in so far as that interest is already safeguarded by the rules', referring to the Standards Directive, which, however, does not foresee any interoperability obligations.[730]

Interoperability between Interactive Digital Services and the API or Operating System in the Consumer Device

Of the different ways of promoting API interoperability,[731] the most widely propagated is probably the use of an open standard. This is amplified by Article 18 (1) of the Framework Directive, which speaks in favour of an open API standard. For the API, the DVB consortium promoted the MHP standard.[732] The MHP standard defines a generic middleware, namely an interface between interactive or enhanced digital applications and the terminals on which those applications are executed. It decouples the providers' applications from the hardware and software used for the set top box and the conditional access system. In so doing, it introduces an additional open operational layer that is principally independent of the underlying hardware. Because this additional layer is open to everyone, services can be run in different application environments.[733] This means that a service provider who wants to write applications for a particular set top box environment does not need to know the specifications of the set top box, which operating system it runs or the programming language used. Instead, the provider only has to know the API specifications. Migration to the MHP standard, however, is still far from becoming reality. A number of important facility controllers do not support the MHP standard

---

[727] Royal Decree-Law No. 1 of 1997 and Royal Decree No. 136 of 1997 transposing into Spanish law the Standards Directive, Boletin Oficial del Estado No. 28, of 1 February 1997, p. 3174.

[728] European Court of Justice, Judgment of the Court of 22 January 2002, Canal Satélite Digital SL v Adminstración General del Estado, and Distribuidora de Televisión Digital SA (DTS), Case C-390/99, European Court reports 2002, p. I-607 [hereinafter 'Canal Satélite Digital'], paragraph 43.

[729] European Court of Justice, Canal Satellite Digital, paragraphs 33-43.

[730] European Court of Justice, Canal Satellite Digital, paragraph 38.

[731] For an in-depth discussion of different solutions to realize API interoperability, see the study prepared for the European Commission by Oxera 2003.

[732] To view the progress of the MHP negotiations, visit <www.mhp.org> (last visited on 20 March 2005).

[733] See for more details <www.mhp.org> (last visited on 20 March 2005).

and prefer their own standards. Moreover, MHP is not the only standard that is negotiated by industry players.

An interesting alternative, at least in terms of compatibility between applications and the middleware used in the hardware elements in a conditional access system's set top box, is open source software.[734] Providing the licensing conditions allow for an API to be developed as open source software, each broadcaster, application provider, etc., would have access to the source code and would be free to adapt applications and even write its own applications.[735] In addition, open source software eliminates licence fees, which would lower the costs of market entry and service development/provision. Ideally, applications written with open source software would also be compatible with each other and enable the realization of indirect network effects.

API interoperability is high on the political agenda in the European Union. Recital 31 of the Framework Directive leaves no doubt that the interoperability of digital interactive television services and enhanced digital television equipment should be encouraged to ensure the 'free flow of information, media pluralism and cultural diversity'. This is also explained by the need for technological neutrality and technological progress, and the need to promote digital television and competition in the digital television service markets.[736] Accordingly, Article 18 (1) of the Framework Directive states:

> 'In order to promote the free flow of information, media pluralism and cultural diversity, Member States shall encourage ... providers of digital interactive TV services for distribution to the public in the Community on digital interactive TV service platforms, regardless of the transmission mode, to use an open API; providers of all enhanced digital TV equipment deployed for the reception of digital interactive television services on interactive digital broadcasting platforms to comply with an open API in accordance with the minimum requirements of the relevant standards or specifications'.

Article 18 (2) of the Framework Directive also calls upon Member States to encourage API proprietors to make available on fair, reasonable and non-discriminatory terms, and against appropriate remuneration, all such information as is necessary to enable providers of digital interactive television services to provide services for that particular API.

At the time of writing, the issue of compatibility between digital interactive television services and enhanced digital television equipment had not given rise to more than declaratory, non-obligatory statements such as those mentioned in

---

[734] See also European Commission, Communication on Barriers to Widespread Access, p. 11. Initiatives to design conditional access solutions with open source software can be found with major players, such as Nokia, IBM, Sun, Intel. See also Koelman 2000, 151pp.

[735] The conditional access operator could then have to ensure that the individual applications are compatible with each other, for example by issuing a 'compatibility certificate'.

[736] Framework Directive, Recital 31.

Article 18 (1) and (2) of the Framework Directive. It has been argued that mandating one particular approach to compatibility could hamper technological and market development by imposing common standards prematurely.[737] Instead, the development and implementation of common standards has been left entirely up to industry initiatives, such as the work of the DVD consortium, despite earlier proposals of the European Parliament to oblige operators to use a single open, interoperable API that is standardized by a recognized European standardization body.[738]

Having said that, the Framework Directive clearly states that certain situations may make it necessary to enforce compliance with specified standards to ensure interoperability in the Internal Market and freedom of choice for consumers.[739] More detailed conditions are laid down in Article 17 (2)-(7) of the Framework Directive. For this purpose, the Communications Framework required examining the extent to which compatibility and freedom of choice have been achieved in Member States by no later than July 2004. Had these objectives not been adequately met, the Commission was entitled to enforce a previously published standard after consulting with the public and obtaining the agreement of the Member States as is laid out in Articles 18 (3) and 17 (3) and (4) of the Framework Directive. In this context, the MHP standard was included in the List of Standards and Specifications to be published in the Official Journal under Article 17 of the Framework Directive.[740]

The first consultation on API compatibility took place in the spring of 2004. The European Commission initiated this first consultation with the goal to examine the effectiveness of the Framework and determine whether compatibility and freedom of choice for users had been achieved. The European Commission invited market players and other interested parties to respond to a previously published Commission Staff Working Paper on the Compatibility of Digital Interactive Television Services.[741] The result of the consultation was summarized in a communication from the Commission.[742] Based on the contributions received from more than fifty-one entities, including manufacturers, network operators, broadcasters, API providers and consumer associations, the European Commission concluded that there was at that time no clear case for mandating standards. One reason that has lead to this conclusion was the Member States' belated implementation of the Communications Framework and the resulting lack of

---

737 European Commission, Communication on Barriers to Widespread Access, p. 10. Framework Directive, Recital 30.

738 European Parliament, Recommendation for Second Reading Access Directive, Amendment 25.

739 Framework Directive, Recital 30, Article 17 (3).

740 European Commission, List of Standards and Specifications.

741 SEC(2004)346, Brussels, March 2004.

742 European Commission, Communication from the Commission to the Council, the European Parliament, the European Economic and Social Committee and the Committee of the Regions on interoperability of digital interactive television services, Brussels, 30 July 2004, COM(2004)541 final (hereinafter 'Communication on Interoperability of Digital Television Services'].

practical experience. Another and more fundamental reason, was the difficulty agreeing on what interoperability really means and whether it had been achieved. One group defended the view that interoperability had not been achieved, because interoperability in the sense of Article 18 of the Framework Directive would be best interpreted in the sense of open, non-proprietary standards (for example, the MHP standard). This view was represented by, among others, free-TV, and in particular public broadcasters that had an interest in free market access, unhampered by proprietary standards and incompatibilities. In contrast, a second group interpreted interoperability result-oriented, meaning in the sense of the availability of the same interactive services on different distribution platforms. Due to, for example, technical solutions that support the portability of interactive applications across different platforms, interoperability would have been achieved. Consequently, the European Commission did not have to interfere. The latter opinion was represented by infrastructure operators, among others. The Commission decided that it would not interfere, and only suggested a number of supportive measures, including the establishment of a workgroup, the legal certainty of public subsidies for consumer equipment, the extension of the list of standards published in the Official Journal, and the monitoring of access opportunities to proprietary technologies.

The outcome of this first consultation shows that the challenges of effective API standardization consist of more than just finding ways to encourage the industry to agree on one API standard. Reaching a consensus on what interoperability actually means could postpone the standardization process indefinitely. Moreover, one must also wonder whether it would not have been justified for the Commission to interpret the lack of agreement as an indication of the need to undertake more pro-active measures to promote API standardization. Article 18 (1) of the Framework Directive does not leave much doubt that interoperability in the sense of the directive refers to the proliferation of open API standards and not to proprietary standards and possible portability solutions.[743]

4.4.4. Terms and Conditions

It is principally left up to the parties to negotiate and define the details of access, meaning the conditions under which it is to be granted, including the price, the beginning and the duration of access, the scope and related matters such as the confidentiality and protection of consumer data, the handling of security questions, dispute settlement, etc. The contracting parties in the sense of Article 6 of the Access Directive are the conditional access operator and the broadcaster, not consumers (subscribers).

It is reasonable to assume that equally strong negotiating parties will negotiate terms that both parties believe are fair, reasonable and non-discriminatory. This is different in markets in which there are still big differences in negotiating power

743 See also Framework Directive, Recital 31.

between enterprises. An example is the pay-TV sector, where smaller broadcasters might find it too costly to establish their own conditional access solution, but wish to have access to the first mover's conditional access platform.[744] Even if the latter agreed to the use of its conditional access system, the provision of access on unfair terms and conditions could have the same effect as simply refusing access. Tying practices, unfair pricing, the refusal to supply information or access to ancillary facilities or services, or simply lengthy procedures are typical forms of abuse of monopoly control.[745]

In response, Article 6 of the Access Directive stipulates that access must be provided on a fair, reasonable and non-discriminatory basis, and must be compatible with European competition law. This is as concrete as the directive gets. Article 6 does not provide for any ex ante guidelines that would outline the scope of the actual access obligation, nor does it provide for accompanying ex ante measures (apart from an obligation of accounting separation) that would help make the sector more transparent and facilitate enforcement. Additionally, the Article does not envisage any ex ante price control for access to the conditional access, nor does it regulate the question of how prices are calculated and which principles may legally influence the price calculation. There are a few other points that Article 6 does not touch on, such as the boundaries of the obligation to share one's facilities, and if the notion of 'fair and reasonable terms' leaves room to acknowledge investments, economic risks and limited technical capacities of the conditional access operator. It is therefore up to the Member States to adopt guidelines that are more detailed.[746] The following paragraphs attempt to shed some light on how these guidelines would have to look if they were drafted according to existing European law and principles.

Fair and Reasonable Terms

The appropriateness of access conditions is generally more difficult to prove in newly emerging markets where there is a lack of market information, reference data or comparable products and services. For the time being, the final assessment of the legitimacy of single conditions in access agreements in the context of Article 6 of the Access Directive is left to the courts, and possibly also to NRAs, on an ex post

[744] See sections 1.5.2. and 1.5.3.

[745] See sections 1.4.3. and 1.5.3., as well as 3.4.3. See also Schulz/Kühlers 2000, p. 60, 108, who suggest the adoption of sector-specific anti-tying rules also for conditional access and associated services and facilities.

[746] Ofcom (then: Oftel) was one of the first NRAs to issue elaborate pricing principles for conditional access and publish them as common guidelines. The guidelines are flexible enough to allow pricing at different levels for different categories of broadcasters (for example, free-TV and pay-TV broadcasters or providers of interactive services) and open enough to allow economically efficient price strategies that maximize the usage of the system, while at the same time ensuring that these strategies do not have significant adverse effects on downstream markets, Oftel 1997, 1997a, 1998 and 2000.

basis.[747] It should be noted that this determination could be difficult in practice, in particular when the practical problems of identifying discriminatory practices as described in Chapter 3 are taken into account, and more particularly those that do not amount to a clear denial of access.[748] The situation is rendered even more difficult by the often intransparent competitive environment that is characterized by various horizontal and vertical links between the market players.

Guidelines for the interpretation of the notion of 'fair, reasonable and non-discriminatory' can be found, if at all, only outside the provision of Article 6 of the Access Directive. It was not in the Access Directive, but in the European Commission's Access Notice that the Commission provides an interpretation of the notion of 'fair and reasonable' access.[749] The Access Notice interprets 'fair and reasonable' access to mean that facility providers may not unduly press broadcasters to purchase a bundled package of services nor refuse to provide separate services at less than the cost of the bundled package, hinder the exercise of single functions or even make the conclusion of the agreement subject to the acceptance of services that are not directly linked to the actual service through contractual provisions or discounts.[750]

According to this interpretation, 'fair and reasonable access' refers to unbundled access. For example, a service provider must be able to gain access to the conditional access platform without being bound to using the operator's EPG or to being marketed under the operator's brand name. In a broader sense, conditions may not a) influence or even impede the way in which parties exercise their business or b) serve the interests of the conditional access provider if they are not directly related to the business of providing access to conditional access systems. Ofcom has interpreted this to mean

> 'a broadcaster should not be put at a competitive disadvantage by using the conditional access operator's services. This applies especially where an associated business of the conditional access operator is competing with the broadcaster in a downstream market'.[751]

---

[747] As to the character and tasks of NRAs, see Articles 3 to 13 and Recitals 11 to 18 of the Framework Directive. For telecommunications networks and facilities, national NRAs can also be required to resolve actual access disputes between enterprises in the same Member State and, if no mutual compromise can be found, impose an adequate solution on the parties, Recitals 32, 33 of the Framework Directive. However, Article 6 of the Access Directive does not contain any corresponding explicit authorization for the NRAs to resolve access disputes.

[748] See section 3.4.2.

[749] The Access Notice is not restricted to access problems in the ONP framework. Instead it is intended to be more generally applicable to other types of access issues, and arguably conditional access, European Commission, Access Notice, paragraph 6: 'As this notice is based on the generally applicable competition rules, the principles set out in this Notice will, to the extent that comparable problems arise, be equally applicable in other areas, such as access issues in digital communications sectors generally'.

[750] European Commission, Access Notice, paragraph 103.

[751] Oftel (now: Ofcom) 1997a, paragraph A 41.

The Commission provided a similar explanation in one of its merger cases. Here, the Commission defined 'non-discriminatory access' in the sense that the licensee of the decoder technology is able to conduct business without being influenced by the technology controller.[752]

One major problem in this sector is unfair pricing, which can be facilitated by a high level of vertical integration. Unfair pricing for access to a dominant operator's facilities consists often of excessively high or discriminatory prices.[753] In practice, it is very difficult to determine what appropriate access prices are. Access prices must be determined in a way that it is economically viable for the regulated enterprise to operate the service and invest in its maintenance and innovation. Access prices must also remain at a level that neither prevents nor discourages the regulated enterprise from offering quality services to consumers. On the other hand, the price to access a conditional access platform must remain affordable to competing operators. The price to access the technical platform can also be influenced by the price that is charged to the subscribers of the pay-TV service. The price paid by broadcasters to access the technical platform and the price paid by consumers to access the pay-TV service are both parameters that the platform operator must take into account to define its pricing model.[754] Another question is whether NRAs are entitled to take welfare issues into account, meaning that prices should not be so high that competitors are forced to offer their services to consumers at unaffordable prices.

Potential anti-competitive practices do not necessarily relate only to technical aspects. Where a vertically integrated conditional access operator operates the service platform at the same time, the terms and conditions can also contain conditions that refer to content-related aspects or aspects that are related to the marketing of content. Examples are a contractual condition that requires broadcasters to agree to being carried via a particular programme bouquet or to adapting his programme format, to disclose its customer database, or to agree to an exclusive relationship and refrain from offering the programme to other service platforms. The access-controlled platform operator may wish to adopt a particular format for all programmes and services delivered via its platform, to admit only popular programmes or programmes that fit into its editorial strategy, or structure the offer in different thematic bundles, etc. Are such considerations relevant when assessing whether conditions are 'fair, reasonable and non-discriminatory', and are telecommunications NRAs, whose competency only lies in the telecommunications sector, entitled to decide whether such practices are justified?

There are valid reasons to doubt this. The European Commission repeatedly stresses the need to 'separate the regulation of transmission from the regulation of

---

[752] European Commission, Bertelsmann/Kirch/Premiere, paragraph 111. Interestingly, the Commission also noted in this context that such influence could be exercised if the technology is controlled by enterprises that also have interests as programme suppliers, thereby drawing attention to the problem of vertical integration.

[753] European Commission, Access Notice, paragraphs 105-106.

[754] See section 1.4.3. Evans 2003, 47pp and 64pp.

content'.[755] Services providing packages of sound or television broadcasting content are not covered by the Communications Framework.[756] Content-related questions are, according to Recital 5 of the Framework Directive, covered by the TWF Directive. The Framework Directive only states in very general terms that the 'separation between the regulation of transmission and the regulation of content does not prejudice the taking into account of the links existing between them'.[757]

On the other hand, the European Commission also indicated that the access obligation in Article 6 of the Access Directive should be sufficient to prevent any likely spill-over or leverage effects in the pay-TV market.[758] In practice, this could mean that telecommunications NRAs will, in future, have to consider content-related aspects providing they are relevant for the realization of open access to the conditional access platform. Bearing in mind that the technical conditional access platform is closely integrated with the marketing platform for pay-TV content, this is probably also the interpretation that best takes the realities of the pay-TV sector into account.[759]

One must bear in mind, however, that in most Member States media supervision is still organized in separate divisions and according to a distinction between transport and content aspects. Most Member States still distinguish between broadcasting and telecommunications NRAs. With the advancing economic and technological convergence of the transport and service areas, effective supervision requires that regulatory authorities in both areas step out of their traditional field of expertise and authority. Some Member States, such as the UK (Ofcom), Italy (AGCOM) and Austria (KommAustria), have already drawn the consequences and merged their regulatory authorities for the telecommunications and broadcasting sectors. In other Member States, such as Germany, doing this could raise complex constitutional issues due to the split competences between *Bund* and *Länder*.

Non-Discriminatory Terms

As explained in Recital 11 of the Access Directive, the principle of non-discrimination is also meant to ensure that enterprises with market power do not distort competition, in particular when such enterprises are vertically integrated and supply services to competitors with whom they compete in related markets.[760] Article 10 (2) of the Access Directive defines (for access to telecommunications networks, services and associated facilities) non-discrimination as follows:

> 'Obligations for non-discrimination shall ensure that the undertaking applies similar conditions in similar circumstances to other undertakings providing

---

[755] Framework Directive, Recital 5.
[756] Universal Service Directive, Recital 45. Critical section 4.3.2.
[757] Framework Directive, Recital 5.
[758] European Commission, Recommendation on Relevant Product and Service Markets, Explanatory Memorandum, p. 37.
[759] See sections 1.2. and 1.4.3.
[760] See also European Commission, Bertelsmann/Kirch/Premiere, paragraph 111.

similar services, and provides services and information to others under the same conditions and of the same quality as they provide for their own services, or those of their subsidiaries or partners'.

This somewhat cryptic definition means that it is prohibited to treat third parties differently than one's own or associated services unless there is an objective justification for doing so. For questions of access to a conditional access system, this could mean that a platform operator who offers access to its own or affiliated services must offer access to rivals at the same conditions providing the services in question are comparable. However, a number of practical questions remain unanswered, such as: when are services comparable? Do providers of pay-TV, free-TV, interactive-TV, special interest channels, foreign channels, etc., all fall under the same category? Does the transmission medium play a role? Do services transmitted through an IP protocol fall under a different category than services transmitted through 'traditional' means of broadcasting, such as satellite or cable? And again, are content-related arguments relevant? Are platform operators entitled to base their decision to grant access on the popularity of a third party's programming content or on how well this content fits into their own editorial arrangement? Can operators ask a higher access price for programmes that are probably more difficult to sell to consumers?

Terms and Conditions—Retail

In Chapter 1, we saw that for pay-TV markets to be competitive not only the conditions under which rival broadcasters obtain access to a technical platform matter, but that the terms and conditions in subscription contracts also impact the sector's competitiveness.[761] The way in which access-controlled broadcasting services are marketed and advertised to subscribers, the composition of bundles and the prices charged—all of these are factors that directly influence the competitors' chances of attracting the consumers' attention.[762]

It is worth noting that the Universal Service Directive acknowledges that there is a risk that an enterprise with significant market power can use control over end-users access to a service to impede competition, for example, by charging excessive prices, setting predatory prices, foreseeing the compulsory unbundling of retail services or showing undue preference to certain customers. Accordingly, NRAs are authorized to consider imposing retail conditions on an enterprise with significant market power, even if it is a last resort and only occurs after due consideration. The regulation of retail conditions is, according to the Universal Service Directive, only appropriate if it has been found that wholesale measures are not sufficient to reach what the directive calls 'the twin objectives of promoting effective competition

---

761 See section 1.5.2.

762 Consumer attention as a relevant economic factor is addressed in 1.5.2. See also Larouche 2000, pp. 374-378.

whilst pursuing public interest needs, such as maintaining the affordability of publicly available services for some consumers'.[763]

Providing these conditions are fulfilled, NRAs can impose requirements not to charge excessive prices, inhibit market entry or restrict competition by setting predatory prices, show undue preference to specific end users or unreasonably bundle services (Article 17 (2) of the Universal Service Directive). NRAs can also impose retail price caps and control individual tariffs or measures to orient tariffs towards costs or prices on comparable markets. In addition, the Universal Service Directive includes, as explained further above, provisions on the fairness of consumer contracts and transparency for consumers, as well as on the quality of services.[764] Applied to the example of pay-TV, this could mean that NRAs are entitled to control the conditions in subscription contracts and if predatory pricing or bundling strategies have an anti-competitive effect.

It is, however, very questionable whether the provisions in the Universal Service Directive, which apply to electronic telecommunications service providers, also apply to pay-TV operators. As the argument goes, the Universal Service Directive only applies to 'electronic communications services' meaning services that consist of the transport of signals.[765] In this sense, services that provide or exercise control over content, such as the making available of access-controlled broadcasting services to consumers, are not telecommunications services and do not fall under the Universal Service Directive.[766] Pay-TV operators do both: they provide content services to consumers, and in most cases, also operate the technical platform that is necessary to do so, for example, the conditional access solution. One could argue that because of their control over the technical pay-TV platform that pay-TV operators fall under the scope of the Universal Service Directive. After all, and as was explained in Chapter 1, consumers conclude contracts with pay-TV operators about the provision of services, or signals, in intelligible form.[767] One indication that this view is not shared by the Commission is the fact that the Commission refrained from defining a retail market for broadcasting services in general, and pay-TV services in particular. This is again a consequence of traditional understanding of broadcasting as 'once-sent-free-access-for-all'.[768]

---

763 Universal Service Directive, Recital 26.

764 Universal Service Directive, Article 20 (Information in consumer contracts, right to withdraw), Article 21 (Transparency), Article 22 (Quality), Article 34 (Out-of-court dispute resolution).

765 Framework Directive, Article 2 (c); Universal Service Directive, Recital 45

766 European Commission, Recommendation on Relevant Product and Service Markets, Explanatory Memorandum, p. 36. Framework Directive, Article 2 (c); Universal Service Directive, Article 1 (1).

767 See sections 1.4.2. and 1.5.2.

768 See sections 2.1.

4.4.5. SUMMARY

Article 6 and Annex I of the Access Directive adopted an absolute approach towards the regulation of access questions whereby all operators of a conditional access facility are obliged to provide access. The precise scope of the access obligation is unclear and remains open for interpretation by the courts. Article 6 and Annex I work with narrowly defined bottlenecks and remain restricted to conditional access devices that provide access to digital radio and broadcasting services. The emphasis is on guaranteeing open access to the related service level, not on actively encouraging competition between different conditional access platforms. More generally, the regulation of conditional access is still focussed on the traditional broadcasting sector, where it is still based on the 'once-sent-free-access-for-all' concept. This is done despite the fact that, with the arrival of electronic access control in broadcasting, the distribution pattern resembles that of other telecommunications services. Consequently, many of the problems are similar, such as the effects of control over retail conditions for functioning competition and general public interest objectives.

4.5. Access to Other Bottlenecks in Digital Television: Article 5 (1)b

Under certain circumstances, NRAs may choose to expand the access obligation to EPGs and APIs if this is necessary to ensure end-user accessibility to digital radio and television broadcasting services.[769] According to Article 5 (1)b of the Access Directive, Member States can authorize their national NRAs to go beyond the present scope of the Access Directive and impose obligations on operators to provide access to EPGs and APIs. NRAs can do so irrespective of the finding of significant market power. Moreover, Article 5 (1)b, Annex I, Part 2 of the Access Directive could be the gateway to extending the authority of national NRAs to monitor other facilities of the technical platform in digital broadcasting, namely in all the cases in which it is necessary to ensure that end users can access digital broadcasting services.[770] For the time being, Annex 1 of the Access Directive refers to the API and the EPG, but the Annex could be extended to cover other facilities or services.

Having said that, the details of Article 5 (1)b of the Access Directive are still very unclear. In particular, the directive leaves open whether obligations under this provision involve only the provision of access itself or whether NRAs can also impose additional obligations, such as the disclosure of technical information and specifications or mandated interoperability. Two options are possible: one would be that the imposition of access to facilities that fall under Article 5 (1)b of the Access Directive follows the previously described concept of access regulation under

---

[769] Article 5 (1) b Access Directive.
[770] In this sense Gibbons 2004, 63pp.

Article 6 of the directive. In this case, national NRAs would probably have little room to choose a remedy other than an access obligation. The other option would be to have Article 5 (1)b follow the concept of Articles 8 to 13 of the Access Directive. This concept, as explained below, differs considerably from the regulation in Article 6 of the Access Directive, and it would give national NRAs far more flexibility to choose how they remedy anti-competitive situations.

One argument that speaks for the first option is its proximity to Article 6 of the Access Directive: both provisions deal with access issues in digital broadcasting. Like Article 6 of the Access Directive, Article 5 (1)b does not require significant market power, and both Articles are treated together in Annex 1. And in the earlier draft versions of the Access Directive, access to associated facilities other than conditional access, such as EPGs and APIs, was still regulated under the umbrella of Article 6 of the Access Directive.[771] An argument in favour of the second option is that access to the API and EPG is now treated under Article 5 and not under Article 6 of the Access Directive. In its first paragraph, Article 5 stipulates that NRAs shall

> 'encourage and where appropriate ensure, in accordance with the provisions of this Directive, adequate access and ... interoperability of services, exercising their responsibility in a way that promotes efficiency, sustainable competition, and gives the maximum benefit to end-users'.

Article 5 (1) of the Access Directive calls on national NRAs to interfere where they consider necessary and in a way that they consider best-suited to achieve maximum benefit for end users, be it in the form of access or interoperability obligations. Article 6 of the Access Directive does not provide national NRAs with this flexibility, but Articles 8 to 13 of the Access Directive do.

With the increasing sophistication of software and middleware solutions such as EPGs and APIs, ensuring end-user accessibility can be far more complicated than simply granting someone access to a facility. Accessibility can also include the provision of technical information, format compatibility, transparency, etc. From a functional point of view, such an argument would favour an interpretation of Article 5 (1)b in the context of Articles 8 to 13. This interpretation also seems to be supported by Article 5 (2) of the Access Directive, which explicitly refers to Article 12 of the Access Directive (obligations of access to and the use of specific network facilities). It does not explicitly exclude EPGs or APIs.

Ofcom put forward a third approach, according to which

> 'it may be most appropriate to use Article 8 of the Access Directive, rather than Article 5 of the Directive, given that this route would allow the

[771] European Parliament, Recommendation for Second Reading Access Directive, Amendment 14.

> imposition of the obligations set out in Articles 8-13 of the Access Directive on operators with SMP [significant market power]'.[772]

This view has its merits, especially as far as the application of the significant-market-power criterion is concerned (see below). On the other hand, it contradicts the wording of Article 5.[773] Moreover, this route seems rather superfluous considering that Article 5 (1)b of the Access Directive could be interpreted to allow the application of the principles set out in Articles 8 to 13 of the Access Directive.

What does not fall under Article 5 (1)b of the Access Directive are content-related aspects, such as the way information services are listed or presented in an EPG. This can be concluded from Article 6 (4) of the Access Directive, which states that the provision is without prejudice to the ability of Member States to impose obligations in relation to the presentational aspects of electronic programme guides and similar listings and navigation facilities. Having said that, it is worth noting that the presentational aspects of an EPG design are crucial in determining if and how services are accessible to end users. In practice, it will therefore be very difficult to make a distinction between, on the one hand, access to the EPG, and, on the other hand, the presentational arrangement. This aspect will probably play a major role mainly for the practical realization of the Access Directive at the enforcement level.

Another question is whether access obligations for EPGs (as suggested in Article 5 (1)b of the Access Directive) will solve the information problem described in Chapter 1.[774] This depends on the effect an access obligation would have. In the best case, the obligation to grant access to an EPG would ensure that EPGs include information about competing services on the same service platform. However, an access obligation still does not provide a guarantee that consumers are adequately informed about services and conditions that are offered on other platforms; access obligations do not guarantee that consumers will be able to learn about service providers who operate their own EPG or conditional access system. Similarly, access obligations are not a means to guarantee that comparable service information will be available for all services, including free-TV services and services from other Member States. In any case, access obligations are still no guarantee for the reliability, quality and accuracy of the service information.

The importance of access to and the quality of transparent and comparable service information for functioning competition can already be assessed by the fact that the Universal Service Directive ascribes some service information a universal service status.[775] According to the Universal Service Directive, Member States must

---

772 Oftel (now: Ofcom) 2002, paragraph 47.

773 European Commission, Recommendation on Relevant Product and Service Markets, Explanatory Memorandum, p. 37. See also Ofcom' Statement on Code on Electronic Programme Guides, Ofcom 2004, paragraphs 15-16; but see also paragraph 19.

774 See section 1.5.2.

775 Universal Service Directive, Recital 11, Article 5. See also Schulz 1998, pinpointing the importance of transparency in a multi-channel environment, p. 190.

make sure that at least one comprehensive telephone directory is available to end users in a form approved by the relevant authority, whether printed, electronic or both, and is updated on a regular basis and at least once a year.[776] Arguably, the telephone directory is a kind of search engine because it helps consumers find the service they are looking for, namely to connect to a particular person. Like the EPG, directory information and directory enquiry services constitute an essential tool in realizing the accessibility of services, be it telephony or broadcasting services.[777] Those directory services that provide an overview of all of the available services, including those offered by rivals, are tools that facilitate overall accessibility. Directory services that only provide information about the subscribers to one particular network cannot fulfil this function. The same can be said of the EPG.

From recalling the social and economic importance of the availability of comparable and comprehensive service information in a multi-channel environment,[778] the step towards considering extending universal service obligations to other 'directory services', such as EPGs is not big. Universal service obligations are a construct used to respond to social and political demands. Consequently, the universal service concept is open to respond to changed perceptions of the relevancy of a medium or of access to particular services for social and economic life.[779] On the other hand, decisions about the extension of universal service obligations should not be undertaken without carefully considering the impact they can have on innovation and investment, and whether the market itself is able to comply with consumer needs.[780] Successful examples, such as programme journals and online search engines, demonstrate that directory services and the provision of information about information can be a profitable market, and one that serves consumer demand. Moreover, Articles 21 and 22 of the Universal Service Directive already provide a framework for NRAs to make sure that consumers have enough information available so that they can compare existing services in terms of quality and pricing. According to Article 21 (1) and Annex II of the Universal Service Directive, Member States shall ensure that transparent and up-to-date information on prices and tariffs as well as on standard terms and conditions are available to consumers. Interesting is also the provision in Article 21 (2) of the Universal Service Directive that suggests providing consumers with interactive guides so they can independently evaluate the costs of alternative services. In other words, the directive suggests to give consumers electronic tools so they can learn not only about the terms and conditions of each particular service separately, but obtain an overview of the different services available to them and compare them. Article 22 of the Universal Service Directive deals with information about the quality of services. It stipulates that Member States shall ensure that NRAs are able to require

[776] Universal Service Directive, Article 5 (1)a.
[777] Universal Service Directive, Recital 11.
[778] See section 1.5.2.
[779] Universal Service Directive, Article 15 (1), (2). Xavier 1997, pp. 830-831.
[780] See Universal Service Directive, Article 5, Annex V; extensively Xavier 1997, 833pp.

undertakings that provide publicly available electronic telecommunications services to publish comparable, adequate and up-to-date information for end users on the quality of their services. Having said that, although potentially useful in the pay-TV case, these provisions are only applicable to electronic telecommunications services, or, more narrowly even, to telephony services (Article 21 of the Universal Service Directive). They do not apply to the broadcasting sector.

4.6. The Rule: Access to Telecommunications Infrastructures and Facilities—A Flexible Approach

Access to telecommunications infrastructures and facilities that do not fall under Article 6 of the Access Directive can fall under Articles 8 to 13 of the Access Directive. With the exception of conditional access, the API and the EPG for digital broadcasting services, and unlike the former ONP concept, the Access Directive does not distinguish between specific bottlenecks. Instead, the Access Directive extends access and interconnection rules to all electronic telecommunications networks and associated facilities. In other words, open access regulation is no longer restricted to selected elements of the telecommunications network; instead, a more general approach was adopted with the goal of establishing throughout Europe a common, harmonized framework for access questions at the transport level.

4.6.1. Scope of Articles 8 to 13 of the Access Directive

Articles 8 to 13 of the Access Directive do not deal with predefined bottlenecks; instead, a flexible approach has been adopted by which the NRAs are entitled to determine the circumstances under which facilities are considered potential bottlenecks to market entry and competition. Conceptually, this means that Articles 8 to 13 of the Access Directive do not automatically label certain facilities as bottleneck facilities, unlike the current approach in Article 6 of the Access Directive. Instead, NRAs evaluate the question of bottleneck control within the context of the market situation. It makes the final assessment dependent on the effect denial of access has on competition or end-user interests.

It remains to be seen whether national NRAs will decide to apply Articles 8 to 13 of the Access Directive to bottlenecks in digital broadcasting that do not fall under Articles 5 (1)b and 6 of the Access Directive. In section 4.5., it was argued that Article 5 (1)b of the Access Directive could leave some room to be interpreted within the context of Articles 8 to 13 of the Access Directive. Were this to be done, such facilities would fall under the below described flexible approach.

The applicability of Articles 8 to 13 of the Access Directive, however, depends on whether a relevant market for ancillary digital broadcasting facilities has been defined. Market definitions mark the limits for the regulatory activity of NRAs. So far, the Commission has not yet defined a wholesale market for ancillary services in digital broadcasting, basically because it was convinced that Article 5 (1)b of the Access Directive would prove to be sufficient. The Commission has also not defined a wholesale market for the pay-TV service platform.[781] The pay-TV subscription service itself, as this is a content service that, falls outside of the scope of the Communications Framework. What the European Commission did define, was a wholesale market for broadcasting transmission services and distribution networks providing they supply the means to deliver broadcast content to end users. It will only be possible for national NRAs to monitor and regulate markets that differ from those identified in the Commission's Recommendation where this is justified by national circumstances and where the Commission does not raise any objections. According to Article 15 (3) of the Framework Directive, NRAs can define relevant markets as 'appropriate to national circumstances' providing they follow the strict procedures set out in Articles 6, 7 and 15 of the Framework Directive.

4.6.2. SIGNIFICANT MARKET POWER

The precondition for any ex ante obligation under the flexible approach is that the enterprise in question must be designated as having significant market power for the market in question. The Commission indicated that the definition of significant market power used in the Communications Framework was equivalent to the concept of dominance as defined in the case law of the European Court of Justice.[782] One important difference, however, is that in the framework of the Access Directive, significant market power is identified from an ex ante perspective. In practice, this often means that the market analysis is based on a purely prospective assessment due to the lack of evidence and records of past behaviour or conduct.[783] The accuracy of the market analysis carried out by NRAs will thus depend on information and data that exist at the time of the adoption of the relevant decision. Further details are specified in the Commission's Guidelines on market analysis and the assessment of significant market power under the Community Regulatory Framework for electronic telecommunications networks and services.[784]

The concept of a market-power-oriented threshold is based on the assumption that bottleneck control is not necessarily harmful to competition, but that only enterprises with a particular degree of market power can efficiently influence

---

[781] European Commission, Recommendation on Relevant Product and Service Markets, Explanatory Memorandum, pp. 36-38.

[782] For a critical discussion, see Larouche 2002, p. 137.

[783] Accordingly, the outcome of the analysis by NRAs can differ from the outcome of general competition law procedures.

[784] European Commission, Commission Guidelines on Market Analysis and the Assessment of Significant Market Power.

competition to their own advantage.[785] As the Commission indicated, 'mandated access is not appropriate in all markets and can have the disadvantage of discouraging investment in innovation by a platform operator'.[786] This statement makes it all the more surprising that the Commission continues to maintain the absolute access obligation in Article 6 of the Access Directive rather than imposing access obligations only on enterprises with significant market power.[787]

National NRAs must justify their decisions based on an analysis of the competition in the market in question and an assessment of the market position of the operator of any bottleneck facility. For this purpose, NRAs are bound to observing a specific market analysis procedure and market definitions as laid down in Articles 14, 15 and 16 of the Framework Directive. Additionally, the Commission issued, after consultation with the NRAs, the aforementioned Recommendation on Relevant Products and Service Markets as well as the Guidelines on Market Analysis and the Assessment of Significant Market Power. The Commission's recommendation should identify markets 'whose characteristics may be such as to justify the imposition of regulatory obligations set out in the specific Directives'.[788] According to the Commission's guidelines, the finding of significant market power is not only a matter of market share.[789] NRAs must also take into account other criteria, such as the ease with which control over the infrastructure can be duplicated, the degree of product or service diversification, economies of scale and scope, the likelihood of new market entries that could constrain the market power of the established undertaking and the existence of barriers to market entry.[790] In particular, the guidelines also address the issue of newly emerging or fast developing markets in which the market leader is likely to have a substantial market share but should be spared from overregulation.[791] According to Article 15(1) of the Framework Directive, NRAs are required to take 'the utmost account' of the Commission's Guidelines.

Where a national NRA determines that the relevant market is not effectively competitive, it should identify enterprises with significant market power in that market in accordance with Article 14 of the Framework Directive and impose

---

785 See also section 1.5.3. Access Directive, Recital 6: 'In markets where there continues to be large differences in negotiating power between enterprises, and where some enterprises rely on infrastructure provided by others for delivery of their services, it is appropriate to establish a framework of ex-ante rules to ensure that the market functions effectively'.

786 European Commission, Commission Staff Working Document on Barriers to Widespread Access, p. 9. Nihoul 1998, p. 213, also the paper by Yoo 2002.

787 See the critical discussion of Article 6 in section 4.4.2.

788 Framework Directive, Article 15 (1).

789 According to the European Commission, an enterprise with no more than 25% market share is not likely to enjoy a (single) dominant position, European Commission, Commission Guidelines on Market Analysis and the Assessment of Significant Market Power, paragraph 75.

790 European Commission, Commission Guidelines on Market Analysis and the Assessment of Significant Market Power, paragraphs 78-80.

791 European Commission, Commission Guidelines on Market Analysis and the Assessment of Significant Market Power, paragraph 80.

appropriate obligations from a catalogue of possible initiatives, as provided for in Articles 9 to 13 of the Access Directive.[792] However, in areas in which NRAs conclude that the market is effectively competitive, they should refrain from imposing sector-specific initiatives and withdraw any existing obligations (Article 16 (3) of the Framework Directive).

Vertical Integration

significant market power at the transport level is not the only motive for NRA intervention. In response to trends in information markets, NRAs can also interfere if they find that the specific position of an enterprise in a related market poses a particular threat to functioning competition:

> 'Where an undertaking has significant market power on a specific market, it may also be deemed to have significant market power on a closely related market, where the link between the two markets are such as to allow the market power held in one market to be leveraged into the other market, thereby strengthening the market power of the undertaking'.[793]

In other words, where an undertaking has a dominant position in one market and a leading, not necessarily dominant position, in a related market, NRAs are entitled to assume that the enterprise holds a dominant position in both markets taken as a whole.[794] Related markets can be horizontal or vertical markets. NRAs may also restrict activities of that enterprise in the related market, impose access obligations in any form and/or accompanying measures that provide for greater transparency and controllability, such as obligations concerning non-discrimination, price control and cost accounting. The Communications Framework might also give NRAs an important and effective means to meet the challenges of strongly vertically integrated markets and to prevent enterprises from abusing their economic strength by leveraging market power from one market into another.

NRAs, however, must not automatically jump to the conclusion that interference is needed in a related market. One must bear in mind, as was explained in Chapter 1, that leverage is likely to occur only in exceptional situations.[795] Even if an enterprise has sufficient market power in both markets, this does not say anything about the profitability of leveraging. Consequently, NRAs should be careful before restricting the activities of an enterprise in the related market. And, as the European Commission claims, there are many cases in which interference in the related market is not even necessary. According to the Commission, NRAs will normally be in a position to prevent any likely spill-over or leverage effects from flowing into the related retail or service market by imposing on that enterprise any of the

---

[792] Access Directive, Article 8 (1).

[793] Articles 8 to 13 of the Access Directive and Article 14 (3) of the Framework Directive

[794] European Commission, Commission Guidelines on Market Analysis and the Assessment of Significant Market Power, paragraph 84.

[795] See section 1.5.3.

obligations provided for in the Access Directive.[796] For example, NRAs may require a vertically integrated enterprise to make transparent its wholesale prices and its internal transfer prices to, among other things, ensure compliance with the non-discrimination requirement or to prevent unfair cross-subsidies.[797] It should also be noted, however, that the abilities of NRAs to remedy spill-over or leverage effects is limited due to the limited competences of telecommunications NRAs to monitor content-related questions.[798]

4.6.3. ACCESS OBLIGATION

Contrary to the regulation of access to conditional access systems, Articles 8 to 13 of the Access Directive are based on the principle of negotiated access. This means that it is expected that the concerned parties negotiate appropriate access agreements and conditions. In the case of interconnection agreements, the Access Directive even explicitly states that operators of public telecommunications networks have a right and, when requested by other enterprises, an obligation to negotiate interconnection with each other (Article 4 (1) of the Access Directive). Only when negotiations fail, are NRAs requested to impose adequate access, interconnection or interoperability obligations.

Flexible Approach

The 'flexible' approach under Articles 8 to 13 of the Access Directive stipulates that national NRAs are to impose specific ex ante obligations where these are necessary to ensure adequate access and interoperability in the market. Unlike under general competition law, access obligations do not require a preceding situation of market power abuse; the infliction of ex ante obligations is based on a prospective assessment.[799]

The nature of the obligation will again depend on the requirements of the market situation. The final access obligation, thus, will be the result of case-by-case decisions, made under the authority of the national NRAs, which take the circumstances of each case into account. Such circumstances include the anticipated effect of an access denial on the overall competition, the market position and protection-worthy interests of the facility operator, the so-called 'essential requirements' (see section 4.6.5.).

The Toolbox

Once the NRA has identified a possible bottleneck situation that was caused by a provider of telecommunications networks or facilities with significant market

---

[796] European Commission, Commission Guidelines on Market Analysis and the Assessment of Significant Market Power, paragraph 84.

[797] Access Directive, Article 11 (1).

[798] See section 4.4.4

[799] See also Dommering/Van Eijk/Theeuwes/Vogelaar 2001, p. 59.

power, the NRA can choose from a list of possible responses the answer that is the most likely to restore the market balance and that has a proportional goal-remedy relationship. This is the so-called 'toolbox' approach.[800] Remedies in the toolbox range from the ability to impose duties that are related to the provision of access and/or interoperability (see section 4.6.4.) to other measures that are aimed at preparing the ground for a sufficiently competitive and transparent environment so that market participants can negotiate access agreements on fair, reasonable and non-discriminatory terms. Examples are the obligation to make public accounting information, technical specifications, network characteristics, terms and conditions for supply and use, and prices (Article 9 (1) of the Access Directive as well as to publish reference offers (Article 9 (2) of the Access Directive). The 'tools' in the toolbox enable NRAs to create the conditions that make monopoly-controlled facilities accessible to rivals. Some of the tools are also aimed at encouraging the emergence of different platforms. Switching between different platforms is facilitated by interoperability arrangements and provisions requiring transparency, non-discrimination, accounting separation and price control. Not included in the toolbox is the ability to enforce structural separation measures, for example, in vertically integrated enterprises.[801]

As for the obligation to allow access to one's own facilities, Article 12 (1) of the Access Directive leaves it up to the NRAs to determine which initiatives are needed to ensure the openness of a particular facility. Different situations may require different initiatives to realize open access. This applies in particular to the more software-oriented elements of the telecommunications network, where accessibility depends on a complex interaction between different standards and interfaces. The set of optional initiatives clearly extends beyond the scope of Article 6 of the Access Directive as it is not restricted to the access to the facility itself, but also covers access to technical interfaces or operational support systems and initiatives that actively promote the interoperability of competing facilities and services. According to Article 12 of the Access Directive, NRAs can also be authorized to monitor the contractual relationship between the facility operator and the access requester even after access has been granted (Article 12 (c) of the Access Directive), for example, to ensure that granted access is not withdrawn; that access to operational support systems, such as customer and information management

---

800 See Dommering/Van Eijk/Theeuwes/Vogelaar 2000, p. 97.

801 Note: As opposed, the Council Regulation on the Implementation of the Rules on Competition Laid Down in Articles 81 and 82 ECT entitles competition NRAs, here: the European Commission, to also adopt structural measures as of mid-2004, European Council, Regulation on the implementation of the Rules on Competition, Article 7: 'any behavioural or structural remedies which are proportionate to the infringement committed and necessary to bring the infringement effectively to an end. Structural remedies can only be imposed either where there is no equally effective behavioural remedy or where any equally effective behavioural remedy would be more burdensome for the undertaking concerned than the structural remedy'. But see also the critical discussion of this distinction in section 3.3.2. For a more general overview of possible remedies: Larouche 2000, pp. 325-329.

systems (Article 12 (h) of the Access Directive) is granted; that access to technical interfaces is granted, and that information, such as accounting information, technical specifications, network characteristics, terms and conditions for the supply and use of the system, and prices (Article 9 (3) of the Access Directive) are made available.

4.6.4. INTEROPERABILITY

Unlike the conditional access regulation, European telecommunications law has had a two-tier approach from the very beginning: it addresses (a) the vertical relationship between network operators and telecommunications service providers with the rules on fair, non-discriminatory access, and (b) the horizontal relationship between competing facility operators with the rules on interconnection and interoperability. The principle of open non-discriminatory treatment applies to both levels.[802] The Access Directive adopted the approach of the former ONP Framework and modernized it in order to respond to the increased intelligence and complexity of telecommunications networks and associated facilities. In addition, it extends the Framework to cover other aspects such as the interoperability of services with middleware or software elements, and the provision of information and specifications needed to run an application or use a facility. In this context, interoperability obligations often work at the interface between the infrastructure and the service level because they guarantee that the services provided by means of a particular infrastructure element are interoperable with the facility itself as well as with other services using the same technical platform. Accordingly, the flexible approach gives NRAs room to oblige operators to actively promote the interoperability of services with the technical application environment, for example, by providing access to the operating systems or converter, or by requiring the installation of a common interface or other interoperability solutions. More importantly, NRAs can also require operators to make available information that is necessary to make systems interoperable, etc. (Articles 12 (e) and (g) of the Access Directive).

In addition, Article 5 (1)a of the Access Directive stipulates more generally that NRAs shall actively encourage, and where necessary and appropriate, also take more pro-active measures to ensure the interoperability of services. The NRAs' task is to ensure end-to-end connectivity for end users where enterprises control access to end users, according to Article 5 (1)a. This provision mirrors the previously described Article 5 (1)b of the Access Directive, which applies to digital broadcasting, and here more specifically to the obligation to provide access to the API and/or EPG. In contrast, Article 5 (1)a provides a more general approach to

---

[802] Directive 97/33/EC of the European Parliament and of the Council of 30 June 1997 on interconnection in telecommunications with regard to ensuring universal service and interoperability through application of the principles of Open Network Provision (ONP) , 26 July 1997, OJ L 199, p. 32 [hereinafter `Interconnection Directive'], Recital 1.

addressing technical lock-in situations by imposing obligations to make systems interoperable or interconnected. For this purpose, NRAs can also impose conditions that refer to the implementation of specific technical standards or specifications (Article 5 (2) of the Access Directive). A precondition, however, is that they respect Article 17 of the Framework Directive, which stipulates the authorities of the European Commission in the field of standardization.

4.6.5. Essential Requirements

In areas in which NRAs have clearly identified the anti-competitive effect of potential access denials, the legal consequence is not necessarily the imposition of an access obligation or the application of another remedy. The Access Directive continues the proportionality approach ('essential requirement') of the former ONP Framework and limits access obligations explicitly to what is practically and technically possible and economically feasible.[803] The obligation to grant access does not apply to situations where a third party's access might cause disproportional technical or economic losses. According to Article 12 (2) of the Access Directive, when imposing access obligations on a case-by-case basis, NRAs must balance all of the interests involved and consider not only the technical aspects, such as system integrity and security, interoperability and capacities, but also the economic and wider competition policy aspects. Examples of such policy aspects are the need to allow enterprises to recoup initial investments, the long-term effects of access denial on competition, the need to consider the economic risks an enterprise runs when setting up certain facilities, and any of the facility provider's property interests (Article 12 (2) of the Access Directive).

4.6.6. Terms and Conditions

In areas in which terms and conditions of access and interconnection are not negotiated for an individual case, it is primarily the task of national NRAs to determine what fair, reasonable and non-discriminatory terms and conditions are and to impose adequate ex ante obligations.[804] The NRAs' task is facilitated by the ability to impose various transparency obligations (Articles 9 and 11 of the Access Directive). As far as the adequacy of terms and conditions is concerned, particular emphasis is placed on the aspect of price discrimination. Facility operators may be subject to tight price controls, including possible obligations regarding cost

[803] See also European Commission, Access Notice, paragraphs 87–98.

[804] Article 12 (1) of the Access Directive reads: 'National regulatory authorities may … impose obligations on operators to meet reasonable requests for access to, and use of, specific network elements and associated facilities, among other things, in situations where the national regulatory authority considers that denial of access or unreasonable terms and conditions having a similar effect would hinder the emergence of a sustainable competitive market at the retail level, or would not be in the end-user's interest'. Article 10 (2) of the Access Directive addresses discriminatory behaviour.

orientation and accounting systems (Articles 9, 11 and 13 of the Access Directive). The burden of proof that charges are derived from costs, including a reasonable rate of return on investment, lies with the facility operator (Article 13 (3) of the Access Directive).

Moreover, Article 17 of the Universal Service Directive applies. Article 17 of the Universal Service Directive acknowledges the consequences of monopolistic control over the consumer base in the form of contractual lock-ins and other forms of anti-competitive behaviour on competition and consumer welfare. The provision authorizes NRAs to monitor retail conditions. Where NRAs find that a given retail market is not competitive and access obligations under the Access Directive are not sufficient to make the market competitive, NRAs can impose additional obligations, such as terms and conditions that define how services are marketed to consumers. It is worth noting that NRAs are entitled to pursue competition and general public interests, such as maintaining the affordability and public availability of certain services for consumers.[805]

Cost-control and transparency play major roles in this context. The Universal Service Directive recognizes that, for reasons of efficiency and to encourage effective competition, it is important that the services provided by an enterprise with significant market power reflect consumer demand conditions and costs. Enterprises should not use their control over access to consumers to impose excessive or predatory prices on them.[806] NRAs can also step in where enterprises show undue preference for certain consumers or unreasonably bundle services. According to Article 17 (2) of the Universal Service Directive, NRAs may therefore impose retail price cap measures, measures to control individual tariffs, or measures to orient tariffs towards costs or prices in comparable markets. According to Articles 21 (2) and 22 of the Universal Service Directive, non-regulatory measures such as the transparency initiatives mentioned in section 4.5. can also be used to make publicly available comparisons on retail tariffs, the quality of services and terms and conditions of access. These measures enable consumers to compare services and exercise their choice, and thereby realize the previously mentioned 'twin objectives'.[807]

4.6.7. SUMMARY

Articles 8 to 13 of the Access Directive implement a flexible concept for the regulation of access to technical bottlenecks. Instead of pre-defined bottlenecks and obligations, it is the task of the NRAs to identify critical bottlenecks within the context of the market situation and choose effective and proportionate remedies. NRAs dispose over a toolbox of possible remedies, including access, transparency

[805] Universal Service Directive, Recital 26.
[806] Universal Service Directive, Recital 26.
[807] Universal Service Directive, Recital 26.

and interoperability obligations as well as the monitoring of retail conditions to prevent abuse of monopolistic control over the consumer base in form of technical/contractual lock-ins or lack of comparable service information.

4.7. Two Conflicting Access Regimes

4.7.1. Ex Ante/Ex Post Control

The two access regimes outlined above could result in very different outcomes for telecommunications markets. The flexible access concept under Articles 8 to 13 of the Access Directive approaches the issue of bottleneck control within the context of the reality of market structures. Its efficacy depends on the choice of the appropriate remedy and the ability to adopt timely procedures. Articles 8 to 13 of the Access Directive establish a system of ex ante market control by which national NRAs regularly monitor market developments and identify sector-specific bottleneck situations. The flexible concept necessarily involves an element of legal uncertainty and subjective assessment due to the uncertainties of any ex ante assessment of market power and the fact that operators cannot necessarily foresee future obligations that might be imposed on them. This also means that the legal situation in national telecommunications markets could change quickly and in line with changes to the economic structure of the respective markets. As telecommunications markets will differ from Member State to Member State, it is also likely that the concept will result in different legal situations in the different national markets.

Conditional access facility operators that fall under Article 6 of the Access Directive will find a less dynamic and, at first glance, more continuous legal environment. The price for continuity and stability, however, is less flexibility to react to newly emerging bottleneck situations in the converging digital service sector in a timely manner and before lasting harm is done. Furthermore, upon closer scrutiny, the approach of Article 6 of the Access Directive gives rise to some legal uncertainty, although of another kind than under the flexible approach. The possible causes for the uncertainty are threefold: the reference to the somewhat outdated definition of broadcasting, the split supervision of NRAs, competition authorities and courts, and, finally, the impossibility of predicting the outcome of judgements in access conflicts. It remains to be seen whether the concept of judicial ex post control is efficient and cost-effective. As opposed to sector-specific NRAs, courts will generally lack the information and experience required to adequately decide cases of access refusal. The overall costs of obtaining a decision, the consequences of a prolonged situation of uncertainty (due to lengthy judicial proceedings) and the possible costs of enforcing decisions can make the absolute approach more inefficient than the flexible approach. In dynamic technology markets such as the pay-TV market, the time aspect plays a crucial role in effective regulation. As Dommering, Van Eijk, Theeuwes and Vogelaar observe, timing is one of the most underestimated aspects in the regulation of the telecommunications sector.[808] The success of the absolute approach and its ability to effectively guarantee fair access to conditional access facilities, EPGS and APIs will depend largely on whether legislators will succeed in predicting the relevant bottlenecks or defining them in a manner that is flexible enough to respond to technological changes or market conditions. It will also depend on the ease with which the law can be amended in response to changing technological and market conditions, and whether the responsible judicial bodies have sufficient competence and access to the necessary information to judge the fairness and adequacy of access conditions.

4.7.2. The Power of the NRAs

Undoubtedly, the flexible concept gives NRAs sufficient opportunity to evaluate the market situation, and in particular to identify bottleneck situations, the facility in question, the relevant market, and whether an enterprise with significant market power is active. It also gives NRAs some flexibility to choose the most appropriate remedy. Unlike court decisions, NRA decisions are also economically and politically motivated decisions that take market factors, competition policy and general public policy interests into account. This could allow NRAs to shape future market structures and conditions of competition for all bottleneck facilities (except conditional access and other technical facilities for broadcasting services). The NRAs' powers, on the other hand, are limited due to the concept of strict separation between content and infrastructure regulation, which is not only immanent to the Communications Framework, but is also reflected in the way supervision over the service and transport sectors is organized. In response, there is a growing acknowledgement in the Member States that a strict separation between the content and the service level is increasingly difficult to maintain. This is why in some Member States, such as the UK and Italy, the regulatory authorities for both sectors were merged into one entity, while other countries, such as Germany, require the NRAs in the different sectors to closely cooperate.

The sector that falls under Article 6 of the Access Directive does not leave any or only little flexibility for NRAs to take competition or general public interest objectives into account and adapt their choice of remedies accordingly; this area is reserved for the legislator.[809] Article 6 of the Access Directive mandates an absolute access obligation, principally irrespective of the market situation and whether other, more effective remedies are available. The goal of Article 6 of the Access Directive is to guarantee the openness of the conditional access platform. This is done with a

---

[808] Dommering/Van Eijk/Theeuwes/Vogelaar 2001, p. 98. 'Timing is een van de meest onderschatte aspecten bij de regulering en het toezicht in de communicatiesector' [Timing is one of the most underestimated aspects in the regulation and supervision of the communications sector. Translation by the author]. Neumann 1998, pp. 33-35 on the particularly high financial risks and difficulties during the entrance phase.

[809] Different opinion Cave/Cowie 1998, section 5.

view towards protecting the interests of competing broadcasters, consumers and general public interest objectives by ensuring that the market for access-controlled services is not impeded by undue bottleneck control. It is up to the courts to evaluate the adequacy of access obligations. The courts' decisions are made on a case-by-case basis, they will focus less on medium or long-term market or public information policy considerations.

4.7.3. STIMULATING VS. DISCIPLINING

Articles 8 to 13 of the Access Directive do not predetermine the remedies that may be imposed on operators because the choice of remedies depends on the case being treated and the 'the nature of the problem'.[810] For facility operators that fall under the flexible approach, this means being exposed to a level of uncertainty in terms of the applicable obligations and their scope. On the other hand, NRAs are explicitly encouraged to take the valid economic interests of the facility operator into account, notably the investment made by the operator and its chances of a reasonable return on investment. The NRAs are also encouraged to take the facility operator's risks into account, as well as the technical and economic viability and the feasibility of imposing access or other obligations. Less economically powerful operators, for example, those without significant market power, will only be burdened with specific obligations in exceptional circumstances.

If correctly exercised, the flexible approach could be a powerful tool for stimulating investment and innovation in the respective markets because it leaves room for investment-friendly and market-policy-oriented choices. In contrast, the absolute access obligation is primarily a tool for disciplining facility operators and prosecuting the abuse of market power. It leaves little room for political market considerations or initiatives to mitigate the deterrent effect an absolute access obligation might have on investment and innovation. The conceptual difference between the two approaches is likely to be reflected in the outcome of decisions in access conflict cases.

4.7.4. NEGOTIATED ACCESS VS. MANDATED ACCESS

Another difference is that Articles 8 to 13 of the Access Directive apply the concept of 'negotiated' access rather than that of 'mandated' access under Article 6 of the Access Directive. Since one of the objectives of the flexible approach is to create a transparent negotiation-friendly environment, facility operators that fall under Articles 8 to 13 are very likely to experience additional transparency obligations regarding the conditions of supply and the obligation to publish reference offers, etc. Another important difference between the flexible approach and the absolute access concept is that the former mandates the imposition of elaborate obligations

[810] Access Directive, Article 8 (4).

concerning pricing and accounting. The flexible approach refers to the economic adequacy and reasonableness of prices, including retail prices. In addition, under the flexible concept NRAs are also entitled to monitor retail conditions in terms of the way services are marketed to consumers. As a result, under the flexible concept, enterprises initially enjoy more negotiating power than under the absolute concept. However, under the flexible approach, NRAs have considerable power to monitor and influence the terms and conditions of access agreements—probably even more than under the absolute approach.

4.7.5. STIMULATING COMPETITION

Both models aim at opening markets to competition and both pursue, to varying degrees, the idea of deregulation. However, there are two major differences in the way they approach this goal. One difference is the way both concepts approach questions of interoperability. The other difference is the role that individual consumer-service-provider relationships play in the NRAs' monitoring activities.

As to the first difference, the strategy of the flexible approach is twofold. First, regulation aims at preventing the anti-competitive use of market power in the form of bottleneck control by monitoring the behaviour of major players in a national market and intervening where necessary. The idea is to guarantee unprejudiced access to the service level by mandating access to a technical platform in order to prevent facility operators from leveraging economic power from the transport level to the service level. Second, the flexible approach strongly mandates interoperability solutions to encourage competition at the transport level. This further reduces the chances of leverage and of a technical service or facility of becoming a lasting bottleneck facility. Ideally, this concept will create a number of alternative facilities that compete 'within the [facility] market' rather than 'for the market'.[811] From the consumer perspective, competition would occur at levels such as price, quality, product features and support services, rather than be a race to establish a dominant standard. Programme and application providers can benefit from this situation, as they will probably be able to choose from alternative technical facilities that compete in terms of quality and service conditions.

In contrast, Article 6 of the Access Directive focuses on opening up the technical platform to competing broadcasters by imposing behavioural rules on the conditional access operator and controlling ex post the adequacy of access conditions. Article 6 of the Access Directive does not promote competition between the different technical platforms but access to the dominant platform. It is worth noting that this is in contrast to the findings in some of the Commission's merger decisions that ascribed major importance to inter-platform competition, including competition between different conditional access standards.[812] As a result of the

[811] See also Shapiro 1998, p. 9.
[812] See section 3.3.1.

application of Article 6 of the Access Directive, and from the broadcasters' point of view, there would be principally no need to establish a second conditional access system since the standard of the first-comer is firmly established and access to this technical platform can be enforced. It is left up to the industry to develop and apply interoperability solutions for the technical pay-TV platform, be it with a view towards conditional access or API interoperability. If industry negotiations fail, competition in the pay-TV sector will continue to develop mainly as a competition 'for the market' in which competitors strive to establish the dominant standard and secure their market position. Presumably, the first mover—or alternatively the operator of the most popular service platform—will establish a de facto standard that does not even have to be the most technologically advanced or consumer-friendliest.

Arguably, Article 6 can also stimulate competition in the pay-TV sector if it encourages the market entry of smaller service providers that cannot afford to operate their own system. It is questionable to which extent these smaller operators will indeed by able to discipline the behaviour of a dominant pay-TV operator. Much depends on whether Article 6 of the Access Directive is an efficient tool to protect the economic and editorial independence of service providers that are carried via this platform. On the other hand, the absolute concept could have a deterrent effect on larger platform operators, meaning potential competitors at the service level who intend to operate their own conditional access system. This is because the lack of mandatory interoperability solutions can generate incalculable risks for launching their own system.

As far as the aspect of monopolization of the installed consumer base is concerned, accessibility for consumers to digital broadcasting services is one of the goals declared in Articles 5 and 6 of the Access Directive. However, the provisions do not address issues of individual consumer access to pay-TV platforms. They focus instead on creating conditions that allow consumers that are subscribed to one pay-TV platform or use one particular decoder technology, to access services from other service providers using that particular technology.

More generally, the Access Directive is not concerned with questions regarding the consumer-service-provider relationship, such as the issues of technical and contractual lock-ins or the information problem that were discussed in Chapter 1. As far as access to telephony, the internet or other telecommunications services is concerned, the provisions of the Universal Service Directive apply and address such issues. This is done in form of provisions dealing with unjustified bundling, unfair pricing, the availability of comparable service information and matters of interoperability. Insofar, social and general public interests can be taken into account when assessing the adequacy of the terms, conditions and prices of services at the retail level. This does not apply, however, for broadcasting services.

	Article 6 Access Directive	Articles 8 to 13 Access Directive
Competent authority	Mixed competencies of national legislators, courts, NRAs and competition authorities	Wide scope of judgement and interference for NRAs
Character of control	Ex post control	Ex ante obligations
Definition of bottlenecks	Predefined bottlenecks	Flexible definition of bottlenecks, dependent on market structure and subject to timely technological change
Access obligation	Absolute access obligation for all CA providers, irrespective of degree of market power and level of vertical integration	Specific initiatives which focus on targeting vertical concentrated structures
Applicable to	Applicable to all operators of CA who produce and market access services for digital television	Applicable only to operators with considerable market power (exception: Article 8 (2))
Focus	Principle of mandated access prevails	Principle of negotiated access prevails
Remedy	Predefined access obligation, while further definition of conditions left to the interpretation of the general notion of 'fair, reasonable and non-discriminatory'	Actual obligations depend on 'nature of the problem'
Interoperability	Questions of interoperability left in the first place to market players	Possibility of NRAs to impose obligations for ensuring interoperability and compatibility
Accompanying measures	Accompanying measures restricted to the obligation to keep separate financial accounts regarding activity as CA provider, Licensing of CA to decoder manufactures at fair, reasonable and non-discriminatory conditions	Catalogue of possible ex ante obligations exceeding the actual access provision and extending to initiatives with the intention of preparing the ground for fair access negotiations in a competitive environment (e.g. transparency obligations and price control, obligation to accounting separation, cost accounting obligations)
Limits	-	Essential requirements, Article 12 (2)
Goals promoted	Competition at service level Continuity and stability with respect to existing systems at the CA level (Deregulation) Open access to established CA system (intra-platform competition)	Competition at service level Competition at transport level Innovation, investment at transport level Deregulation Stimulate inter- and intra-platform competition

Table 2–Comparison of Article 6 and Articles 8 to 13 of the Access Directive. Table 2 compares Article 6 to Articles 8 to 13 of the Access Directive using different criteria.

4.7.6. SUMMARY

The existence of two such disparate approaches to bottleneck control in the broadcasting and non-broadcasting sector is likely to lead to considerable inconsistencies in the market structure for both the digital TV broadcasting and non-broadcasting service markets. Conceptually, one system promotes full competition at both levels (the transport and the content service level), while the other might have the opposite effect of consolidating the dominance of one conditional access platform that is commonly also associated with the dominant service platform. Both concepts will have very different effects on competition, innovation and consumers. Very likely, only a flexible approach will leave enough room to actively stimulate innovation and investment in digital broadcasting markets and encourage larger service providers and conditional access operators to enter the market.[813]

The uncertainty is very prominent in areas in which the two regulations overlap, as is the case with advanced conditional access systems that provide access to both broadcasting and information society services. By distinguishing between providers of access-controlled broadcasting and non-broadcasting services, the Access Directive submits them to very different legal regimes. At the same time, the operators will be the target of contradictory economic impulses, as one regulation can aim at competition between alternative systems, while the other aims at stability and continuity with respect to existing systems. The resulting incoherence in market structure and policy might seriously impede the development of the market for advanced access-controlled content services, such as interactive services that are offered at pay-TV platforms as well as broadcasting services that are streamed via the internet.

4.8. Absolute Approach versus Flexible Approach

Section, 4.7. illustrated that the Access Directive maintains two different and even contradictory access regimes for technical facilities, and that this could lead to considerable legal uncertainty and inconsistency. In section 4.8., arguments in favour and against both approaches are weighed to determine which approach is better-suited to deal with bottleneck situations in pay-TV.

The need for and the character of public intervention depends to a large extent on the structure of the markets in question and on their openness. Obviously, there is a delicate balance between the advantages of an access obligation and the possible

[813] See, for example, Shelanski/Sidak 2001, who argue that the drivers of innovation are the major players rather than small newcomers, p. 124. However, examples such as the case of joint control of German Telekom over the cable and the telephone network illustrate that the opposite can also be true, whereby internal conflicts of interests and structural inflexibility can block innovation (in this case, investment in broadband technology).

adverse effect on competition and investment. An 'access-at-any-price' approach may not only be detrimental to its goals, but may also be questionable from a constitutional point of view. To be justifiable, any approach to access regulation must be proportionate and sufficiently balanced to adequately consider the legitimate interests and constitutional freedoms of all of the parties involved, meaning the parties requesting access as well as the facility operator.

Imposing an absolute obligation on conditional access operators to grant access to their system might facilitate the entry for new access-controlled services. It could be a viable instrument for addressing the high entrance risks and costs that could deter competitors from entering the market. Newcomers would not be under pressure to invest in new technical facilities because they would have access to existing facilities. Conditional access controllers would be widely deprived of possibilities to fight market entry with access refusals since they would be legally obligated to provide access to their systems.

On the other hand, are all conditional access systems by definition obstacles to market entry? Chapter 1 explained why it is so difficult to identify critical bottlenecks. The definition of critical bottlenecks is a process that must take particular market situations and a sector's competition and general public policy goals into account.[814] The greatest difficulty with any ex ante access obligation is probably the identification of critical bottlenecks. The price for legal certainty in dealing with bottleneck situations is that a predefined access obligation can quickly become outdated and even harmful in a fast moving technology and economic environment, and thus fail to achieve its goal.

4.8.1. When is a 'Bottleneck' a Bottleneck?

Enterprises that specialize in the development and installation of conditional access systems do not automatically have an incentive to reduce the availability of the conditional access system to competitors or to discriminate against access requesters.[815] In the absence of additional circumstances, why would an economically minded enterprise refuse to sell its technology to as many users as possible and profit from the resulting profits and economies of scale?[816] It is often argued that the degree of vertical integration in the pay-TV sector would not only create the possibility but also the incentive for pay-TV operators to leverage market power using bottleneck control.[817] But as was explained in Chapter 1, those views too easily ignore that leverage is not necessarily a profitable, and therefore likely, strategy.[818] The existence of a bottleneck situation and the occurrence of leverage

[814] See section 1.5.3.
[815] See section 1.5.3.
[816] See also Besen/Farrell 1994, 122pp.
[817] See König 1997, p. 89; Neumann 1998, p. 239; the paper of Lemley/Lessig 2000 for the broadband sector. More differentiated, Owen 1975, 888pp.
[818] See section 1.5.3.

depends on, among others, the competitive strength of the enterprise in question, the degree of concentration in the markets for conditional access systems and access-controlled services, and the existence of obstacles to market entry.[819] Even in vertically integrated structures, pay-TV operators may have valid economic incentives to admit competing operators to their conditional access system and service platform. A rich array of content could enhance the attractiveness of their own platform. Hence, it seems that it is not the control over the conditional access facility itself that threatens to block market entry, but the existence of additional structural factors that accompany this control. Consequently, simple control over facilities does not justify limiting the operators' property rights and contractual freedom by imposing access obligations.[820] Such overregulation can interfere with legitimate business practices, render a service inefficient and harm competition and consumer welfare. It can also create a disincentive to innovate and invest in the development and maintenance of an own technical platform. Accordingly, at least the European Court of Justice subjects the analysis of alleged bottleneck facilities in competition law cases to a strict and critical assessment[821] With this in mind, and against the background of the intended deregulation of the telecommunications sector, it is difficult to see why facilities in digital broadcasting should be more strictly regulated under Articles 5 (1)b and 6 of the Access Directive than they would be under general competition law, unless there are other reasons that would require the stricter approach (see section 4.8.5.).

In order to adopt effective, innovation-friendly and proportionate regulation, it is necessary to specify the circumstances that make a facility or service a bottleneck. Factors that can favour bottleneck situations can be, for example, the existence of indirect network effects, individual and collective adaptation costs as well as the lack of adequate interoperability solutions.[822] This is particularly true for bottleneck situations at the upper levels of the telecommunications model, such as the conditional access system. The conditional access bears little resemblance to the natural monopoly situations that the former ONP framework sought to address. Natural monopolies describe situations in which demand for a particular service or facility can be best and most efficiently served by a single operator.[823] Usually, this concerns resources whose capacities can be extended only with difficulties, due to the technical and economic particularities of a sector. Examples for sectors that are often considered susceptible to natural monopolies are the energy and telecom networks. The existence of natural monopolies is an extreme case of market

---

[819] See section 1.5.3.

[820] Fritsch/Wein/Ewers1999, p. 238; Knieps 1998, p. 276.

[821] See section 3.4.1. As far as access to the conditional access system is concerned, this is already reflected in Article 6, Annex I, Part 1 (b) of the Access Directive. This is different for the facilities in Annex I, Part 2, notably the API and the EPG. But both Article 6 and 5 (1)b of the Access Directive allow the imposition of access obligations even in the absence of significant market power.

[822] See section 1.5.2.

[823] See Herdzina 1999, p. 41.

development. Usually, the technological and economic conditions will restrict the number of potential competitors (oligopoly) without necessarily excluding any chance of competition.

Bottleneck situations in pay-TV are often the result of the economic strength and the control of a proprietary standard. The co-existence of several competing conditional access standards in larger markets such as France, Italy, the UK, Germany or The Netherlands, seems to suggest that access-controlled service providers do not necessarily have to depend on one particular conditional access operator, but that alternative systems can live alongside each other. In other words, often it will be the proprietary control over a dominant conditional access or API standard that provides enterprises with sufficient market power to effectively exclude third parties from accessing a conditional access platform or its underlying content platform.

In contrast, the flexible approach under Articles 8 to 13 of the Access Directive makes regulatory interference in potential bottleneck situations dependent on a finding of significant market power as well as on a more comprehensive assessment of the actual market situation. It has to be acknowledged, however, that Article 6 (3) of the Access Directive leaves NRAs some room to perform a market analysis for conditional access facilities and to restrict the imposition of access obligations, at least, to operators with significant market power, too. Within the context of the market analysis and according to Articles 16, 14 (3) of the Framework Directive,[824] NRAs are also entitled to take into consideration the involvement in related markets, such as a market for access-controlled services, the possible interaction between both markets, and the likelihood of new market entries. On the other hand, section 4.4.2. has been demonstrated that the provision is difficult to apply in practice.

Moreover, according to Article 5 (1)b of the Access Directive, such possibilities are not provided for in the regulation of EPGs and APIs. Bearing in mind that Article 5 (1)b, in combination with Annex 1, Part 2 of the Access Directive, could become the main tool used to deal with unforeseen bottlenecks in digital television, the lack of a more comprehensive requirement to analysis the economic conditions is deplorable and potentially harmful to innovation and competition

4.8.2. CONDITIONAL ACCESS IS NOT RESTRICTED TO DIGITAL BROADCASTING

While the scope of Articles 5 (1)b and 6 of the Access Directive appears to be too broad in some respects, it is certainly too narrow in others. As far as bottlenecks in digital broadcasting are concerned, the Access Directive and the relevant provisions in the Framework Directive and the Universal Service Directive have a strongly technology-dependent approach. The Communications Framework still

[824] European Commission, Commission Guidelines on Market Analysis and the Assessment of Significant Market Power, paragraphs 72-78.

distinguishes between broadcasting and non-broadcasting services. To this extent, it clearly fails to live up to its promise to create the conditions for an effective, horizontal framework for the regulation of convergent service markets.

To begin with, conditional access, APIs and EPGs are not the only potential bottlenecks in pay-TV,[825] and it cannot be excluded that, with further technological development, new and unheard of facilities will emerge and become bottlenecks to market entry. As far as telecommunications services and facilities in general are concerned, the flexible bottleneck concept in Articles 8 to 13 of the Access Directive is open and sufficiently technology-independent to respond to newly emerging bottleneck situations. This is not the case for newly emerging bottlenecks in pay-TV. To cover them will require amending at least Annex I, Part 2 of the Access Directive.

Second, conditional access is not a problem that is restricted to broadcasting. The use of conditional access is no longer (and probably never was) restricted to the distribution of digital broadcasting content. With progressing convergence, providers of other non-broadcasting services require access to the technical pay-TV platform.[826] Modern pay-TV platforms also offer e-commerce services, access to the internet, and internet-based services. Likewise, conditional access solutions can, as explained in Chapter 1, appear in the non-broadcasting environment.[827] 'Pay-TV' services can be received via the internet or mobile platforms. Conditional access can also be a part of technical solutions that have nothing to do with broadcasting or broadcasting-like services at all, for example, when it is implemented in DRM solutions. The same is true for EPGs and the API. The equivalent of an EPG can be an electronic directory service, a search engine or a web browser. And APIs can be found in set top boxes as well as in portable devices, in computers and mobile phone devices. It is difficult to understand why the operators of such facilities are treated differently based on whether they are active in what was traditionally considered the broadcasting sector or the non-broadcasting sector.

Third, by distinguishing between providers of access-controlled broadcasting and non-broadcasting services (for example, providers of on-demand services that are delivered via the internet), the Access Directive creates artificial distinctions between different kinds of services that are potentially received via the same set top box. More importantly, the directive creates a considerable level of legal uncertainty as to whether and under which provision(s) the following cases fall under the Communications Framework:

- Access by service providers to conditional access solutions for an internet or mobile platform environment.
- Access by service providers to conditional access solutions that control access to both broadcasting and IS services.

---

[825] See section 1.4.1.
[826] See section 1.4.3.
[827] See section 1.2.

- Access by service providers to bottlenecks in digital television other than the conditional access, the EPG or the API (for example, access to the billing system).
- Access by service providers that do not qualify as broadcasters in the traditional sense.

Finally, operators of convergent access-controlled platforms will be left with considerable legal uncertainty if they have to open their facilities to third parties, and the conditions and supervisory regime they should be subject to.

The conflicts in the current approach will therefore particularly affect providers of digital services and/or facilities that do not clearly fall under Articles 5 (1)b, 6 or Articles 8 to 13 of the Access Directive, or where the two legal regimes overlap. The consequence can be either a regulatory gap or overregulation. One might argue that none of the situations listed above fall under the scope of Articles 5 (1)b or 6 and that they therefore fall under the more general framework of Articles 8 to 13 of the Access Directive. The applicability of Articles 8 to 13 of the Access Directive to conditional access and other facilities that do not fall under Articles 5 (1)b and 6 of the Access Directive would require, however, the definition of a relevant market. As far as facilities other than transmission networks in digital broadcasting are concerned, no such market has been defined yet, at least not at the European level.

4.8.3. Pitfalls of Absolute Solutions

Providing that NRAs succeed in identifying a bottleneck situation, the issue of finding the best remedy to bottleneck control still remains. Obliging the operator of a technical pay-TV platform to grant access to the conditional access system, the EPG or the API is certainly an option. However, with the increasing sophistication of the technical distribution process, and depending on the market structure, other remedies might prove to be more effective. In areas in which applications are written in different formats, access to the API may be less crucial than knowledge of its technical specifications and standards. If the conditional access operator is willing to grant access to the API but makes only limited capacities in the set top box memory available, the obligation to provide for sufficient capacity might be more appropriate. Moreover, the way in which services are marketed to consumers, the size of the bundles and the prices requested could form no less of an obstacle to market entry and functioning competition, even if access to the conditional access system were granted. Here, it might be necessary to monitor the retail conditions in pay-TV markets.

On the other hand, any legal framework should give the NRAs sufficient room to take the legitimate interests of the facility operator into account. A few examples of legitimate interests are capacity constraints, security concerns or the need to recoup investments. The point made here is that it is difficult to formulate, as Articles 5 (1)b and 6 of the Access Directive do, a predefined one-fits-all response to bottleneck control. This is the reason why the Communications Framework in

Articles 8 to 13 of the Access Directive opted for a more flexible concept of bottleneck definition and regulation. It remains to be seen if the approach is indeed more flexible or whether procedural obstacles, difficulties in identifying and enforcing the adequate remedies, and time-consuming market definition procedures will prevent effective and timely initiatives. One thing, however, is certain: remedies are proportional and constitutional only where they are necessary and effective in achieving their goals and where no other, less stringent measure is available to achieve the same goal.[828] A more flexible approach is more likely to accommodate these concerns than the absolute approach.

4.8.4. INTEROPERABILITY

So far, the discussion focussed on the openness of a technical pay-TV platform for competitors. There may be situations in which service providers are not even interested in using a third party's platform, but wish to operate their own technical platform, for example, for security reasons, strategic marketing reasons, or because another technical system is superior. The presence of a dominant, proprietary standard can have a discouraging effect on these operators. Arguably, mandating interoperability could result in more competition in the facility and service market, because it would encourage the market entry of parties that prefer to operate their own technical conditional access platform, instead of using an existing one. An intra-platform scenario could, for consumers, very well lead to lower prices, more choice, broader availability of services across different platforms, diversity, etc.[829] The example of the Global System for Mobile Communications (GSM) standard showed that mandating one common standard could substantially increase consumer welfare and acceptance. Other examples of successful standardization are the digital compression standard DVB- Moving Picture Experts Group (MPEG) 2 and the standards for digital transmission (DVB-S, DVB-T, DVB-T) that paved the way for the proliferation of digital television, but also for the MPEG Audio Layer 3 (MP3) and the DVD standard.

The Access Directive recognizes for all facilities that fall outside of the scope of Articles 5 (1)b and 6 how important some form of interoperability between telecommunications services and/or other facilities can be as a precondition for functioning competition and consumer freedom of choice.[830] According to Articles 5 (1)a and 12 of the Access Directive, NRAs have far-reaching authority to mandate interoperability for telecommunications facilities and services at all levels of the telecommunications infrastructure.[831] In contrast, as far as facilities that are affiliated

828 Stern/Dietlein 1999, p. 4.
829 See only Besen/Farrell 1994, p. 120.
830 Access Directive, Recital 9.
831 Meaning at the level of physical infrastructure (interconnection between networks), network and carrier services (for example, roaming, switching) and teleservices (compatibility of operating systems), see section 1.3.

with the technical pay-TV platform are concerned, the European Union has adopted a more cautious approach and only encourages Member States to promote the MHP standard for the API. Having said that, the difficult and fruitless discussion in the European pay-TV market on achieving a common industry standard provides an example of the obstacles that must be overcome before an industry-driven standard can be achieved. Moreover, APIs are only one of many facilities that can cause interoperability issues. Similarly, and no less important, obstacles to the widespread availability of digital services at different platforms include the interoperability between rival conditional access systems, electronic payment systems, DRM systems, media players and plug-ins, and hardware devices.

This is not to say that interoperability solutions are always beneficial.[832] A negative example was the DH-MAC experiment.[833] Here, neither the industry nor consumers were prepared to accept a standard that was officially mandated but soon perceived as outdated and inferior. In such situations, mandated interoperability solutions can backfire and actually freeze standards that are technically not optimal, but are the result of a political decision. Publicly mandating one specific interoperability solution not only risks promoting an inferior standard, it can even be in conflict with the EC Treaty, as Spain experienced with the way it promoted the Multicrypt standard.[834]

Moreover, the experiences gathered in telecommunications regulation demonstrate that the practical difficulties of enforcing interoperability can be considerable. This became evident when the telecommunications networks were interconnected as a means of achieving interoperability in the fixed telephony market. One aspect that makes enforcing interoperability so difficult, is the fact that interoperability solutions are often enforced against the will of the incumbent who has chosen a particular (proprietary) standard for strategic reasons. For example, in the case of conditional access, control over a dominant conditional access standard can be an effective means of monopolizing large parts of the consumer base.[835] As a consequence, enforcement procedures can be complicated and cost-intensive. Forcing conditional access operators to make their system interoperable with other conditional access systems, and compatible with competing applications, could also mean discouraging competition. For example, Playstation owners certainly benefit from the fact that Playstation competes with the X-Box in terms of hardware and by producing more and more attractive games. One might wonder whether Microsoft and Sony would invest the same effort if their game consoles were compatible. In other words, the interoperability approach could remove important incentives to invest in ever more advanced technologies and new applications.

---

[832] A concise overview of the different economic arguments in favour and against mandating interoperability is given in Van Geffen/De Nooij/ Theeuwes 2002, pp. 55-64.

[833] See section 4.2.1.

[834] See section 4.4.3.

[835] See section 1.5.2.

On the other hand, if the devices were compatible, Playstation consumers could buy more easily games by Microsoft and vice versa, thereby stimulating the demand for more and better games from different sources.[836] Also, more third-party game writers might feel inclined to write applications that are compatible with both platforms. It is also unclear how much innovation consumers are willing to accept. Will consumers constantly buy the newest hardware or software updates? Or will the majority of consumers be content with durable and user-friendly devices? The answer to this question will probably also depend on the kind of services that are transmitted through the set top box—simple broadcasting services or video games and other more sophisticated applications.[837] The decision of whether or not to impose interoperability is, thus, also a question of policy objectives for each sector,[838] for example, whether to support competition in or between platforms.[839]

In any case, any decision to intervene should be the result of careful consideration about the need for intervention, the costs and the expected benefits.[840] This can also be the general obligation to make systems or services interoperable or compatible, without actually deciding on one particular standard. Again, the transport level cannot be seen separately, and regulatory goals for the service level can influence the final decision. Arguably, from a public information policy point of view, interests of consumers of films and other broadcasting content might be better served if consumers are able to access a broad range of content without having to invest in ever newer and better hardware because this could stimulate broad and affordable access opportunities and prevent social exclusion. This could be another argument in favour of giving NRAs the power to impose interoperability obligations in the digital broadcasting sector.

---

836 See also Besen/Farrell 1994, 119pp.

837 IPTS 2001, with reference to the Betamax/VHS case, 54pp. An interesting parallel that may demonstrate the importance and the controversial character of this discussion is the AOL Messenger problem and the question if AOL should be obliged to make its Messenger Network compatible for third parties. For an extensive discussion of the different arguments, see the case before the US Federal Communications Commission, decision from 11 January 2001 in the matter of applications for consent to the transfer of control of licenses and section 214 authorizations by Time Warner, Inc. and America Online, Inc., transferors, to AOL Time Warner, Inc., transferee, Memorandum Opinion and order, 22 January, 2001, CS Docket No. 00-30, paragraphs 128-190, available at <http://www.fcc.gov/Bureaus/Cable/Orders/2001/fcc01012.pdf> (last visited on 20 March 2005). See also the statement of Commissioner Michael K. Powell, concurring in part and dissenting in part, from 22 January 2001, available at <http://www.fcc.gov/Speeches/Powell/Statements/2001/stmkp104.doc> (last visited on 20 March 2005).

838 See also the comprehensive discussion in Larouche 2000, pp. 382-388 and 388-398; Van Geffen/De Nooij/ Theeuwes 2002, pp. 55-64.

839 Besen/Farrell, pp. 120 and 121pp., see also the papers by Farrell/Saloner 1985; Shapiro 2000, 7pp and 13pp., Larouche 2000, 383pp. See also sections 3.3.1., 3.3.2. and 4.4.2.

840 In this sense also e.g. Larouche 2000, p. 397.

4.8.5. REASONS TO MAINTAIN ARTICLE 6 OF THE ACCESS DIRECTIVE

The previous analysis begs the question of why the absolute access obligation in Articles 5 (1)b and 6 of the Access Directive should be maintained. Why not move to a more flexible approach, such as that provided for in Articles 8 to 13 of the Access Directive? Not only do Articles 8 to 13 of the Access Directive allow for a more flexible, technology-independent definition of bottleneck facilities that also takes the actual market power and structure into account. They also allow for a more flexible 'toolbox' of instruments to remedy bottleneck control, which may be better suited for the dynamics of information markets.

One important argument that favours treating bottleneck situations in digital broadcasting separately, concerns the purportedly substantial differences in industry structure between the broadcasting and telecommunications sectors.[841] However, control over access to the conditional access system and/or other elements of the transport level raises similar questions and concerns in the broadcasting and non-broadcasting sectors:

- In both sectors, a third party seeks access to the distribution infrastructure or parts thereof to generate profit from the consumer base the access provider controls.
- Likewise, monopolistic use of and refusal to grant access to the facility can block access to related markets and hinder potential competition and new market entries.
- The governing regulatory approaches to bottlenecks in the digital broadcasting and telecommunications sectors in general aim at market openness and deregulation.
- The focus of each set of rules, be it the regulation of bottlenecks in pay-TV or the regulation of bottlenecks in telecommunications markets, is the obligation to grant access to networks and facilities on fair, reasonable and non-discriminatory terms.
- Both sets of rules deal with arguments of scarcity and the balance of conflicting economic interests of the facility operator, competitors and consumers.

Still, one could argue that an absolute access obligation such as that in Article 6 of the Access Directive would ensure more legal certainty and should hence be maintained. On the other hand, as the analysis shows, it is still far from clear whether Article 6 of the Access Directive really provides for more legal certainty than, for example, Articles 8 to 13 of the Access Directive. Although this seems to be the case at first glance, this form of certainty is unreliable because it is based on parameters that may cease to be crucial. The most obvious example is the technology-dependent bottleneck definition at a time when it is becoming increasingly difficult to identify what broadcasting is. Other causes for legal

[841] For a concise overview of the industry arguments for and against a common treatment, see Ovum 2001, pp. 14-15.

uncertainty are the vague scope of the access obligation according to Article 6 of the Access Directive, the lack of guidelines of what adequate, non-discriminatory and fair terms and conditions are, and the unpredictable outcome of court decisions in access conflicts.

It is probably true that the process of determining market power up front brings an element of uncertainty with it. This applies in particular to the digital broadcasting sector, which has vertically integrated structures and fragmented audiences. However, ease of application should not be at the expense of an appropriate and proportionate solution. In areas in which the problem is less the control over the technical facility itself than that its use within the context of the market structure, the market structure should be tested for bottlenecks.

It is also argued that the regulation of broadcasting has traditionally required different and stricter rules because of a particularly strong social and public interest in the regulation of broadcasting.[842] This study is not the place to demonstrate whether this general policy argument is still valid. But even without such a demonstration, the argument is not very convincing within the context of the regulation of access to the technical platform. The Communications Framework claims that it is technologically and content neutral. This is also true for the technical pay-TV platform.[843] As discussed earlier, the European Parliament's suggestion to apply must-carry rules to pay-TV platforms was dismissed because the content-neutral Article 6 of the Access Directive was said to be sufficient when dealing with this kind of bottleneck problem.[844] Moreover, as explained in Chapter 2, the fact alone *that* some access-controlled broadcasting services are distributed and marketed through access-controlled intermediary platforms does not necessarily endanger the realization of public information policy objectives for this sector. Rather, it is the way access-controlled services are marketed to consumers that triggers public information policy concerns.[845] This again is a question that principally falls under the Communications Framework, even if the relevant rules have not yet been extended to the broadcasting sector.[846]

Finally, one reason to maintain Article 6 of the Access Directive could be the existence of urgent consumer interests. This is the need to protect consumers from so-called decoder towers, meaning a situation in which consumers have to install more than one set top box to receive broadcasting programmes from competing providers. Article 6 of the Access Directive could bring peace to the set top box war. The obligation of the established pay-TV operators to grant other broadcasters access to their conditional access platform may clear the way for the joint use of a single set top box. Consumers could then choose from competing offers without

[842] See Access Directive, Recital 10. See also section 2.2.
[843] See Access Directive, Article 6 (4). The act of providing broadcasting content still falls under the regime of broadcasting law.
[844] See section 2.3.3.
[845] See section 2.2.1.
[846] See also Access Directive, Recital 10.

having to face the costs of acquiring additional set top boxes. Ideally, the absolute access obligation would generate a pluralistic offering of competing broadcasting programmes that can be received through the same conditional access platform.

On the other hand, one may wonder if the decoder-towers argument is still valid, or, if it is, if decoder towers are the lesser evil. As previously explained, Article 6 of the Access Directive does not necessarily guarantee that consumers get the best and most innovative decoder standard; the provision supports the standard of the first mover. Moreover, the prices for set top boxes keep falling and many operators subsidize the provision of set top boxes for consumers. Finally, a regulation that promotes the consumers' ability to switch between different platforms and protects them against technical lock-ins would be equally, if not more, suitable to obviate abuse of monopoly power and encourage consumers to invest in hardware. If an adequate interoperability regime existed, there would be no standards war.

4.8.6. ARE ACCESS OBLIGATIONS THE OPTIMAL REMEDY?

A different question altogether is if access obligations are at all the optimal answer to monopoly control over technical facilities in pay-TV. The answer to this question is not only a matter of telecommunications regulation, but an assessment of the wider public information policy goals for the pay-TV sector, which were discussed in Chapter 2.

Mandated access regimes can be very questionable from the standpoint of static and dynamic efficiencies and consumer welfare. Access obligations could hamper investment by cutting down incentives to invest in technical innovation and improvement, and by discouraging other enterprises from doing so. As a consequence, so says Yoo, mandated access 'preempts the only solution to the bottleneck problem that is viable in the long run (i.e. the development of a viable alternative to the bottleneck facility)'.[847] McGowan further differentiates by saying that compelling access might be adequate in a natural monopoly market if it will not be profitable for competitors to replicate existing systems as here access obligations could increase competition, efficiency and consumer welfare. However, as McGowan continues, 'in situations not involving natural monopoly the market will support more than one firm... So long as replication is possible, the claim that a competing facility would be too costly to build is entitled to no weight'.[848]

As already mentioned, the bottleneck character of a conditional access system is more likely to be the result of control over a particular (proprietary) standard, than that the facility is a bottleneck facility as such. In such a situation, access obligations encourage the use of a particular conditional access standard. Because of the right of access, there is no need to develop costly alternatives. Encouraging

---

847 Yoo 2002, 246-247. In this sense also McGowan 1996, p. 806; Bittlingmayer/Hazlett 2002, 299pp and 308pp.; Veljanovski 2001, p. 119; Nihoul 1998, p. 213; differentiating Waterman, 528pp.
848 McGowan 1996, p. 805. See also section 3.4.1.

service providers to use one particular conditional access standard is likely to further increase the strategic and economic importance of this conditional access solution. This is the result of indirect network effects and the need to generate efficiencies and economies of scale and scope. The larger the number of services carried via a particular technical platform, the more attractive the platform will be in the eyes of consumers and, finally, of other service providers. Access to the first conditional access platform and its particular standard will become even more important for market entry, with the result that its controller will have even more influence on market developments. The stronger a particular standard becomes, the more likely is it that the market will 'tip' towards this particular standard and that this will result, finally, in a monopoly position. This was, for example, one reason why Netscape finally lost the competition against Microsoft Explorer. In the end, access obligations can contribute to a situation in which a sector is dominated by one proprietary standard—even when in theory different, competing technical solutions would have been possible. The dominant standard is very likely the standard of the first mover or the most popular pay-TV platform, even if it is not necessarily the most technically advanced or most consumer friendly. In other words, access obligations risk 'freezing' a certain technical standard. The question is who are the drivers behind innovation and the development of new, diverse digital content services—the large established players or highly motivated newcomers?[849]

If access obligations reduce demand for alternative systems or standards, this will not only affect competition between different conditional access technologies, but also competition between different pay-TV platforms. In pay-TV, the technical and the service platforms are closely integrated, which can make it attractive or even indispensable for major operators to operate their own conditional access system for strategic, security and efficiency reasons. As far as the entry of competitors is concerned, only those that can expect to generate similar popularity and scale in a relatively short time will consider entering the market. Others will rely on being able to use the established pay-TV platform. Access obligations might therefore favour further monopolization of the markets for conditional access *and* access-controlled services rather than discourage it.[850]

To make the situation more difficult, access obligations still leave ample room for the monopolist to influence competition in its favour at both the technical and the service level.[851] Much will depend on how effectively NRAs can ensure that the terms and conditions of access to a conditional access solution are fair, reasonable and non-discriminatory for rivals. The principle of strict separation between the transport and the service level in regulation and supervision does not make this task

---

[849] See Shelanski/Sidak 2001, p. 124.

[850] In this sense also Veljanovski 2001, pointing out that the Communications Framework was biased towards promoting short-term service competition a the expense of network competition, Veljanovski 2001, p. 120.

[851] See section 3.3.2.

any easier, at least not in areas in which activities at both levels are tightly integrated. In practice, the efficiency of access remedies will also depend on the way the broadcasting and telecommunications markets are supervised, and the extent to which telecommunications NRAs are entitled to take content-related aspects into account and vice versa. The example of electronic access control makes very clear that with the advancing economic and technological convergence between the transport and service levels, regulatory authorities must step out of their traditional field of expertise and authority. They can no longer restrict their activities to one level and ignore the other.

Access obligations are also controversial from a public information policy point of view. On the one hand, one could even argue from the public information point of view that a dominant pay-TV platform might have advantages. Arguably, a powerful privately controlled service platform could be an invaluable partner for media regulators in realizing public information policy goals, such as the digital switchover or broadband rollout. States could place burdensome tasks on the broad shoulders of a national media giant. Powerful pay-TV operators are major drivers of and catalysts for digitization; they invest in campaigns that convince reluctant consumers of the merits of digitization and develop the necessary equipment and attractive services. Finally, the access obligation could stimulate intra-platform competition because providing access-controlled services through the existing infrastructure would be less costly and more attractive to smaller operators in particular. The result might very well be a kind of 'internal pluralism' and more choice for consumers.

On the other hand, one may already wonder whether access obligations will indeed stimulate internal pluralism within a pay-TV platform. A reduction in economic freedom often comes hand in hand with a reduction of initiative and responsibility, in this case journalistic responsibility.[852] In such situations, access obligations can strengthen not only the economic, but also the journalistic influence of a platform operator. This is to say, even in areas in which a monopoly position of a pay-TV platform may still be acceptable from a competition point of view, the requirements of pluralism and diversity of sources raise serious concerns about whether it is desirable to actively promote the creation and strengthening of one dominant access-controlled pay-TV platform. Monopoly control over the pay-TV market can challenge fundamental objectives in broadcasting regulation, namely to prevent one private player from exercising excessive influence on large parts of the audience. In addition, the effect of one large and comprehensive pay-TV platform on the position of free-TV providers in general and public broadcasting in particular, must be taken into consideration, as well as the extent to which a pay-TV platform can affect their position when negotiating programme rights and

---

[852] See Wentzel, p. 172; Bullinger (1997), speaking of 'Massenprogrammhaltung' and the influence on the 'personality' of the individual programme, p. 763. Reduction of individual editorial freedoms is a necessary part of the strategic concept of service platforms, as otherwise it could not develop its own, distinguishable profile.

competing for audience attention. To conclude, access obligations could be counterproductive, not only from a competition policy point of view, but also from a public information policy point of view.

4.8.7. REFORM PROPOSAL

Having said that access obligations are not the best remedy to realize competition policy and public information policy objectives in the pay-TV market, the next question is what the better remedy is. Access obligations are a remedy for the symptoms of a lack of competition in the market for technical pay-TV facilities. That such a standard can prevail and influence competition in the facility market is due to a lack of competition in the technical facilities market, as well as a lack of competition in the pay-TV market itself. Providing consumers can choose from a number of equally attractive pay-TV offers, trying to dominate a share of the consumer base by means of a proprietary conditional access standard would be a risky and probably not very profitable strategy. The operator of that intermediary platform would risk loosing those parts of the audience that do not value its services highly enough to agree to subscribing to a platform that is incompatible with other services.

A better approach to tackle bottleneck situations in pay-TV could be to stimulate competition at the technical facility level, and more importantly, the service level. Creating the conditions that make the associated service market sufficiently competitive could decrease the prospects of anti-competitive leverage as a result from control over technical bottlenecks. A further advantage of stimulating competition at the service level is that it would prevent the existence of one powerful pay-TV platform from conflicting with guiding principles behind broadcasting regulation—the promotion of pluralism, diversity and fair access opportunities.

Of course, whether a pay-TV market becomes competitive or not depends on whether service providers develop sufficiently attractive service offers and applications. This is particularly true where pay-TV providers compete with an attractive free-TV offer. Experience will show if pay-TV will turn into a popular third form of broadcasting alongside advertising and fee-financed services, whether several competing access-controlled platforms can exist alongside each other in a given market and whether consumers will accept pay-TV. The author argued that digitization and convergence could have a stimulating effect on the take-off of and competition between access-controlled services. Competition in the pay-TV market, however, also depends on whether consumers are free to choose between competing offers. In other words, competition also depends on whether the regulatory framework foresees remedies to prevent that single operators monopolize the consumer base.

The existing regulatory framework under the Access Directive is very focused on the supply-side of the market. The access obligations in the Access Directive focus on the access of competitors to access-controlled platforms, ignoring entirely the impact that the contractual and technical relationship between service provider and consumer has for competition. To remedy the monopolization of the consumer base means to create conditions that enable consumers to access diverse services of their choice at fair, affordable and non-discriminatory conditions. Hence, an alternative approach to tackle a monopolization of the consumer base would be to shift the regulatory focus to the other side of the market—the retail side—and to create the conditions that enable consumers to choose and give them the freedom to choose by lowering their switching costs.[853] Adequate tools would guarantee the mobility of consumers and stimulate their willingness to switch between services. Suggestions of what these tools could be will be developed in the following paragraphs.

Contractual Lock-ins and Fairness of Contractual Conditions

As far as consumer contracts are concerned, the Universal Service Directive provides a legal framework that gives NRAs instruments to ensure that service providers offer services to consumers at adequate, non-discriminatory conditions and fair prices, and refrain from unjustified bundling strategies.[854] The Universal Service Directive acknowledges that for 'reasons of efficiency and social reasons, end-user tariffs should reflect demand conditions as well as cost conditions, provided that this does not result in distortions of competition'.[855] The Universal Service Directive also stipulates that service providers must provide consumers with a minimum level of legal certainty in subscriber contracts, concerning contractual terms and conditions, service quality, contract and service termination conditions, compensation measures and dispute resolution. Contracts must include information on prices, tariffs and terms and conditions in order to increase the consumers' ability to optimize their choices and thus fully benefit from competition.[856] These provisions provide tools to tackle contractual lock-ins and unfair provisions in consumer contracts. It remains to be seen whether these provisions are sufficient to meet consumer concerns in practice.

---

[853] Klemperer 1987, p. 391. See also the G7 Summit, Conclusion of G7 Summit 'Information Society Conference', Doc/95/2/, Brussels, 26 February 1995: 'The regulatory framework should put the user first and meet a variety of complementary societal objectives. It must be designed to allow choice, high quality services and affordable prices. It will therefore have to be based on an environment that encourages dynamic competition, ensures the separation of operating and regulatory functions as well as promotes interconnectivity and interoperability. Such an environment will maximize consumer choice by stimulating the creation and flow of information and other content supplied by a wide range of service and content providers'.

[854] Universal Service Directive, Article 17 (2).

[855] Universal Service Directive, Recital 26.

[856] Universal Service Directive, Recital 30, Articles 20 (1), (2).

Technical Lock-ins

Regarding the problem of technical lock-ins, the Universal Service Directive offers a solution regarding the interoperability of consumer equipment. According to Article 24 of the Universal Service Directive, Member States must ensure that any analogue television set has at least one open interface socket to connect additional devices and ensure interoperability. The provision, however, only refers to digital television equipment, not to set top boxes.

Questions concerning decoder interoperability are dealt with in the Access Directive and the Framework Directive. However, neither of them obliges operators in the pay-TV sector to provide for adequate interoperability solutions. In addition, under Articles 5 (1)b and 6 of the Access Directive NRAs do not have as many possibilities to impose interoperability obligations on pay-TV providers as they do for all other telecommunications service and facilities providers according to Articles 8 to 13 of the Access Directive. It is true that the discussion about the adequacy and desirability of mandated interoperability can be very controversial. Nevertheless, the author concludes that there are valid and important arguments in favour of mandated interoperability solutions in areas in which mobility and fair access opportunities would otherwise be at risk. It is difficult to see why the (un)willingness of major industry players towards standardization in pay-TV would change in the future and why necessary initiatives in this field should be further postponed. Without imposing a particular standard, initiatives could focus on making consumer equipment interoperable through open interfaces and open standards for software and middleware. The approach of Article 24 of the Universal Service Directive to mandate a common interface could serve as a model.

Information Problem

In terms of the information problem, the importance of comparable service information has already been acknowledged in the Communications Framework. The Universal Service Directive, and in particular the provisions on directory services and transparency obligations in Articles 21 and 22 of the Universal Service Directive, demonstrates that transparency, as a precondition for functioning competition and the realization of public information policy objectives, is taken very seriously in Europe. Access to telephony directory services, a kind of search agent, is even subject to an universal service obligation. The study argued in section 4.5. that the examples of programme journals for the broadcasting sector or search engines and browsers for the internet demonstrate that there can be a market for independent information agents and that competition between such services is generally possible. NRAs could seek to stimulate such competition. Article 21 (2) of the Universal Service Directive, a provision that mandates providing consumers with comprehensive, comparable and user-friendly service information, could serve as a model. So far, the provision does not apply to EPGs.

Pay-TV Subscribers Excluded from Rules on Consumer Protection

Most of the aforementioned provisions on consumer protection in the Universal Service Directive, however, such as the provisions on transparency and the publication of service information in Article 21 of the Universal Service Directive, are only applicable to providers of telephony services. Others only apply to electronic telecommunications services and associated facilities. Services providing content, such as the offer for sale of a package of sound or television 'broadcasting' content have been deliberatively excluded from the Universal Service Directive, except in Article 31, which is the provision on must-carry.[857] Again, this is an indication that, as far as broadcasting services are concerned, the Communications Framework handles a fundamentally different idea of the consumer-service-provider relationship than it does for the rest of the telecommunications sector. In the case of digital broadcasting, consumers are still not considered active market participants—which they are in pay-TV—but passive receivers.

Hence, pay-TV is a good example to illustrate why the technology-dependent concept of the regulation of the transport level is no longer adequate in media markets. Technology-dependent regulation is also not in line with European policies to promote multi-platform access to all kinds of digital services. Pay-TV services are distributed to consumers, similarly to telephony and other telecommunications services, on an individualized basis. Consequently, in pay-TV too, technical or contractual lock-ins, or the lack of transparency can be a means to monopolize the consumer base. With ongoing convergence, the differentiation between broadcasting and non-broadcasting services is not any longer justified as far as the modalities of the way services are marketed to consumers are concerned. It is difficult to see why subscribers to digital broadcasting services should be in a weaker legal position than subscribers to telephone networks or internet service providers.

4.8.8. SUMMARY

To conclude, the flexible concept under Articles 8 to 13 of the Access Directive is better-suited to react to bottleneck situations in digital broadcasting. There are no viable reasons to maintain Article 6 of the Access Directive. Moreover, the present top-down approach of dealing with electronic access control in European law fails to address the monopolization of the consumer base and the effect this has on competition and consumer access to content. To effectively address these concerns, this study concludes that a bottom-up approach that focuses on access conditions for consumers is needed. Consumers and their demands are key to stimulating competition and to disciplining the behaviour of the market parties.

---

[857] Universal Service Directive, Recital 45, Framework Directive, Article 2 (c).

4.9. Conclusion

Despite the formal common framework, the way the European Communications Framework approaches access questions is strikingly incoherent. The Access Directive, the Framework Directive and the Universal Service Directive are far from achieving the declared goal of realizing a horizontal, transmission-technology-independent approach when dealing with technical bottlenecks in general and conditional access in specific. Instead, the regulations still maintain a distinction between 'broadcasting' and 'non-broadcasting' services. In an age of convergence, this is an increasingly questionable distinction. Moreover, the approach towards bottleneck regulation that is provided for in Articles 5 (1)b and 6 of the Access Directive seems too simplistic and too inflexible to identify critical bottleneck situations in digital broadcasting and allow NRAs to respond with adequate remedies. Finally, Article 6 of the Access Directive ignores, as does the Communications Framework in general, that there is a retail relationship between pay-TV operators and consumers, and that the way this relationship is designed is relevant for the realization of competition and general public policy interests.

We have seen that there are no urgent economic or legal reasons that necessitate treating technical facilities in pay-TV differently from technical facilities in the telecommunications sector in general. The technology-dependent concept of Article 6 of the Access Directive conflicts with European policies for the digital service sector to promote a multi-platform' environment in which consumers have access to services irrespective of the technical distribution platform.

If regulators opt for access obligations as a regulatory tool, a more flexible technology-independent approach seems better-suited to respond to the challenges of technical bottleneck control. Articles 8 to 13 of the Access Directive, together with the relevant provisions of the Universal Service Directive, are a more promising way of addressing bottleneck issues and the monopolization of the consumer base than Article 6 of the Access Directive. The Access Directive should be adapted during a review to reflect this and abandon Article 6 of the directive. Until then, however, in order to prevent the obstruction of market development, an adaptation at the enforcement level is probably the best preliminary option. NRAs could narrowly interpret Article 6 of the Access Directive, with the effect that all conditional access systems that provide access to convergent services (which may soon be the majority) would fall under the flexible concept of Articles 8 to 13 of the Access Directive. Moreover, NRAs could interpret Article 5 (1)b of the Access Directive in the spirit of Articles 8 to 13 at least for EPGs and APIs.

Access obligations are not necessarily the optimal remedy in the pay-TV sector. Providing market conditions allow for it, a better approach would be to make the associated service market sufficiently competitive by tackling the monopolization of the consumer base. Giving NRAs the flexibility to impose interoperability obligations can be a means to remedy technical lock-ins. The provisions on transparency for consumers in Articles 21 and 22 of the Universal Service Directive respond at least in part to the previously identified information problem. The provisions on consumer contracts and retail conditions in Article 17 of the same directive provide remedies for contractual lock-ins.

One problem that is common to the access obligations as provided for in the present form of Article 6 and Articles 8 to 13 of the Access Directive, is the formal distinction between the transport and the service level when dealing with technical bottlenecks. In practice, this distinction is increasingly artificial and difficult to maintain in pay-TV markets, and all the more so as the nature of the problems involved is often the same for the transport and the service level. This has to do with the functional relationship between the technical conditional access platform and the service platform, meaning the subscription service itself. The technical pay-TV platform is an integrated part of the pay-TV operators' business model, whose core business is still the provision of access-controlled services through the service platform. Consequently, access decisions must also take into account content-related aspects, such as the content of the competitor's service, the form in which and the price for which it is offered to subscribers, etc. It is very questionable whether the Access Directive leaves room to take the functional and strategic unity between the service and transport levels into account. It is also not clear to which extent the notion of 'fair, reasonable and non-discriminatory' conditions under the Access Directive is to be interpreted within the context of content-related conditions. A more comprehensive assessment of pricing, foreclosure and interoperability strategies requires that NRAs are able to take both levels of a pay-TV operator's activities into account.

Chapter 5

Summary and Conclusion

5.1. Summary

Pay-TV is an example of a sophisticated content management and distribution infrastructure designed to sell digital content to consumers. In pay-TV, access to services is controlled electronically. The role that technology plays in this context is prominent—conditional access is a means of identification, authorization and control. Conditional access also enables distribution channels to be secured and lasting commercial relationships to be established with consumers.

This study provides a critical analysis of how current European broadcasting law, competition law and telecommunications law have responded to the challenges from electronic access control for competition and consumer access to access-controlled content. It examined how access-controlled platforms in digital broadcasting are regulated, and if the existing regulatory framework is adequate to realize the policy goals for this sector. The purpose of this chapter is to summarize the main findings, and provide some conclusions about the current framework, its inadequacy to deal with electronic access control in digital broadcasting and how it could be reformed. In a last section, Chapter 5 explains why electronic access control is a phenomenon that is not reserved to the digital broadcasting sector.

5.1.1. Changes and Regulatory Implications

As an element of a business model for digital broadcasting, electronic access control changes the sector's traditional distribution patterns in different ways. These changes impact the applicability and adequacy of current pay-TV regulation. They also trigger the need to rethink existing concepts and have led to a number of regulatory initiatives at the European level.

The most obvious change is that electronic access control allows for fine-grained private control over individual access to broadcasting content. Conditional access allows for a high degree of individualization and differentiation. This is why conditional access is well-suited for business models that are based on strategic exclusion, meaning, on the one hand, the exclusion of consumers from access to the content they are not willing to pay for, and, on the other hand, the exclusion of competitors from access to consumers. One can imagine conditional access as an intelligent, strong wrapping paper that is able to decide who may unwrap which particular content under which conditions. Unlike traditional parts of the transmission infrastructure, conditional access is not blind to the content of the service distributed; instead, the content and the conditions attached to it are made an integral part of the technical routine. This is why the use of conditional access plays a role for transport-related and content-related questions, notably the conditions under which consumers can access content.

Secondly, electronic access control solutions are often not operated by the broadcasters themselves, but by intermediary platform operators, such as BSkyB or Canal+. The intermediary platform takes on a middle-function between consumers, operators at the service level and operators at the technical distribution or, as the study calls it, transport level. Such platform operators are 'information brokers' that bring third-party content providers and consumers together and present consumers with content services, content à la carte or in pre-selected packages. Like business-to-consumer (B2C) platforms, intermediary platforms can and will engage in marketing strategies other than single broadcasters would or could do. Traditional free-TV broadcasters sell consumer attention to advertisers and/or finance themselves through public fees. Pay-TV platforms can also offer advertising-financed content, but their main business strategy is to sell content directly to consumers. The economic success of intermediary platforms depends on their ability to get, on the one hand, attractive service providers, and, on the other hand, large subscriber numbers on board and keep them. Pay-TV platforms can attract consumers by offering them a whole 'world' of services. Such platforms concentrate on the commercial side of packaging and selling content to consumers. In so doing, they can generate efficiency advantages individual broadcasters would not be able to generate in the form of (large-scale) bundling and price differentiation models.

This also means, however, that control over a dominant conditional access standard, a popular programme package, or an Electronic Programme Guide can result in bottleneck situations for competition. In areas in which a platform operator succeeds in monopolizing its consumer relationships, it can be more difficult for consumers to switch platforms and for rivals to gain access to this part of the consumer base.

Regulatory Implications

Once consumers have subscribed to one pay-TV platform, they are entangled in a web of contractual, technical and informational relationships with its operator. The privately dictated terms and conditions of access in subscriber contracts overrule one of the basic concepts on which broadcasting law is based, namely that once a service has been sent, it is principally 'freely' available. This also means, however, that the supremacy of broadcasting law to shape the broadcasting offer and to guarantee that the latter complies with a number of public information policy rationales is eroded by the exercise of private control over consumer access to broadcasting content. With pay-TV, the controller of the intermediary pay-TV platform decides which content citizens can watch under which conditions. Broadcasting regulators are confronted with new issues, such as bottleneck control, access issues, and how to prevent platform operators from abusing their commercial relationship with consumers to exercise excessive influence over choice and the accessibility and availability of digital broadcasting content.

Restricting choice and participation can conflict with the realization of competition, the Internal Market and the public information policy interests that govern this sector. Chapter 1 and 3 described the competition problems involved, and why conditional access can become a bottleneck to market entry. Chapter 2 explained why electronic access control in broadcasting can raise considerable concerns regarding the fragmentation of society into subscribed and unsubscribed citizens. It discusses the exclusion of citizens who cannot, for technical or financial reasons, have a subscription. Pay-TV can bring inequalities into the 'broadcasting' world as we know it. Finally, controlling access to access-controlled platforms can also conflict with the interests of consumers in exercising choice, accessing services from other Member States and fair dealing when 'purchasing' digital broadcasting content.

Traditionally, issues of access to the distribution platform were a matter for European telecommunications law. The question is whether the access rules in telecommunications law, which deal with conditional access, are able to effectively address access issues for both competitors and consumers in the pay-TV sector.

5.1.2. Electronic Access Control as a Bottleneck to Market Entry

'Bottleneck' control refers to a situation of monopoly control over a scarce resource, be it an element of the technical level or the service level. Scarcity can have legal (subject to exclusive licenses), practical (not easy to duplicate) or market-related reasons, for example, when the controller of a resource has economic reasons to exclude rivals from accessing that resource. Bottleneck control becomes a competition issue when it prevents competitors from entering a market.

Bottleneck control poses several challenges for regulators. The first challenge consists of finding a formula to determine which facilities are potential bottleneck facilities and which conditions must be fulfilled before refusing access to this facility has an anti-competitive effect. This is because, as long as monopoly control is not accompanied by abusive behaviour, it is very likely to be accepted as the result of a successful business strategy, at least according to competition law principles. From a public information policy point of view, there can be other public policy reasons why monopoly control, even without abusive behaviour, is not desirable. An example is that bottleneck control would result in excessive private influence in the broadcasting sector. Another challenge consists of finding adequate and effective remedies.

There are no Bottlenecks as such in Pay-TV

In the general competition law analysis in Chapter 3, we saw that bottleneck facilities are identified on a case-by-case basis. Within the framework of merger analysis, the European Commission sought to avoid ex ante the creation of anti-competitive structures that could result in bottleneck situations. Under the heading 'Essential Facilities Doctrine', the European Court of Justice and the European Commission specified the conditions under which competition law authorities can remedy existing bottleneck control ex post. These conditions are dominant market power, the fact that the facility is essential for market entry and the creation of a new product and, last but not least, the fact that it cannot be reasonably expected that a rival duplicates the facility.

Articles 5 (1)b and 6 of the Access Directive, the relevant provisions in European telecommunications law that deal with bottlenecks in pay-TV, pursue another approach. The provisions list specific facilities—conditional access, Electronic Programme Guides and Application Programme Interfaces—and declare them as bottlenecks as such, principally irrespective of their controller's degree of market power or the competitive situation in general.

The study criticized the sector-specific approach towards bottleneck definition under Articles 5 (1)b and 6 of the Access Directive for a number of reasons. First, the bottleneck character of control over a conditional access system, an Electronic Programme Guide or an Application Programme Interface, is often not so much the result of control over the facility, but of specific market conditions, notably control over a proprietary de facto standard—possibly strengthened by the existence of intellectual property rights—together with a lack of adequate interoperability solutions. This is closely related to the existence of strong indirect network effects, the degree of market power of the bottleneck controller and individual/collective adaptation costs in the pay-TV sector. It is true that Article 6 (3) of the Access Directive leaves room for NRAs to make the imposition of access obligations subject to a market test. However, we also saw in Chapter 4 that Article 6 (3) of the Access Directive is subject to a number of conditions that could restrict its application in practice. In addition, as Chapter 3 and the analysis of the European Court of Justice's interpretation of the Essential Facilities Doctrine and the European Commission's decisions in the pay-TV sector have demonstrated, the existence of a controller with significant market power alone does not make a facility a bottleneck facility, unless the controller has incentives to refuse access to that facility because it is profitable. Whether refusal of access to a facility is profitable or not depends on different factors, such as the degree of vertical integration, the core business of the enterprise and the competitiveness of related markets. Imposing access obligations on facilities that are not true bottlenecks risks creating overregulation and negative investment incentives. Articles 5 (1)b and 6 of the Access Directive do not leave sufficient room for National Regulatory Authorities to take these considerations into account as Articles 8 to 13 of the Access Directive do. Articles 8 to 13 of the Access Directive apply to bottleneck situations concerning all telecommunications infrastructures and facilities that do not fall under Articles 5 (1)b and 6 of the Access Directive. Instead of pre-defined bottlenecks and obligations, telecommunications National Regulatory Authorities identify critical bottlenecks within the context of the market situation.

No valid reasons were found that would justify discriminating between broadcasters and non-broadcasters and between bottleneck facilities in pay-TV and other technical bottlenecks. The study rejects the assumption underlying Articles 5 (1)b and 6 of the Access Directive that the conditional access system, Application Programme Interface or Electronic Programme Guide can only be bottlenecks for broadcasters and not for providers of interactive or telecommunications services. Second, the Application Programme Interface, the Electronic Programme Guide and the conditional access system are not the only potential bottlenecks in pay-TV.

By distinguishing between providers of access-controlled broadcasting and non-broadcasting services, for example, providers of on-demand services that are delivered via the internet, Articles 5 (1)b and 6 of the Access Directive create artificial barriers between different kinds of services that are eventually received via the same set top box. This outcome threatens to obstruct the realization of the so-called multi-platform approach, a guiding rationale of European policies in the converging media environment. The flexible approach under Articles 8 to 13 of the Access Directive better reflects market reality.

Access Obligations—Not the Optimal Remedy in Pay-TV

Secondly, with the increasing sophistication and diversity of the digital service market in general and the underlying transmission processes in particular, the question of what is needed to respond to bottleneck problems and obstacles to market entry is becoming increasingly difficult to answer. In this respect, too, Articles 5 (1)b and 6 of the Access Directive leave National Regulatory Authorities little flexibility to decide what the most efficient remedy against anti-competitive bottleneck control is. Moreover, Articles 5 (1)b and 6 of the Access Directive leave little room for telecommunications National Regulatory Authorities to differentiate between anti-competitive and non-anti-competitive control over technical facilities in pay-TV. As opposed, according to Articles 9 to 13 of the Access Directive National Regulatory Authorities can choose from a toolbox of different instruments the most effective remedy. Possible regulatory measures include access obligations, transparency enhancing measures and obligations to publish technical specifications. Bearing in mind that the long-term goal of the Communications Framework is to deregulate and prevent overregulation, an orientation towards a more flexible approach and general economic and competition law principles would seem the better route to follow.

Access obligations, meaning the obligation to share one's facility with competitors, are a traditional remedy for bottleneck situations in general competition law and telecommunications law. They are, however, not by default the adequate remedy for bottleneck situations. Access obligations are a means of controlling the behaviour of the dominant bottleneck controller. This kind of remedy might be necessary and appropriate in mature markets in which new entries seem unlikely without forcing existing players to share crucial resources. Access obligations may be necessary in situations in which a bottleneck facility is a natural monopoly or an essential facility, namely when the duplication of the facility is not an option and when the refusal to provide access would prevent market entry. Here, forced access to the facility may be the only viable way to stimulate competition in a related market.

In the case of pay-TV, it is already very questionable whether there are facilities that cannot be duplicated, or for which no alternative is available. The development of conditional access systems, set top boxes, Application Programme Interfaces, Electronic Programme Guides, etc. is no any longer driven only by a small number of pay-TV providers. Major software and hardware manufacturers, such as Microsoft and Nokia, are already in the business of developing conditional access systems and even complete technical solutions for the distribution of pay-TV. With digitization and the development of more differentiated service offerings and alternative forms of payment, the demand for electronic access control systems will be further stimulated by broadcasters and non-broadcasters. It follows that the conditional access system, the Application Programme Interface and the Electronic Programme Guide are probably all facilities for which the duplication of and competition between alternative systems in a market is possible and, arguably, desirable.

The study explained that from a competition policy point of view access obligations are controversial. After examining general competition law and sector-specific access obligations, it is apparent that access remedies that were successfully applied to traditional bottleneck situations are not necessarily the best or most effective way of dealing with ‘new’ technical bottleneck situations, such as control over conditional access. It may be true that there are situations in which access obligations are necessary and helpful, notably in settled and established markets. As far as the pay-TV sector is concerned, one should be careful not to take this assumption for granted, as much will depend on the time-frame. In the short term, at least in smaller pay-TV markets, competition between competing platforms in one market might seem unlikely. In the medium and long term, however, the broadcasting sector will face stimulating effects from digitization, which is a declared policy goal of the European Union as well as the European Member States. Access obligations should be imposed only in exceptional situations in order to avoid overregulation and negative investment impulses. Furthermore, the possible side-effects of access obligations—to stifle demand for alternative facilities and increase the power of one particular standard—should not be overlooked. Relying on access obligations as a remedy to bottleneck situations risks promoting a static position of the established players in the pay-TV sector at the cost of promoting a dynamic competitive environment in the future.

Finally, the study concluded that for similar reasons access obligations could be counterproductive, not only from a competition policy point of view, but also from a public information policy point of view. Monopoly control over a pay-TV platform may still be acceptable from a competition point of view. However, the requirements of pluralism and diversity of sources raise serious concerns about whether it is desirable to actively promote the creation and strengthening of one dominant access-controlled pay-TV platform. The risk is that one player can exercise excessive influence on large parts of the audience as well as influence the position of free-TV providers in general and public broadcasting in particular.

5.1.3. Consumer Access to Access-Controlled Platforms

The European Communications Framework leaves questions about the distribution of pay-TV services to consumers, that is questions concerning the retail market, outside of its scope. The broadcasting retail market has traditionally been a matter of broadcasting regulation and general competition law. Main objectives behind broadcasting law are the realization of public information policy principles, such as the realization of freedom of expression, fair access opportunities, democratic principles, the broad availability of information of general importance for the public, and the free flow of services in the Internal Market. The realization of these goals pivots around the question of how to ensure that a range of plural and diverse content is broadly available and accessible to consumers, alias citizens.

Electronic access control triggers valid concerns about the general availability and accessibility of content for consumers, including access to transborder services. This is particularly true for content of public interest—often that is the content that is most likely to disappear behind the electronic fences of pay-TV. Other concerns are the quality of the content offering, including the principles of diversity and pluralism, as well as the future of public broadcasting as a qualitatively high-standing programme offering. More generally, the latter is a question of how electronic access control impacts the popularity of free-TV programmes, including public broadcasting programmes and competition between free-TV and pay-TV. Less outspoken but no less pressing, are distinct policy concerns of national governments about their ability to use the broadcasting medium directly or indirectly for the realization of public policy objectives. Both the general availability of and the choice between information sources become a question of access and slips away into private control.

Broadcasting law does not seek to accommodate these concerns by granting consumers access rights to access-controlled platforms. European broadcasting law has taken another route. The approach of current European broadcasting law that deals with pay-TV is to make control over access to particular content non-exclusive. The underlying idea is that certain content should also be available outside access-controlled platforms. Moreover, this is the idea that it is the role of the media to act as an intermediary between the audience and the information source and, in so doing, to realize the 'right of the public to being properly informed'. The role of the audience or of individual members of the audience is best described as the role of passive information receivers, not as active information seekers who negotiate access to content they wish to receive.

This approach is reflected in the two main instruments that European broadcasting law has developed within the context of pay-TV. These are the right to short reporting and the list-of-important-events concept.

List of Important Events

Article 3a of the Television Without Frontiers Directive recognizes a right of Member States to ensure that particular popular sports or cultural events can be transmitted in full or in part, in parallel or deferred on free-TV. Member States are entitled to draw up so-called lists of important events, meaning events that should not be exclusively transmitted on pay-TV. Hence, the list-of-important-events concept mandates the non-exclusivity of particular content that is considered of major public interest. The role of free-TV in this concept is to act as a carrier of general interest content and make it publicly available.

Right to Short Reporting

The right to short reporting enables all broadcasters to inform their audience about particular news, sports and cultural or social events, even if a third party holds the exclusive rights to the transmission of that event. A precondition is that the broadcaster considers the event to be of high public interest and newsworthy, and that the transmission complies with the requirements of a 'short report'. Unlike the list-of-important-events concept, the right to short reporting does not touch on the exclusivity of a transmission right. Instead, the right to short reporting, or rather its effect, is better compared to an exception in copyright law. It gives broadcasters the right to make certain use of material to which another party holds exclusive rights. In contrast to the list-of-important-events concept, the right to short reporting does not aim so much at ensuring consumer access to a particular event, but at consumers being informed about the essence of the event to the extent that it is newsworthy and of public interest. Another difference is that, for the right to short reporting, the broadcasters determine which information is of public interest, whereas this task is left to the governments for the list-of-important-events concept.

Helpful, but not Helpful Enough

Chapter 2 examined the list-of-important-events concept and the right to short reporting in depth and concluded that, in combination, both provisions might provide some guarantee that those parts of the audience that are unsubscribed do not remain ignorant if particular events are shown through access-controlled broadcasting. The list-of-important-events concept aims at the broad accessibility of events of public importance, free from exclusive electronic control. The right to short reporting addresses the impact of exclusive electronic control over content for the media's mission to keep the audience properly informed. Both instruments are valuable tools that can stimulate the broad accessibility of content in free-TV. They are also regulations that can stimulate competition between services. Both regulations restrict the exclusive control over particular content thereby stimulating journalistic and economic competition between free-TV and pay-TV providers. Having said that, both instruments were drafted to 'protect' consumers from pay-TV, meaning to remedy the effects of electronic exclusion. They do not tackle the question of the consumers' situation regarding content that is offered on pay-TV.

One can doubt whether the impact of electronic access control on the accessibility and availability of digital content can be addressed effectively by simply relying on free-TV as a solution. One reason to doubt this is based on the valid assumption that when digitization is completed, electronic access control in digital broadcasting will gain in popularity and importance as a business model. If digitization progresses and electronic access control becomes a common form of collecting remuneration, pay-TV will more than ever be part of the media landscape. It is not unlikely that, for consumers, pay-TV could become a common and perhaps even particularly popular way of consuming digital content. It is already very questionable whether the notion of a broadly available free-TV programme will, under these circumstances, still carry the same value as before. Most importantly, however, protecting consumers in such a situation 'from' pay-TV instead of ensuring that pay-TV services are offered in a way that respects their legitimate interests is like holding on to an idea from the past with all one's might.

It is also worth noting, that, despite what the pay-TV critics say, pay-TV does not necessarily have to be a development for the worse. Conditional access can also offer new and attractive opportunities. Conditional access explores new ways of collecting remuneration, and the technology offers ample opportunities for innovative, interactive and differentiated service offerings. Now, content offered on pay-TV is mostly entertainment content, such as films and sports. The question is if this must necessarily stay this way, or if electronic access control could be used to finance other, high-quality content. Conditional access could be a means of improving media responsiveness. As Brittan puts it:

> 'If the state is to pay the piper, or regulate the piper's activities, it will eventually seek to call the tune—to the amazed indignation of the broadcasting fraternity which thinks it can have the benefits of state finance or regulation without paying the costs'. [858]

Individualization and interactivity can provide consumers with a way to pay the piper and express their preferences through a means the market understands: Euros, dollars, etc. Maybe the time is ripe for the audience to use technical and market developments to actively exercise choices in digital broadcasting. Pay-TV could give the audience the opportunity to become more than 'eyeballs', namely active players. A necessary precondition is that consumers have a choice and the freedom to exercise it. Without claiming completeness, this study has identified some factors

[858] Brittan 1989, p. 32.

that explain why the position of consumers who wish to access content on access-controlled platforms and exercise choice can be weak.

Concern No. 1: Fairness of Contractual Conditions
For consumers to benefit from pay-TV, the services offered to them must be acceptable. The adequacy and fairness of terms and conditions under which pay-TV services are offered will influence the consumers' willingness to accept access-controlled services. In a study that was performed in the UK for the former British telecommunications regulator Oftel, consumers indicated that they tend to feel overpowered by pay-TV operators, which they described as 'strong global players that ... could, ... and would, do exactly what they want'.[859] Examples stated were the changing of the configuration of packages, deleting key channels from packages consumers had subscribed to, moving the best-quality programmes to pay-per-view programmes (particularly sports and films), and pricing issues, such as the demand for fees in addition to the monthly subscription.[860]

More generally, the arrival of individual consumer-service-provider contracts in the broadcasting sector calls for more attention to be paid to the way subscriber contracts are drafted and to ensure that terms and conditions of access are fair. Common not only to the broadcasting sector are examples of potentially unfair conditions, such as the demand for unreasonably high prices and the imposition of conditions that are in no way related to the request for content. This can be the requirement to provide personal information on age, number of children, education, profession, etc. when subscribing to a pay-TV service, or the condition that the consumer must accept information mail. Subscription contracts can force consumers to subscribe to more channels than they wish to receive as the result of extensive bundling strategies. Conditions can directly or indirectly conflict with established rights and principles: privacy concerns and concerns about far-reaching control to the access and usage of digital content are further issues. The latter fall, in part, under privacy and copyright regulations, fields of law in which it might be worth conducting further research.

To examine the legitimacy and adequacy of contractual conditions in pay-TV subscription contracts could be the subject of a study on its own. Relevant for the context of this study is its more abstract observation that the fairness and adequacy of contractual conditions in the commercial relationship between consumers and pay-TV providers is a matter that falls outside traditional broadcasting law. Fairness of contractual conditions is a matter that is better known from consumer protection law. This is another field that needs further research. Traditional broadcasting law does not have provisions for consumer protection. It does not acknowledge the existence of a commercial relationship between the consumer and the service

[859] Counterpoint Research 2001, p. 21. See also the submission by Voice of Listeners & Viewers 2001.
[860] Counterpoint Research 2001, 9pp. and pp. 23-38.

provider. This deficiency can be explained by the 'once-sent-accessible-for-all' character of traditional broadcasting services.

*Concern No. 2: Contractual Lock-ins/*Lock-out*s*
Nobody would seriously argue that the choice between a basic channel, a family package and a premium channel would allow consumers to express any specific preferences, not to mention the ability to influence the structure of broadcasting markets—particularly if all three bundles came from the same provider. Competition between different access-controlled services, consequently, would be true only in an environment in which consumers have the choice between a considerable number of affordable competing offers and sufficiently small or differentiated programme units, and in which they are in a position to actively exercise this choice. A further major obstacle for the realization of fair access opportunities and the possibility to exercise choice could be the effect of bundling strategies. Three particularly critical examples in this context are large-scale programme bundling, the bundling of premium and basic services and making the subscription to a pay-TV platform conditional on the purchase of a particular conditional access technology. Other no less critical forms, are forms of bundling in time, notably long-term subscription contracts and other such terms that restrict consumer mobility.

The potential anti-competitive and welfare decreasing effects of such strategies require careful scrutiny. To some extent, bundling strategies can be addressed applying competition law. Chapter 3, however, demonstrated why the tying prohibition in general competition law is subject to major uncertainties, vagueness and timely delays and why, furthermore, this is not a provision that can be easily invoked by consumers.

*Concern No. 3: Technical Lock-ins/*Lock-out*s*
Of major importance is the aspect of the lack of adequate interoperability solutions. The lack of interoperability solutions can result in technical lock-in or lock-out situations that deprive consumers of the possibility to access services of their choice. Interoperability in information markets in general and in pay-TV markets in particular is a topic that has already received much attention. And rightly so. This study concluded that choice is also a matter of the quality of consumer equipment and whether it is able to process rival services, including services form other Member States. This is also why interoperability is a major factor for consumer acceptance of pay-TV services on a large scale.

Concern No. 4: The Information Problem
The last major factor in this list, the importance of which cannot be emphasized enough and whose dimension in the broadcasting sector is still often not properly understood, is what the study calls the 'information problem'. Behind the information problem lies the realization that access and choice are, among other

things, a matter of access to trustworthy, comprehensive and comparable information about the available services from different providers. The ability to find and compare such information can be a major problem in a digital multi-channel environment. The importance of search engines for access to content on the internet illustrates this situation well. The search engine example also illustrates the value of transparency in the form of reliable unbiased information as a means to overcome informational lock-ins.

The regulation concerning Electronic Programme Guides in European law lack vision in this respect. The potential impact of the Electronic Programme Guide on economic and ideological competition, democracy and culture is still widely underestimated. For the time being, the public information policy discussion concerning Electronic Programme Guides still pivots around the question whether public broadcasters should receive a prominent place or not. But the potential of information agents, such as the Electronic Programme Guide, goes far beyond ensuring that consumers have access to particular programmes of particular public interest. Looking towards the future, information agents have the potential of becoming platform-independent gateways for access to content irrespective of national borders and language obstacles or of the technical transport platform used.

5.1.4. Reform Proposal: A Bottom-up approach

To conclude, this study suggests that tackling electronic access control in digital broadcasting is tackling at least four issues—fairness of access conditions, contractual lock-in or lock-out situations, technical lock-in or lock-out situations and the information problem. A future goal for the regulation of pay-TV should be to ensure that pay-TV platforms are publicly accessible, meaning that the terms and conditions of access are such that all members of the public are not arbitrarily excluded for technical, financial or transparency reasons. In the information society, each citizen should be able to benefit from new services being made available by means of advanced communications: 'The information society is not only affecting the way people interact but it is also requiring the traditional organisational structures to be more flexible, more participatory and more decentralised'.[861] Likewise, subscribers to one particular platform should not be unreasonably impeded from benefiting from pluralism between different platforms because of technical, contractual or informational lock-ins. The fairness and openness of the individual commercial relationship between service provider and consumer is key to preventing that electronic access control is used to the detriment of competition, consumers and public information policy.

There is no need to reinvent the wheel. Technical and contractual lock-ins and information problems are not new issues, and they have already been answered for

[861] G7 Summit, Conclusion of G7 Summit 'Information Society Conference', Doc/95/2/, Brussels, 26 February 1995.

other sectors. For all media sectors other than the broadcasting sector, the key role that the fairness of the conditions of access for consumers plays in guaranteeing the accessibility of electronic services has already been widely acknowledged. This is why the Universal Service Directive aims to

> 'ensure the availability throughout the Community of good quality publicly available services through effective competition and choice and to deal with circumstances in which the needs of end-users are not satisfactorily met by the market' (Article 1 (1) of the Universal Service Directive).

The Universal Service Directive, part of the Communications Framework, acknowledges that functioning competition and the broad availability of a range of different telecommunications services for consumers is not only a matter of access for service providers to telecommunications networks and facilities, but also a matter of access for consumers to services. A study of the provisions of the Universal Service Directive revealed that it would address the four aforementioned factors, namely fairness of contractual conditions, contractual lock-ins, technical lock-ins and the information problem, were it not for the pitfalls of technology-dependent regulation.

The existing European regulatory framework might have been a good starting point to address urgent access issues in the digital broadcasting sector if the Communications Framework had kept its promise to realize a more technology-independent approach to regulation. None of the potentially relevant provisions in the Universal Service Directive (in particular Articles 17, 20, 21, 22 and 32), with the exception of Article 24 (interoperability of television equipment), apply to the broadcasting sector. It is difficult to see why consumers of digital broadcasting services should receive less legal protection than consumers of other electronic services.

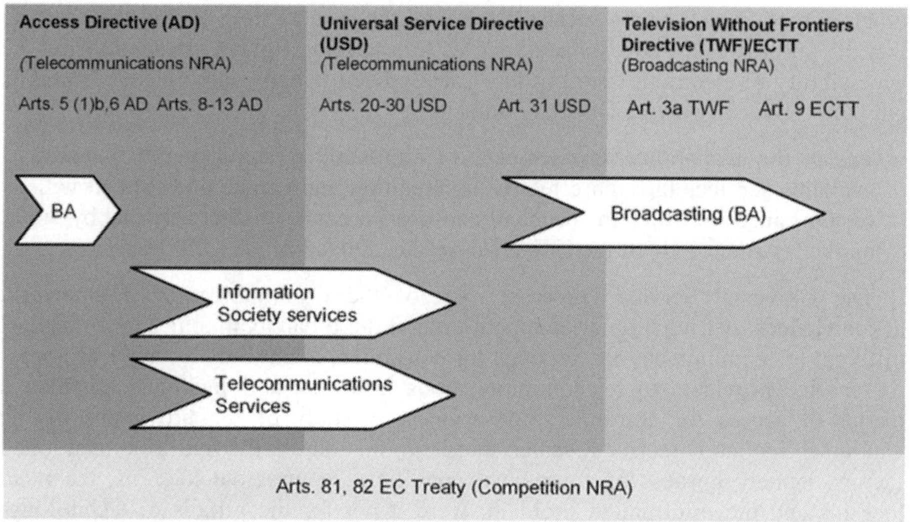

Figure 10—Different regulatory frameworks applicable to broadcasting, telecommunications and information society services. Figure 10 provides a schematic overview of the different regulatory frameworks that apply to broadcasting, telecommunications and information society services offered via one and the same pay-TV platform. The figure illustrates the aspects of access to the technical infrastructure (Access Directive: Articles 5 (1)b and 6 (access to the conditional access platform), and Articles 8 to 13 (access to telecommunications services and facilities). It also illustrates how differently the position of consumers (Universal Service Directive: Articles 17 to 30 (consumer access to services) and Article 32 (must-carry)) and the accessibility of content (TWF: Article 3a (list of important events), ECTT: Article 9 (the right to short reporting)) is regulated at the European level. This is done despite the converging nature of pay-TV platforms and the alleged horizontal approach of regulation under the Communications Framework.

5.2. Conclusion

The study concludes that there is no convincing reason to treat electronic access control in the digital broadcasting sector different from other technical facilities in the telecommunications sector. The above figure 10 illustrates the different regulatory environments that apply to broadcasting and non-broadcasting services. It shows that the broadcasting sector falls under a differing access regime, and that the rules on consumer protection and control of the retail market do not apply to broadcasting services. The present approach creates unnecessary obstacles for the realization of competitive and convergent service markets and the so-called multi-platform approach. Furthermore, the figure illustrates the strict separation of content and transport-related aspects. This is a further obstacle to the effective regulation and supervision of access-controlled service platforms. Because of the often tight integration of the technical and content platforms as elements of one and the same service platform, a number of interactions take place at both levels. Regulating and supervising the technical platform while ignoring content-related aspects or the retail market in general is difficult, if not impossible. As the figure shows, the approach leaves a regulatory gap: consumer access to access-controlled services.

The study further concludes that access obligations, such as provided for in Article 6 of the Access Directive (access to the conditional access system), are not always the best tool to resolve bottleneck situations in the pay-TV sector. Access obligations are also not sufficient to address valid competition and public information policy concerns that arise in conjunction with conditional access. What is needed are tools to remedy the abuse of the commercial relationship between consumers and pay-TV providers. Most importantly, a stronger focus on the position of consumers is needed. Without claiming completeness, the study suggests to give National Regulatory Authorities four tools to deal with conditional access:

- The ability to mandate interoperability solutions.
- The ability to prohibit anti-competitive bundling practices in the retail market.
- Transparency enhancing measures, including the public availability of comparable service information.
- The ability to monitor the adequacy of contractual conditions, pricing, etc. in pay-TV provider-consumer contracts also in the digital broadcasting sector.

In addition, National Regulatory Authorities should have sufficient flexibility to consider market-related factors, such as the likelihood of new market entries, economies of scale and scope, the degree of product diversification, the existence of market entry barriers, the need to recoup investments as well as security and capacity considerations.

These tools already exist for the non-broadcasting sector, which is why this study concludes that the various distinctions that still rule the Communications Framework are not helpful at all. The study advocates abandoning Article 6 of the Access Directive, which is still based on technological distinctions and which does not give NRAs a lot of flexibility. Instead, the more flexible approach in Articles 8 to 13 of the Access Directive should apply. Second, it suggests interpreting Article 5 (1)b of the Access Directive (access to Application Programme Interfaces and Electronic Programme Guides) in the sense of the flexible concept under Articles 8 to 13 of the Access Directive. Third, it recommends putting an end to the exclusion of the broadcasting sector from the application of the Universal Service Directive, and in particular Articles 17, 20, 21, 22 and 32, to ensure that consumers can have access to access-controlled services at fair, affordable and non-discriminatory conditions. Discrimination on grounds of residence should also be acknowledged as a form of discrimination. Doing this would be a first step in ensuring that consumers within the European Union have fair access opportunities when confronted with electronic access control. Strengthening the position of consumers is in the interest of functioning competition in the digital service sector. Strengthening the position of consumers is also in the interest of consumers, alias citizens, in an access-controlled broadcasting environment. Finally, strengthening the position of consumers is in the interest of public information policy where it is directed at utilizing new technical and economic developments and creating the conditions that enable consumers to optimally benefit from the possibilities of digital media.

Further research should explore the question of exactly how helpful existing consumer protection law can be to address all consumer concerns for the pay-TV case. This study, furthermore, repeatedly pointed to the fact that subscribers are not only consumers but also citizens. General consumer protection law applies to all kinds of services and is probably not designed to reflect the idea of information as a product of particular social and democratic relevance, continuous and reliable access to which should be available at affordable prices, with good quality and on user-friendly terms. Further research is also needed to assess to what extent consumer protection law can realize public information policy goals, such as pluralism, the realization of freedom of expression and democratic principles. The Universal Service Directive could be a good starting point provided its scope is being extended to also cover broadcasting services. Unlike general consumer protection law, the Universal Service Directive leaves room to combine consumer protection with the realization of general competition and public information policy objectives. This is to promote what the directive calls 'the twin objectives of promoting effective competition whilst pursuing public interest needs, such as maintaining the affordability of publicly available services for some consumers'.[862]

Another aspect is that electronic access control should neither hamper competition between free-TV and pay-TV nor the functioning of the media. The same is true for initiatives to stimulate the creation and distribution of content regardless of consumer demand. Electronic access control must not obstruct competing media in complying with their task to inform and to criticize. European broadcasting law has developed two potentially useful instruments, the list of important events and the right to short reporting, which, however, need further improvement to be effective.

5.3. Conditional Access is not a Phenomenon Reserved to Pay-TV

Electronic access control and its effect on competition and individual access to content are not reserved to the broadcasting domain. The electronic management and enforcement of exclusive content rights and the controlled distribution of

[862] Universal Service Directive, Recital 26.

digital content to individualized customers is not a strategy that is specific to broadcasting. The possibility to target niche areas, collect remuneration, control content and maintain individualized commercial relationships with consumers are aspects that make the use of electronic access control techniques equally attractive for, for example, the internet sector. In the internet sector too, access-controlled service platforms, such as portals and B2C platforms, form the integrating link between various service providers and consumers.

The similarities do not end with the way services are marketed. The provision of exclusive content and controlled access to such content becomes a means to attract subscribers and to monopolize profitable distribution channels. The iTunes example from Chapter 3 demonstrates that, correspondingly, many of the concerns that the study listed for the realization of competition and public information policy objectives in the pay-TV case have already been made for other sectors. It was repeatedly argued that the combination of exclusive control over content and the portal, and control over Apple's proprietary FairPlay Digital Rights Management solution together with the iPod, could be a means to monopolize the consumer base, influence competition in one's favour and exclude consumers who do not comply with the operator's business rules. Similar to the pay-TV case, private electronic control over digital content raises for the internet issues of consumer freedom to access and choose between different services.

The iTunes case triggered a new West Coast crusade against electronic content control, this time led by competitor RealNetworks who also operates a music download store. RealNetworks launched the 'Freedom of Choice' campaign to help consumers 'break the chains that tie their music device [iPod] to proprietary music downloads'. And, according to RealNetworks, 'We are here to inform AND motivate'.[863] Note how a traditional public policy argument—freedom of choice—is again made with the intention of mobilizing consumers against a rival's business methods. RealNetworks understood very well that strategies to monopolize the consumer base can have an effect on competition and that a means to remedy monopoly control is to 'break the chains', meaning to inform and to remove technical and contractual lock-ins.

Not only competitors complain about proprietary content-control schemes on, for example, the internet; consumers complain too. Again, the complaints from consumers, consumer representatives and scholars repeat arguments that are eerily familiar to the case of electronic access control in digital broadcasting. For example, the English Consumers' Association complained to the Office of Fair Trade about price discrimination between FairPlay-protected content that is sold in the UK and tracks that are sold in France and Germany. The Consumers' Association claimed that this was in conflict with the European Union's Internal

[863] http://www.musicfreedomofchoice.org/

Market principles.[864] Furthermore, and also within the internet context, access to content becomes a matter of complying with the service provider's contractual terms and conditions. Here too, private control over access to content triggers public information policy concerns. One example is the on-going discussion about the relationship between technological content protection measures and the objectives of intellectual property law, meaning the effect of electronic control on access to and the use of creative works. Copyright law is characterized by a well-developed ideology of sharing in return for the assignment of certain exclusive exploitation rights. This idea is reflected in the attempt of copyright law to achieve a balance between the assignment of exclusive exploitation rights and the need to limit exclusive control over creative works where the interest in the wide usage and the dissemination of them is preponderant. Here too, there are concerns about the negotiating position of consumers and the compatibility of consumer-content-provider contracts with legitimate consumer rights and interests.

Are there lessons to be learned from the pay-TV case? Obviously, one would first have to analyze exactly how far electronic access control in the digital broadcasting sector and other cases of electronic control, for example Digital Rights Management use on the internet, are comparable. While some rationales are broadcasting specific, others may be not. The purpose of this last section is to take the pay-TV example further in order to stimulate this discussion.

A lesson that could be learned from the pay-TV case is that there can be a certain tension between electronic control of access to content and public information policy objectives. This conflict is particularly apparent in the case of broadcasting. It might be worth, however, conducting further research to determine the extent to which the public accessibility and availability of content is an important rationale behind regulatory policies for the internet. One factor that could bring the policies for the different sectors closer to each other is convergence and the European policy objective of promoting multiple access to services through digital television, mobile and internet platforms alike. Another factor could be the role that intellectual property law plays. Finally, access to electronic services and electronic exclusion are issues that are also discussed in context with European policies for the internet domain.

Providing the finding was that there are also tensions in the internet sector between electronic access control and public information policy objectives, it could be worth discussing the approach of broadcasting law and building on the experiences gained. European broadcasting law responds to the fact that it is the combination of electronic access control of and exclusive rights to content that affects the accessibility and broad availability of content. It advocates to impose

[864] The Office of Fair Trade (OFT) has decided to refer the matter further to the European Commission, as this was an issue that touched upon broader Internal Market issues, namely how the online exploitation of music is licensed across Europe, OFT Press Release, available at <www.oft.gov.uk/news/press+releases/statements/2004/itunes.htm> (last visited on 20 March 2005).

limitations on the electronic enforcement and scope of exclusive exploitation rights, at least regarding content of public interest.

Another lesson that could be learned from access control in digital broadcasting is that Articles 8 to 13 of the Access Directive and the Universal Service Directive provide tools for dealing with bottleneck situations and anti-competitive behaviour that is targeted at monopolizing the consumer base. The pay-TV case also ascertains that caution is in place concerning access obligations as a remedy to technical bottlenecks that are the result of dominant standards. Articles 8 to 13 of the Access Directive and the Universal Service Directive, however, provide viable alternatives in form of tools that stimulate competition at the facility and service levels, providing this is economically possible. In order to do this, the pay-TV case demonstrated that the effective regulation of electronic access control needs to take both the competitors' and the consumers' position into account. Guarantees concerning the fairness, affordability and adequacy of retail conditions are needed, as is market transparency. Making sure that the terms and conditions of access are fair, affordable and non-discriminatory is not 'just' a matter of fairness and consumer protection, it can be an important element of realizing competition and access to content.

Bibliography

Aghion/Bolton 1987
P. Aghion and P. Bolton, 'Contracts as a Barrier to Entry' (1987) 77 *The American Economic Review* 388

Andersen Consulting 2002
Andersen Consulting, *Outlook of the Development of Technologies and Markets for the European Audiovisual Sector up to 2010*, Study (Report to the European Commission Directorate-General for Education and Culture, Brussels, 2002), available at <europa.eu.int/comm/avpolicy> (last visited on 15 March 2005)

Areeda 1990
P. Areeda, 'Essential Facilities: An Epithet in Need of Limiting Principles' (1990) 58 *Antitrust Law Journal* 841

Armstrong 1999
M. Armstrong, 'Competition in the Pay-TV Market' [1999] 13 *Journal of the Japanese and International Economics* 257

Arnbak/Van Cuilenburg/Dommering 1990
J.C. Arnbak, J.J. van Cuilenburg and E.J. Dommering, *Verbinding en Ontvlechting: een Studie naar Toekomstig Overheidsbeleid voor de Informatievoorziening* (Otto Cramwinckel, Amsterdam, 1990)

Article 19 1999
Article 19, *The Public's Right to Know. Principles on Freedom of Information Legislation* (International Standards Series, London, 1999), available at <www.article19.org> (last visited on 14 March 2005)

Asscher 2002
L. Ascher, *Communicatiegrondrechten* (Otto Cramwinckel, Amsterdam, 2002)

Baccaro 2003
V. Baccaro, 'The Commission Closes Probe into Pay-TV Industry in Italy Approving Newscorp/Telepiù merger deal' [2003] *Competition Policy Newsletter* 8

Badura 1989
P. Badura, 'Das Recht auf freie Kurzberichterstattung' [1989] *Zeitschrift für Urheber- und Medienrecht* 317

Baker 2002
E. Baker, 'First Amendment Limits on Copyright' (2002) 55 *Vanderbilt University Law Review* 891

Bakos/Brynjolfsson 2000
Y. Bakos and E. Brynjolfsson, 'Aggregation and Disaggregation of Information Goods: Implications for Bundling, Site Licensing, and Micropayment Systems', in D. Hurley, B. Kahin and H. Varian (eds.), *Internet Publishing and Beyond: The Economics of Digital Information and Intellectual Property* (MIT Press, Cambridge, 2000)

Bakos/Brynjolfsson 2000a

Y. Bakos and E. Brynjolfsson, 'Bundling and Competition on the Internet' [2000] 19 *Marketing Science* 63

Bakos/Brynfolfsson 1999

Y. Bakos and E. Brynfolfsson, 'Bundling Information Goods: Pricing, Profits and Efficiency' [1999] 45 *Management Science* 1613

Barendt 1993

E. Barendt, *Broadcasting Law* (Clarendon Press, Oxford, 1993)

Barendt 1992

E. Barendt, 'Fundamental Freedoms', in W.F. Korthals Altes, E.J. Dommering, P.B. Hugenholtz and J.J.C. Kabel (eds.), *Information Law Towards the 21st Century* (Kluwer Law International, Information Law Series, 1992, Deventer)

Bavasso 2003

A. Bavasso, *Communications in EU Antitrust Law. Market Power and Public Interest* (Kluwer Law International, The Hague, 2003)

Beers 1992

T.A.L. Beers, 'Public Access to Government Information', in W.F. Korthals Altes, E.J. Dommering, P.B. Hugenholtz and J.J.C. Kabel (eds.), *Information Law Towards the 21st Century* (Kluwer Law International, Information Law Series, 1992, Deventer)

Benkler 1999

Y. Benkler, 'Free as the Air to Common Use: First Amendment Constraints on Enclosure of the Public Domain' (1999) 74 *New York University Law Review* 354

Besen/Farrell 1994

S. Besen and J. Farrell, 'Choosing How to Compete: Strategies and Tactics in Standardization' (1994) 8 *Journal of Economic Perspectives* 117

Van Den Beukel/Nieuwenhuis 2000

J. van den Beukel and A. Nieuwenhuis, 'Pluriformiteit in het Mededingingsrecht' [2000] 4 *Mediaforum* 116

Bird/Bird 2002

Bird and Bird, *Market Definition in the Media Sector. Comparative Legal Analysis*, Study (Report to the Commission Directorate-General Competition, Brussels, 2002), available at <europa.eu.int/comm/competition/publications/studies/legal_analysis.pdf> (last visited on 15 March 2005)

Birnhack 2004

M.D. Birnhack, 'Copyrighting Speech: A Trans-Atlantic View', in P.L.C. Torremans (ed.), *Copyright and Human Rights, Freedom of Expression - Intellectual Property - Privacy,* (Kluwer Law International, Information Law Series, The Hague, 2004)

Bishop/Walker 1999

S. Bishop and M. Walker, *The Economics of EC Competition Law: Concepts, Application and Measurement* (Sweet & Maxwell, London, 1999)

Bittlingmayer/Hazlett 2002

G. Bittlingmayer and W. Hazlett, '"Open Access:" the Ideal and the Real', (2002) 26 *Telecommuncinations Policy* 295

Breuning 2000

C. Breuning, 'Programmbouquets im Digitalen Fernsehen' [2000] *Media Perspektiven* 378

Brittan 1989

S. Brittan, 'The Case for the Consumer Market', in C. Veljanovski (ed.), *Freedom in Broadcasting* (IEA-Institute of Economic Affairs, London, 1989)

Bullinger 1997

M. Bullinger, 'Verbreitung entgeltlicher Rundfunkprogramme und Mediendienste in Paketen' [1997] *Archiv für Presserecht* 761

Bullinger 1980

M. Bullinger, *Kommunikationsfreiheit im Strukturwandel der Telekommunikation* (Nomos, Baden-Baden, 1980)

Caffarra/Coscelli 2003

C. Caffarra and A. Coscelli, 'Merger to Monopoly: NewsCorp/Telepiù' (2003) 24 *European Competition Law Review* 625

Capiau 2001

J. Capiau, 'EC Must-Carry Rules on the Brink of a Lost Opportunity: Harmonisation and Free Movement of TV Broadcasts within the Communications Review' [2001] *Journal of Network Industries* 227

Cave/Cowie 1998

M. Cave and C. Cowie, 'Not Only Conditional Access. Towards a Better Regulatory Approach to Digital TV', Paper presented at the Euro-CPR '98 Conference, Venice, March 1998, available at <www.cordis.lu/infowin/acts/ienm/newsclips/arch1998/980402at.html> (last visited on 15 March 2005)

Cave/Cowie 1996

M. Cave and C. Cowie, 'Regulating Conditional Access in European Pay Broadcasting' [1996] *Communications & Strategies* 119

Cave/Crandall 2001

M. Cave and R.W. Crandall, 'Sports Rights and the Broadcast Industry' (2001) 111 *The Economic Journal* F4

Choi 2004

J.P. Choi, 'Tying and Innovation: A Dynamic Analysis of Tying Arrangements' (2004) 114 *Economic Journal* 83 <ideas.repec.org/e/pch73.html#articles> (visited at 15 March 2005)

Clements 2004

M. Clements, 'Direct and Indirect Network Effects: Are They Equivalent?' Working Paper, 2004, available at <papers.ssrn.com> (last visited on 15 March 2005)

Cohen

J.E. Cohen, 'A Right to Read Anonymously: A Closer Look at Copyright Management in Cyberspace' (1996) 28 *Connecticut Law Review* 981

Cohendet 2003
P. Cohendet, *The Digital Divide in the European Enlarged Economic Scenario : an Assessment of the Socio-Economic Effects*, Report (Report for ESA, Strasbourg, 2003) <telecom.esa.int/telecom/media/document/digdiv220903final.pdf> (last visited on 14 March 2005)
Contest Consultancy 2003
Contest Consultancy, *Standardisation in digital interactive television, strategy and recommendations for a standardisation policy supporting the effective implementation of the Framework Directive 2002/21/EC and the establishment of required interoperability levels in digital interactive television*, Study (Study prepared for the European Commission Directorate-General Enterprise, Brussels, 2003), available at <www.cenelec.org/NR/rdonlyres/CCC0D181-330E-4275-BF5F-95DE06BA2298/0/DigitalTV_Sec0002_DC_Final.pdf> (last visited on 15 March 2005)
Counterpoint Research 2001
Counterpoint Research, *Digital Television – Consumer's Use and Perceptions, a Report on a Research*, Study (Study prepared for OFTEL, London, 2001)
Department of Trade and Industrie
Department of Trade and Industry (DTI), *Communications Whitepaper. A New Future for Communications*, 12 December 2000
Deselaers 1995
W. Deselaers, 'Die "Essential Facilities"- Doktrin im Lichte des Magill-Urteils des EuGH' [1995] *Europäische Zeitschrift für Wirtschaftsrecht* 563
Diesbach 1998
M. Diesbach, *Pay-TV oder Free-TV, Zur Zulässigkeit der verschlüsselten Exclusivübertragung sportlicher Großereignisse* (Nomos, Baden-Baden 1998)
Dietl/Franck 2000
H. Dietl and E. Franck, 'Free-TV, Abo-TV Pay per view – Organisationsformen zur Vermarktung von Unterhaltung' [2000] *Schmallenbachs Zeitschrift für Betriebswirtschaft* 592
Van Dijk/Van Hoof 1998
P. van Dijk and G.J.H. van Hoof, *Theory and Practice of the European Convention of Human Rights* (Kluwer Law International, Den Haag, 1998)
Dommering 2000
E.J. Dommering, *Informatierrecht* (Otto Cramwinckel, Amsterdam, 2000)
Dommering 1990
E.J. Dommering 1990, 'Groppera en Autronic. Maken de Zwitsers Gatenkaas van de Nederlandse Mediawet?', in P.B. Hugenholtz a.o. (ed.), *Recht in de Kijker* (W.E.J. Tjeenk Willink, Zwolle 1990)
Dommering/Hugenholtz/Kabel 2002
E.J. Dommering, P.B. Hugenholtz and J.J.C. Kabel, *De Overheid en het Publiek Domein van Informatie voor Wetenschappelijk Onderzoek* (Institute for Information Law, Amsterdam, 2002)

Dommering/Van Eijk/Theeuwes/Vogelaar 2001
E.J. Dommering, N.A.N.M. van Eijk, J. Theeuwes and F. Vogelaar, *Toezicht en Regulering in de Telecommunicatiemarkt* (Institute for Information Law, Amsterdam, 2001)
Van Eechoud 1998
M. van Eechoud, 'Openbaarheid, Exclusiviteit en Markt: Commercialisering van Overheids-informatie' [1998] *Mediaforum* 177
Van Eik 1992
N.A.N.M. van Eijk, *Omroepvrijheid en Overheidsbemoeienis* (Otto Cramwinckel, Amsterdam, 1992)
Elkin-Koren 2001
N. Elkin-Koren (2001), 'A Public Regarding Approach to Copyrights' in R. Dreyfuss, D.L. Zimmerman and H. First (eds.), *Expanding the Boundaries of Intellectual Property* (Oxford University Press, New York, 2001)
Engel
B. Engel, 'Digitales Fernsehen – Neue Perspektiven für die Zuschauerforschung' [2001] *Media Perspektiven* 480
Enβlin 2000
H. Enβlin, *Kontrahierungszwang für Anbieter von Dienstleistungen für das digitale Fernsehen* (Peter Lang, Frankfurt am Main, 2000)
European Audiovisual Observatory 2004
European Audiovisual Observatory, *Statistical Yearbook 2004* (European Audiovisual Observatory, Strasbourg, 2004)
European Audiovisual Observatory 2003
European Audiovisual Observatory, *Digital Television Glossary* (IRIS Special, European Audiovisual Observatory, Strasbourg, 2003)
Eurostrategies 2003
Eurostrategies, *Assessment of the Member States' measures aimed at fulfilling certain general interest objectives linked to broadcasting imposed on providers of electronic communication networks and services in the context of the new regulatory framework, Final Report - Part 1*, Study (Eurostrategies, Boulogne Billancourt, 2003), available at <europa.eu.int/information_society/topics/telecoms/regulatory/studies> (last visited on 14 March 2004)
Evans 2003
D. Evans, 'The Antitrust Economics of Multi-Sided Platform Markets' (2003) 20/2 *Yale Journal on Regulation* 325
Farrell/Saloner 1985
J. Farrell and G. Saloner, 'Standardization, Compatibility, and Innovation' (1985) 16 *Rand Journal of Economics* 70
Farell/Shapiro 1989
J. Farell and C. Shapiro, 'Optimal Contracts with Lock-in' (1989) 79 *The American Economic Review* 51

Fraser 1998
S. Fraser, The Conflict between the First Amendment and Copyright Law and its Impact on the Internet (1998) 16 *Cardozo Arts & Entertainment Law Journal* 1
Frtisch/Wein/Ewers 1999
M. Fritsch, T. Wein and H.J. Ewers, *Marktversagen und Wirtschaftspolitik* (3rd edition, Vahlen, München, 1999)
Furse 1995
M. Furse, 'The Essential Facilities Doctrine in Community Law' [1995] *European Competition Law Review* 469
Galbiati/Nicita/Nizi 2004
R. Galbiati, A. Nicita and G. Nizi, 'Regulation, Competition, and Institutional Design in Media Markets: The Evolution of pay-TV in UK, Australia and Italy' [2004] *American Law & Economics Association Annual Meetings*, No. 76
Garfield 2001
A. Garfield, 'The First Amendment as a Check on Copyright Rights' [2001] *Hastings Communications and Entertainment Law Journal* 579
Garzaniti 2000
L. Garzaniti, *Telecommunications, Broadcasting and the Internet. EU Competition Law and Regulation* (Sweet& Maxwell, London, 2000)
Gassner 2001
U.M. Gassner, 'Internet-Handelsplattformen im Spiegel des Kartellrechts' [2001] *Multimedia und Recht* 140
Van Geffen 2001
S. van Geffen, *Koppelverkoop van Informatieprodukten en Netwerkelementen. Implicaties van de US v. Microsoft-zaak voor de Regulering van de Informatiemaatschappij*, Master thesis (Universiteit van Amsterdam, Amsterdam, February 2001)
Van Geffen/Nooij/Theeuwes 2002
S. van Geffen, M. de Nooij and J. Theeuwes, *Marktwerking & ICT – Is de Mededingingswet ICT-bestendig?*, Study (SEO-rapport nr. 624, Amsterdam, 2002)
Gersdorf 2000
H. Gersdorf, *Kabeleinspeisung von Programmbouquets - Zugang digitaler Programmbouquets des öffentlich-rechtlichen Rundfunks zum breitbandigen Kommunikationsnetz* (Vistas, Berlin, 2000)
Gersdorf 1998
H. Gersdorf, *Chancengleicher Zugang zum digitalen Fernsehen: eine Untersuchung des verfassungsrechtlichen Regulierungrahmens am Beispiel des Entwurfs zum vierten Rundfunkänderungstaatsvertrages vom 27. Februar 1998*, Expert Opinion (Rechtsgutachten erstellt im Auftrag der in der Arbeitsgemeinschaft der Landesmedienanstalten (ALM) zusammengeschlossenen Landesmedienanstalten, vertreten durch die Hamburgische Anstalt für Neue Medien (Vistas, Berlin, 1998)
Gibault 2002

L. Gibault, *Copyright Limitations and Contract – An Analysis of the Contractual Overridability of Limitations on Copyright* (Kluwer Law International, Information Law Series, London/The Hague/Boston, 2002)

Gibbons 2004

T. Gibbons, 'Control over Technical Bottlenecks - A Case for Mediaownership Law?', in European Audiovisual Observatory (ed.), *Regulating Access to Digital Television. Technical Bottlenecks, Vertically-Integrated Markets and New Forms of Media Concentration* (European Audiovisual Observatory, Strasbourg, 2004)

Gibbons 2000

T. Gibbons, 'Pluralism, Guidance and the New Media' in C. Marsden (ed.), *Regulating the Global Information Society* (Routledge, London, 2000)

Gibbons 1999

T. Gibbons, 'Concentrations of Ownership and Control in a Converging Media Industry' in C. Marsden & S. Verhulst (eds.), *Convergence in European Digital TV Regulation* (Blackstone Press, London, 1999)

Gibbons 1998

T. Gibbons, *Regulating the Media* (2nd edition, Sweet & Maxwell, London, 1998)

Gleich 2001

U. Gleich, 'Poluläre Unterhaltungsformate im Fernsehen und ihre Bedeutung für die Zuschauer' [2001] *Media Perspektiven* 524

Van de Gronden/Mortelmans 2003

H. van de Gronden and K. Mortelmans, 'Mededinging en Algemene Belangen: Welke Signalen geeft de Europese Rechter in de Eurovisiezaak?' [2003] 1 *Mediaforum* 11

Grosheide 2000

F.W. Grosheide (ed.), *Communicatie- & Mediarecht* (Ars Aequi Libri, Nijmegen, 2000)

Gunzerath 2000

D. Gunzerath, '"Darn that Pay-TV!": STV's Challenge to American Television's Dominant Economic Model' [2000] *Journal of Broadcasting & Electronic Media* 655

Habermas 1979

J. Habermas, *Strukturwandel der Öffentlichkeit, Untersuchungen zu einer Kategorie der bürgerlichen Gesellschaft* (10th edition, Luchterhand, Darmstadt, 1979)

Harbord/Szymanski 2005

D. Harbord and S. Szymanski, *Restricted View. The Rights and Wrongs of FA Premier League Broadcasting*, Study, (Consumers' Association, London, 2005)

Harcourt

G. Harcourt, 'Public broadcasting – the Tasmanian Tiger of New Zealand' (2000) 4 ABCzINE 18

Harris/O'Boyle/Warbrick 1995

D.J. Harris, M. O'Boyle and C. Warbrick, *Law of the European Convention on Human Rights* (Butterworths, London, 1995)

Helberger 2002

N. Helberger, 'Brot und Spiele – Die Umsetzung der Listenregelung des Artikel 3a der Fernsehrichtlinie' [2002] *Archiv für Presserecht* 292

Helberger 2001

N. Helberger, A. Scheuer and P. Strothmann, 'Non-Discriminatory Access to Digital Access Control Services', IRIS Plus, Supplement, [2001] *iris - Legal Observations of the European Audiovisual Observatory* 2

Helberger 1999

N. Helberger, *Neighbouring Rights Protection of Broadcasting Organisations: Current Problems and Possible Lines of Action*, Study (Report prepared for the Council of Europe, MM-S-PR (99)9 final, Strasbourg 1999), available at <www.humanrights.coe.int/media/documents/main.htm> (last visited on 15 March 2005)

Helberger 1999a

N. Helberger, 'Hacking BskyB: The Legal Protection of Conditional Access Services Under European Law' [1999] *Entertainment Law Review* 88

Helberger/Van Eijk 2000

N. Helberger and N.A.N.M. van Eijk, *Study on the use of conditional access systems for reasons other than the protection of remuneration, to examine the legal and the economic implications within the Internal Market and the need of introducing specific legal protection*, Study (Report prepared for the European Commission Directorate-General Information Society, Brussels, 2000) <www.ivir.nl/publications/other/ca-report.html> (last visited on 15 March 2005)

Henry

J.B. Henry, 'The Economics of Pay TV Media' in E. Noam (ed.), *Video Media Competition, Regulation, Economics, and Technology* (Columbia University Press New York, 1985)

Herdzina 1999

K. Herdzina, *Wettbewerbspolitik* (5th edition, Lucius & Lucius, Stuttgart 1999)

Hesse 2002

A. Hesse, 'Rechtsprobleme der Verschlüsselung bei der Ausstrahlung von Fernsehprogrammen aus der Sicht des öffentlich-rechtlichen Rundfunks' [2002] *Zeitschrift für Urheber- und Medienrecht* 692

Hesse 1999

A. Hesse, *Rundfunkrecht* (2nd edition, Franz Vahlen, München, 1999)

Hills 1993

J. Hills, 'Universal Service: a Social and Technological Construct' [1993] *Communications & Strategies* 61

Hinloopen 2002

J. Hinloopen, 'The Incentives to Bundle for a Stackelberg leader', paper presented at the University of Amsterdam (and Tinbergen Institute) and Delft University of Technology, 6 March 2002
Hins 1998
A.W. Hins, 'Uitzendrechten voor Belangrijke Evenementen en de EG-Televisierichtlijn' [1998] *Mediaforum* 318
Hins 1991
A.W. Hins, *Ontvangstvrijheid en Buitenlandse Omroep* (Kluwer, Deventer, 1991)
Hoffmann-Riem 1990
W. Hoffmann-Riem, 'Kommunikationsfreiheit und Chancengleichheit', in J. Schwartländer and E. Riedel (eds.), *Neue Medien und Meinungsfreiheit im nationalen und internationalen Kontext* (N. P. Engel, Kehl am Rhein, 1990)
Hugenholtz 2001
P.B. Hugenholtz, 'Copyright and Freedom of Expression in Europe' in R. Dreyfuss, D.L. Zimmerman, H. First (eds.), *Expanding the Boundaries of Intellectual Property* (Oxford University Press, New York, 2001)
Hugenholtz (1989)
P.B. Hugenholtz, *Auteursrecht op Informatie* (Kluwer, Deventer, 1989)
Idate 2001
Institut de l'audiovisuel et des télécommunications en Europe (Idate), *Reference Report 2000 of the Study 'Development of Digital Television in the EU'*, Study (Study prepared for the Commission Directorate-General Information Society, Brussels, 2001), available at <europa.eu.int> (last visited on 15 March 2005)
Idate 2000
Institut de l'audiovisuel et des télécommunications en Europe (Idate), *Reference Report 1999 of the Study 'Development of Digital Television in the EU'*, Study (Study prepared for the Commission Directorate-General Information Society, Brussels, 2000), available at <europa.eu.int> (last visited on 15 March 2005)
Jones/Sufrin 2001
A. Jones and B. Sufrin, *EC Competition Law, Text, Cases, and Materials* (Oxford University Press, Oxford, 2001)
Jürgens 2002
H.W. Jürgens, *Zur Handhabbarkeit von Zugangseinrichtungen und Verfahren zur Nutzung digitaler Medienangebote* (Unabhängige Landesanstalt für das Rundfunkwesen (ULR), Kiel, 2002)
Kaitatzi-Whitlock 1997
S. Kaitatzi-Whitlock, 'The Privatising of Conditional Access Control in the European Union' [1997] *Communications & Strategies* 91
Kaplow 1985
L. Kaplow, 'Extension of Monopoly Power through Leverage' [1985] *Columumbia Law Review* 515
Katz/Shapiro 1998

M.L. Katz and C.C. Shapiro, 'Antitrust in Software Markets', paper prepared for presentation at the Progress and Freedom Foundation conference, Competition, Convergence and the Microsoft Monopoly, February 5, 1998, available at <faculty.haas.berkeley.edu/shapiro/software.pdf> (last visited on 20 March 2005)

Katz/Shapiro 1985

M.L. Katz & C.C. Shapiro, 'Network Externalities, Competition and Compatibility' [1985] 75 *American Economic Revue* 424

KEK 2000

Kommission zur Ermittelung der Konzentration im Medienbereich (KEK), *Fortschreitende Medienkonzentration im Zeichen der Konvergenz, Bericht der Kommission zur Ermittlung der Konzentration im Medienbreich (KEK) über die Entwicklung der Konzentration und über Massnahmen zur Sicherung der Meinungsvielfalt im privaten Rundfunk*, Report (Schriftenreihe der Landesmedienanstalten, Band 17 (Vistas, Berlin, 2000)

Klemperer 1987

P. Klemperer, 'Markets with Consumer Switching Costs' (1987) 102 *The Quarterly Journal of Economics* 375

Knieps 1998

G. Knieps, 'Zugang zu Netzen, Verselbständigung, Nutzung, Vergütung, Eigentumsschutz' [1998] *Multimedia und Recht* 275

Koelman 2000

K.J. Koelman, 'Terug Naar de Bron: Open Source and Copyleft' [2000] *Tijdschrift voor Auteurs-, Media- & Informatierecht* 151

Koelman 2000a

K.J. Koelman, 'A Hard Nut To Crack: The Protection of Technological Measures' [2000] *European Intellectual Property Review* 272

König 1997

M. König, *Die Einführung des digitalen Fernsehens: Neue Probleme und Implikationen für den Wettbewerb und die Medienkonzentration* (Nomos, Baden Baden, 1997)

Lange

A. Lange (ed.), *Developments in Digital TV in the European Union* (European Audiovisual Observatory, Strasbourg, 1999)

Larouche 2002

P. Larouche, 'A Closer Look at Some Assumptions Underlying EC Regulation of Electronic Communications' [2002] 3 *Journal of Network Industries* 129

Larouche 2002a

P. Larouche, 'Communications Convergence and Public Service Broadcasting', Discussion paper 2002, available at <www.tilburguniversity.nl/tilec/publications/discussionpapers/archive/dp2002/> (last visited on 20 March 2005)

Larouche 2000

P. Larouche, *Competition Law and Regulation in European Telecommunications* (Hart, Oxford and Portland Oregon, 2000)

Ledoux Book 2004

C. Ledoux Book, *Digital Television, DTV and the Consumer* (Blackwell, Ames, 2004)

Lemley/Lessig 2000

M.A. Lemley and L. Lessig, 'The End of End-to-End: Preserving the Architecture of the Internet in the Broadband Era' (2001) 48 *University of California Law Review* 925

Lemley/McGowan 1998

M.A. Lemley and D. McGowan, 'Legal Implications of Network Economic Effects' (1998) 86 *California Law Review* 481

Lessig 1999

L. Lessig, *Code and Other Laws of Cyberspace* (Basic Books, New York, 1999)

Leveque 2000

F. Leveque, 'The Controversial Choice of Remedies to Cope with the Anti-Competitive Behavior of Microsoft', [2000] Berkeley Olin Program in Law & Economics, Working Paper Series, Paper 34, available at <repositories.cdlib.org/blewp/34> (last visited on 15 March 2005)

Levy 1997

D.A.L. Levy, 'The Regulation of Digital Conditional Access Systems' [1997] *Telecommunications Policy* 661

Lichtenberg 1990

J. Lichtenberg, 'Foundations and Limits of Freedom of Press', in J. Lichtenberg (ed.), *Democracy and the mass media* (Cambridge University Press, Cambridge, 1990)

Liebowitz 1983

S.J. Liebowitz, 'Tie-In Sales and Price Discrimination' [1983] *Economic Inquiry* 387

Litman 1997

B.R. Litman, 'Vertical Integration and Consumer Welfare in the Cable Industry' (1997) 41 *Journal of Broadcasting & Electronic Media* 231

Llorens-Maluquer 1998

C. Llorens-Maluquer, 'European Responses to Bottlenecks in Digital Pay-TV: Impacts on
Pluralism and Competition policy' (1998) 16 *Cardozo Arts & Entertainment Journal* 557

Van Loon 2001

A. van Loon, *Television and Media Concentration, Regulatory Models on the National and the European Level*, (IRIS Special, European Audiovisual Observatory, Strasbourg, 2001)

Mackaay 1992

E. Mackaay, 'The Public's Right to Information', in W.F. Korthals Altes, E.J. Dommering, P.B. Hugenholtz and J.J.C. Kabel (eds.), *Information Law Towards the 21st Century* (Kluwer Law International, Information Law Series, 1992, Deventer)
Mackaay 1982
E. Mackaay, *Economics of Information and Law* (Kluwer, Boston, 1982)
McGowan 1996
D. McGowan, 'Regulating Competition in the Information Age: Computer Software as an Essential Facility under the Treshman Act' (1996) 18 *Hastings Communications & Entertainment Law Journal* 771
De Meij/Hins/Nieuwenhuis/Schuijt 2000
J.M. de Meij, A.W. Hins, A.J. Nieuwenhuis and G.A.I. Schuijt, *Uitingsvrijheid. De Vrije Informatiestroom in Grondwettelijk perspectief* (3rd edition, Otto Cramwinckel, Amsterdam, 2000)
De Meij 1996
J. de Meij, *Uitingsvrijheid* (2nd edition, Otto Cramwinckel, Amsterdam, 1996)
Meese
A.J. Meese, 'Tying Meets the New Institutional Economics: Farewell to the Chimera of Forcing' (1997) 146 *University of Pennsylvania Law Review* 1
Mendes
P.M. Mendes, 'Recent Consolidation in the European Pay-TV Sector' [2003] *Competition Policy Newsletter* 29
Moglen 1997
E. Moglen, 'The Invisible Barbecue', originally published (1997) 97 *Columbia Law Review* 945, in slightly different form as a comment in 'Symposium: Telecommunications Law: Unscrambling the Signals, Unbundling the Law' available at <old.law.columbia.edu/my_pubs/barbecue.html> (last visited on 20 March 2005)
Mowbray 2004
A. Mowbray, *The Development of Positive obligations under the European Convention on Human Rights by the European Court of Human Rights* (Hart, Oxford, 2004)
Nalebuff 2004
B. Nalebuff, 'Bundling as an Entry Barrier' (2004) 119 *The Quarterly Journal of Economics* 159
Nalebuff 1999
B. Nalebuff, *Bundling* (Yale ICF Working Paper Series No. 99-14, Yale, 1999)
Nauheim 2001
M. Nauheim, *Die Rechtmässigkeit des Must-Carry-Prinzips im Bereich digitalisierten Kabelfernsehen in der Bundesrepublik Deutschland* (Peter Lang, Frankfurt am Main, 2001)
Neumann 1998
I. Neumann, *Pay-TV in Deutschland, Markteintritts- und Wettbewerbsbedingungen für neue Anbieter* (Deutscher Universitätsverlag, Wiesbaden, 1998)

Nihoul 1998
P. Nihoul, 'Competition or Regulation for Multimedia?' [1998] *Telecommunications Policy* 207
Noam 1988
E. Noam, 'Der Einfluß von Markstruktur und Eintrittsschranken auf die Vielfalt der Fernsehprogramme', in Mestmäcker (ed.), *Offene Rundfunkordnung* (Bertlesmann Stiftung, Gütersloh, 1988)
Nocke/Peitz/Stahl 2004
V. Nocke, M. Peitz and K. Stahl, 'Platform Ownership', [2004] *Centre for Economic Policy Research, Discussion Paper Series*, No. 4657, available at <www.cepr.org> (last visited on 15 March 2005)
Nolan 1997
D. Nolan, 'Bottlenecks in Pay Television' [1997] *Telecommunications Policy* 597
O'Driscoll 2000
G. O'Driscoll, *The Essential Guide to Digital Set-top Boxes and Interactive TV* (Prentice Hall PTR, Upper Saddle River, New York, 2000)
Ofcom 2004
Ofcom, *Statement on Code on Electronic Programme Guides, Statement by Ofcom,* London, 26 July 2004, available at <www.ofcom.org.uk> (last visited on 20 March 2005)
Oftel 2002
Office of Telecommunications (Oftel), *Draft Commission Recommendation on relevant product and service markets within the electronic communications sector susceptible to ex ante regulation*, London, July 2002, available at <www.ofcom.org.uk> (last visited on 15 March 2005)
Oftel 2000
Office of Telecommunications (Oftel), *Class Licence for the running of communication systems for the provision of access control services granted by the secretary of State under section 7 of the Telecommunications Act 1984*, London, May 2000, available at <www.ofcom.org.uk> (last visited on 15 March 2005)
Oftel 1998
Office of Telecommunications (Oftel), *Conditional Access Charges for Digital Television, Statement*, London, February 1998, available at <www.ofcom.org.uk> (last visited on 15 March 2005)
Oftel 1997
Office of Telecommunications (Oftel), *The Regulation of Conditional Access for Digital Television, Office of Telecommunications*, London, March 1997, available at <www.ofcom.org.uk> (last visited on 15 March 2005)
Oftel 1997a
Office of Telecommunications (Oftel), *Bundling in the pay television market, Submission by the Director General of Telecommunications to the Independent Television Commission*, London, December 1997, available at <www.ofcom.org.uk> (last visited on 15 March 2005)

Ostrom/Hess 2001
E. Ostrom and C. Hess, 'Artifacts, Facilities, and Content: Information as a Common-Pool Resource', paper presented at the Duke conference on the public domain, Duke University School of Law, 9-11 November 2001
Ovum 2001
OVUM and Squire, Sanders & Dempsey LLP, *Study of the development of competition for electronic conditional access networks and services, together with an Inventory of EU 'Must Carry' regulations*, Study (Final report prepared for the European Commission Directorate-General Information Society, Brussels, 2001)
Owen 1989
B.M. Owen, 'Determining Optimal Access to Regulated Essential Facilities' (1989) 58 *Antitrust Law Journal* 887
Owen 1975
B.M. Owen, *Economics and Freedom of Expression: Media Structure and the First Amendment* (Ballinger, Cambridge, 1975)
Oxera 2003
Oxera, *Study on Interoperability, service Diversity and Business Models in Digital Broadcasting Markets*, Study (Study prepared for the European Commission Directorate-General Information Society, Brussels, 2003), available at <europa.eu.int/information_society/topics/telecoms/index_en.htm> (last visited on 20 March 2005)
Peacock 1989
A. Peacock (1989), 'The Future of Public Service Broadcasting', in C. Veljanovski (ed), *Freedom in Broadcasting* (Institute of Economic Affairs (IEA), London, 1989)
Peacock 1986
A. Peacock (1986), *Report of the Committee on Financing the BBC, Presented to the Parliament by the Secretary of State for the Home Department by Command of Her Majesty*, Report (Her Majesty Stationary Office, London, 1986) (Peacock Report)
Pereira 2003
M. Pereira, 'Recent Consolidation in the European Pay-TV Sector' [2003] *Competition Policy Newsletter* 29
Petrick 2004
P. Petrick, *Why DRM Should be Cause for Concern: an Economic and Legal Analysis of the Effect of Digital Technology on the Music Industry*, Study (2004, Berkman Center for Internet & Society at Harvard Law School Research Publication No. 2004-09)
Poel & Hawkins 2001
M. Poel and R. Hawkins, 'The Evolution of Access Bottlenecks in Europe: Re-locating the Regulatory Issues' [2001] *Communications & Strategies* 71
Posner 1976
R.A. Posner, *Antritrust Law. An Economic Perspective* (The University of Chicago Press, Chicago, 1976)

Rawls 1971
J. Rawls, *A Theory of Justice* (The Belknap Press of Harvard University Press, Cambridge, 1971)
Riegel 1988
R. Riegel, 'Besteht ein Recht der Medien, kostenlos Ausschnitte von Fußballspielen zu übertragen?' [1988] *Zeitschrift für Urheber- und Medienrecht* 527
Rifkin 2000
J. Rifkin, *The Age of Access: The New Culture of Hypercapitalisation. When all of Life is a Paid-For Experience* (Tarcher/Putnam, New York, 2000)
Rochet/Tirole 2003
J.-C. Rochet and J. Tirole, 'Platform Competition in Two-Sided Markets' [2003] *Journal of the European Economic Association* 990
Rubinfeld 1998
D.L. Rubinfeld, 'Competition, Innovation, and Antitrust Enforcement in Dynamic Network Industries', speech before the Software Publishers Association, 1998 Spring Symposium, San Jose, California, 24 March 1998
Rumphorst 2001
W. Rumphorst, 'Sports Broadcasting Rights and EC Competition Law, European Broadcasting Union', 25 July 2001, available at <www.ebu.ch> (last visited on 14 March 2005)
Schellhaaß 1994
H.M. Schellhaaß, *Instrumente zur Sicherung der Meinungsvielfalt im Fernsehen*, Discussion Paper (Diskussionspapier 1/1994, Wirtschaftliche Dokumentation, Fachbereich 14, Technische Universität Berlin, 1994)
Scherer 1999
J. Scherer, 'Das Bronner-Urteil des EuGH und die Essential Facilities Doktrin im TK-Sektor' [1999] *Multimedia und Recht* 315
Schmidt 1981
I. Schmidt, *Wettbewerbstheorie und –politik* (Fischer, Stuttgart, 1981)
Scheuer/Knopp 2003
A. Scheuer and M. Knopp, 'Digital Television Glossary', Complement, in: European Audiovisual Observatory, *Regulating Access to Digital Television* (IRIS Special, European Audiovisual Observatory, 2003)
Schulz 1998
W. Schulz, *Gewährleistung kommunikativer Chancengleicheit als Freiheitsverwirklichung* (Nomos, Baden-Baden/Hamburg, 1998)
Scholz 1995
R. Scholz, 'Zukunft von Rundfunk und Fernsehen: Freiheit der Nachfrage oder reglementiertes Angebot?' [1995] *Archiv für Presserecht* 357
Schulz/Kühlers 2000
W. Schulz and D. Kühlers, *Konzepte der Zugangsregulierung. Schriftenreihe der Landesmedienanstalten* (Vistas, Berlin, 2000)
Schulz/Seufert/Holznagel 1999

W. Schulz, W. Seufert and B. Holznagel, *Digitales Fernsehen: Regulierungskonzepte und – perspektiven* (Schriftenreihe Medienforschung der Landesanstalt für Rundfunk Nordrhein-Westfalen, Opladen, 1999)

Schwartz 1985

I. Schwartz, 'Rundfunk und EWG Vertrag' in J. Schwarze, Jürgen (ed.), *Fernsehen ohne Grenzen: die Errichtung des Gemeinsamen Marktes für den Rundfunk, insbesondere über Satellit und Kabel* (Nomos, Baden-Baden, 1985)

Schwarz-Schilling 1998

C. Schwarz-Schilling, 'Pay-TV und doch kein Rundfunk!' [1998] *Zeitschrift für Urheber- und Medienrecht* 487

Shapiro 2000

C. Shapiro, *Competition Policy in the Information Economy, Foundations of Competition Policy Analysis* (Routledge, London, 2000)

Shapiro 1999

C. Shapiro, 'Exclusivity in Network Industries' [1999] *George Mason Independent Law Review* 3

Shapiro 1998

C. Shapiro, 'Setting Compatibility Standards: Cooperation or Collusion?', paper prepared for the conference of the Engelberg Center on Innovation Law and Policy, La Pieta, Italy, 25-28 June 1998, available at <faculty.haas.berkeley.edu/shapiro/standards.pdf> (last visited on 20 March 2005)

Shapiro/Varian 1999

C. Shapiro and H. R. Varian, *Information Rules. A Strategic Guide to the Network Economy* (Harvard Business School Press, Boston, 1999)

Shelanski/Sidak 2001

H.A. Shelanski and J.G. Sidak, 'Antitrust Divestiture in Network Industries' (2001) 688 *University of Chicago Law Review* 1

De Sola Poel

I. de Sola Poel, *Technologies of Freedom* (Harvard University Press, Cambridge, Massachusetts and London, 1983)

Spence/Owen 1977

M.A. Spence and B.M. Owen, 'Television Programming, Monopolistic Competition and Welfare' (1977) 91 *Quarterly Journal of Economics* 103

Speta 2004

J. Speta, 'Vertical Regulation in Digital Television: Explaining why the United States has no Access Directive', in: European Audiovisual Observatory, *Regulating access to digital television. Technical bottlenecks, vertically-integrated markets and new forms of media concentration* (IRIS Special, European Audiovisual Observatory, Strasbourg, 2004)

Stern/Dietlein 1999

K. Stern and J. Dietlein, 'Netzzugang im Telekommunikationsrecht' [1999] *Zeitschrift für das Gesamte Recht der Telekommunikation* 2

Stettner

R. Stettner, 'Rechtsfragen einer Ausweitung der nationalen Liste frei empfangbarer Ereignisse (§ 5 Rundfunkstaatsvertrag)', Rechtsgutachten erstellt im Auftrag der Kirch Media GmBH & Co KG aA 2002' [2002] *Zeitschrift für Urheber- und Medienrecht* 627

Temple Lang 1997

J. Temple Lang, 'Media, Multimedia and European Community Antitrust Law', in *International Antitrust Law and Policy 1997', Annual Proceedings of the Fordham Corporate Law Institute* (Juris Publishing, Inc., New York 1997), available at <europa.eu.int/comm/competition/speeches/text/sp1997_070_en.pdf> (last visited on 20 March 2005)

Temple Lang 1994

J. Temple Lang, 'Defining Legitimate Competition: Companies' Duties to Supply Competitors, and Access to Essential Facilities' (1994) 18 *Fordham International Law Journal* 441

Thierfelder 1999

J. Thierfelder, *Zugangsfragen digitaler Fernsehverbreitung* (Beck, München, 1999)

Trafkowski 2002

A. Trafkowski, *Medienkartellrecht. Die Sicherung des Wettbewerbs auf den Märkten der elektronischen Medien* (Beck, München, 2002)

Urek 1991

T. Urek, *Grenzen der Zulässigkeit von Exklusievvereinbarungen über die Fernsehberichterstattung* (V. Florentz GmbH, München, 1991)

Valcke 2003

P. Valcke, *Pluralism in een Digitale Interactieve Mediaomgeving. Analyse van Sectorspecifieke Regelgeving & Algemeen Mededingingsrecht*, Doctoral Thesis (Katholieke Universiteit Leuven, Leuven, 2003)

Varian

H.R. Varian, 'Markets for Information Goods', Paper, University of California, Berkeley, 1998, available at <www.sims.berkeley.edu/~hal/Papers/japan/> (last visited on 15 March 2005)

Voice of Listeners & Viewers 2001

Voice of Listeners & Viewers, 'The Pricing of Conditional Access Services and Related Issues', submission by the Voice of the Listeners & Viewers (VLV) to a Consultation Document issued by the Director General of Telecommunications on 20 October 2001', London, January 2002

Wagner

M.A. Wagner, 'Rechtliche Aspekte elektronischer Programmführer. The Legal Dimension of EPGs' [1998] *Multimedia und Recht* 243

Walker

D.M. Walker, *The Oxford Companion to Law* (Oxford University Press, New York, 1980)

Waterman

D. Waterman, 'Vertical Integration and Program Access in the Cable Television Industry' 47 *Federal Communications Law Journal* 511

Waterman/Weiss 1996

D. Waterman and A.W. Weiss, 'The Effects of Vertical Integration between Cable Television Systems and Pay Cable Networks' [1996] 72 *Journal of Econometries* 357

Weiss/Wood

A. Weiss and D. Wood, 'Was elektronische Programmführer leisten sollten. Difficult to be easy – The Electronic Programme Guide' [1998] *Multimedia und Recht* 239

Weisser

R. Weisser, 'Dienstleistungen zum Vertrieb digitaler Pay-TV Angebote' [1997] *Zeitschrift für Urheber- und Medienrecht* 877

Wentzel 2002

D. Wentzel, *Medien im Systemvergleich. Eine ordnungsökonomische Analyse des deutschen und amerikanischen Fernsehmarktes, Schriften zu Ordnungsfragen der Wirtschaft* (Lucius & Lucius, Stuttgart, 2002)

Whinston 1990

M.D. Whinston, 'Tying, Foreclosure, and Exclusion' (1990) 80 *The American Economic Review*, 837

Wirtz 1994

B.W. Wirtz, *Neue Medien, Unternehmensstrategien und Wettbwerb im Medienmarkt* (Peter Lang, Europäischer Verlag der Wissenschaften, Frankfurt am Main, 1994)

Wright 2003

J. Wright, 'One-Sided Logic in Two-Sided Markets', Working Paper 03-10, September 2003, available at <ssrn.com> (last visited on 15 March 2005)

Wyatt/Dashwood 2000

D. Wyatt and A. Dashwood, *European Union Law* (4th edition, Sweet & Maxwell, London, 2000)

Xavier 1997

P. Xavier, 'Universal Service, and Public Access in the Networked Society' [1997] *Telecommunications Policy* 829

Yoo 2002

C.S. Yoo, 'Vertical Integration and Media Regulation in the New Economy' (2002) 19 *Yale Journal on Regulation* 171

INDEX

INFORMATION LAW SERIES

GENERAL EDITOR

Prof P. Bernt Hugenholtz
Institute for Information Law
University of Amsterdam
The Netherlands

TITLES IN THE SERIES

1. Egbert J. Dommering & P. Bernt Hugenholtz (eds.), *Protecting Works of Fact: Copyright, Freedom of Expression and Information Law* (1991), ISBN 90-6544-567-6.
2. Willem F. Korthals Altes, Egbert J. Dommering, P. Bernt Hugenholtz & Jan J.C. Kabel (eds.), *Information Law Towards the 21st Century* (1992), ISBN 90-6544-627-3.
3. Jacqueline Seignette, *Challenges to the Creator Doctrine: Authorship, Copyright, Ownership and the Exploration of Creative Works in the Netherlands, Germany and the United States* (1994), ISBN 90-6544-876-4.
4. P. Bernt Hugenholtz (ed.), *The Future of Copyright in a Digital Environment* (1996), ISBN 90-411-0267-1.
5. Julius C.S. Pinckaers, *From Privacy Toward a New Intellectual Property Right in Persona* (1996), ISBN 90-411-0355-4.
6. Jan J.C. Kabel & Gerard J.H.M. Mom (eds.), *Intellectual Property and Information Law: Essays in Honour of Herman Cohen Jehoram* (1998), ISBN 90-411-9702-8.
7. Ysolde Gendreau, Axel Nordemann & Rainer Oesch (eds.), *Copyright and Photographs: An International Survey* (1999), ISBN 90-411-9722-2.
8. P. Bernt Hugenholtz (ed.), *Copyright and Electronic Commerce: Legal Aspects of Electronic Copyright Management* (2000), ISBN 90-411-9785-0.
9. Lucie M.C.R. Guibault, *Copyright Limitations and Contracts: An Analysis of Contractual Overridability of Limitations on Copyright* (2002), ISBN 90-411-9867-9.
10. Lee A. Bygrave (ed.), *Data Protection Law: Approaching its Rationale, Logic and Limits* (2002), ISBN 90-411-9870-9.
11. Niva Elkin-Koren & Neil Weinstock Netanel (eds.), *The Commodification of Information* (2002), ISBN 90-411-9876-8.

12. Mireille van Eechoud, *Choice of Law in Copyright and Related Rights. Alternatives to the Lex Protectionis* (2003),
ISBN 90-411-2071-8.
13. Martin Senftleben, *Copyright, Limitations and the Three-Step Test. An Analysis of the Three-Step Test in International and EC Copyright Law* (2004),
ISBN 90-411-2267-2.
14. Paul L.C. Torremans (ed.), *Copyright and Human Rights. Freedom of Expression – Intellectual Property – Privacy* (2004),
ISBN 90-411-2278-8.
15. Natali Helberger, *Controlling Access to Content – Regulating Conditional Access in Digital Broadcasting* (2005),
ISBN 90-411-2345-8.